MW01257666

Engineering Optimization

Engineering Optimization

An Introduction with Metaheuristic Applications

Xin-She Yang
University of Cambridge
Department of Engineering
Cambridge, United Kingdom

WILEY
A JOHN WILEY & SONS, INC., PUBLICATION

Copyright © 2010 by John Wiley & Sons, Inc. All rights reserved.

Published by John Wiley & Sons, Inc., Hoboken, New Jersey.

Published simultaneously in Canada.

No part of this publication may be reproduced, stored in a retrieval system, or transmitted in any form or by any means, electronic, mechanical, photocopying, recording, scanning, or otherwise, except as permitted under Section 107 or 108 of the 1976 United States Copyright Act, without either the prior written permission of the Publisher, or authorization through payment of the appropriate per-copy fee to the Copyright Clearance Center, Inc., 222 Rosewood Drive, Danvers, MA 01923, (978) 750-8400, fax (978) 750-4470, or on the web at www.copyright.com. Requests to the Publisher for permission should be addressed to the Permissions Department, John Wiley & Sons, Inc., 111 River Street, Hoboken, NJ 07030, (201) 748-6011, fax (201) 748-6008, or online at http://www.wiley.com/go/permission.

Limit of Liability/Disclaimer of Warranty: While the publisher and author have used their best efforts in preparing this book, they make no representations or warranties with respect to the accuracy or completeness of the contents of this book and specifically disclaim any implied warranties of merchantability or fitness for a particular purpose. No warranty may be created or extended by sales representatives or written sales materials. The advice and strategies contained herein may not be suitable for your situation. You should consult with a professional where appropriate. Neither the publisher nor author shall be liable for any loss of profit or any other commercial damages, including but not limited to special, incidental, consequential, or other damages.

For general information on our other products and services or for technical support, please contact our Customer Care Department within the United States at (800) 762-2974, outside the United States at (317) 572-3993 or fax (317) 572-4002.

Wiley also publishes its books in a variety of electronic formats. Some content that appears in print may not be available in electronic formats. For more information about Wiley products, visit our web site at www.wiley.com.

Library of Congress Cataloging-in-Publication Data:

Yang, Xin-She.
Engineering optimization : an introduction with metaheuristic applications / Xin-She Yang.
p. cm.
Includes bibliographical references and index.
ISBN 978-0-470-58246-6 (cloth)
1. Heuristic programming. 2. Mathematical optimization. 3. Engineering mathematics. I. Title.
T57.84.Y36 2010
620.001'5196—dc22 2010003429

10 9 8 7 6 5 4 3 2 1

CONTENTS

LIST OF FIGURES

3.5 Strong and weak maxima and minima. A is a weak local maximum; point B is a local maximum with discontinuity; C and D are the minimum and maximum, respectively. E is a weak local minimum. Point F corresponds to a strong maximum and also the global maximum, while point G is the strong global minimum. 39

3.6 Convexity: (a) non-convex, and (b) convex. 54

3.7 Convexity: (a) affine set $\boldsymbol{x} = \theta \boldsymbol{x}_1 + (1-\theta)\boldsymbol{x}_2$ where $\theta \in \Re$, (b) convex hull $\boldsymbol{x} = \sum_{i=1}^{k} \theta_i \boldsymbol{x}_i$ with $\sum_{i=1}^{k} \theta_i = 1$ and $\theta_i \geq 0$, and (c) convex cone $\boldsymbol{x} = \theta_1 \boldsymbol{x}_1 + \theta_2 \boldsymbol{x}_2$ with $\theta_1 \geq 0$ and $\theta_2 \geq 0$. 54

3.8 Convexity of a function $f(x)$. Chord AB lies above the curve segment joining A and B. For any point P, we have $L_\alpha = \alpha L$, $L_\beta = \beta L$ and $L = |x_B - x_A|$. 55

4.1 The basic steps of a line search method. 66

4.2 Schematic representation of linear programming. If $\alpha = 2$, $\beta = 3$, $n_1 = 16$, $n_2 = 10$ and $n = 20$, then the optimal solution is at $B(10, 10)$. 69

4.3 Minimization of a function with the two equality constraints. 79

4.4 The feasible region, infeasible solutions (marked with $\circ$) and the optimal point (marked with $\bullet$). 81

5.1 The pseudocode of the BFGS method. 86

5.2 The concept of a simplex: (a) 1-simplex, (b) 2-simplex, and (c) 3-simplex. 87

5.3 Simplex manipulations: (a) reflection with fixed volume (area), (b) expansion or contraction along the line of reflection, (c) reduction. 87

5.4 Pseudocode of Nelder-Mead's downhill simplex method. 89

5.5 Pseudocode of a trust region method. 90

5.6 Procedure of sequential quadratic programming. 93

6.1 Newton's method for the equality constrained optimization. 100

6.2 The barrier method: (a) log barrier near a boundary, and (b) central path for $n = 2$ and $N = 4$. 102

6.3 The procedure of the barrier method for convex optimization. 104

16.5 Yang's standing wave function with the global minimum at $(0, 0)$. 218

17.1 Pseudo code of the firefly algorithm (FA). 223

17.2 Landscape of a function with two equal global maxima. 226

17.3 The initial locations of 25 fireflies (left) and their final locations after 20 iterations (right). 227

17.4 Landscape of Ackley's 2D function with the global minimum 0 at $(0, 0)$. 228

17.5 The initial locations of the 25 fireflies (left) and their final locations after 20 iterations (right). 229

18.1 Non-dominated set, Pareto front and ideal vectors in a minimization problem with two objectives f_1 and f_2. 236

18.2 Three functions reach the global minimum at $x_* = \beta, y_* = \alpha - \gamma$. 238

18.3 Final locations of 40 particles after 5 iterations. The optimal point is at $(1/3, 0)$ marked with $\circ$. 239

18.4 Finding the Pareto solution with maximum utility in a maximization problem with two objectives. 241

18.5 Pareto front is the line connecting $A(5, 0)$ and $B(0, 5/\alpha)$. The Pareto solution with maximum utility is $U_* = 25$ at point A. 242

19.1 The design optimization of a simple spring. 248

19.2 Pressure vessel design and optimization. 249

19.3 The rectangular design domain is divided into N elements. As optimization and material distribution evolve, the shape becomes a truss-style structure (bottom). 250

19.4 Harmonic vibrations. 254

19.5 A rectangular beam with inhomogeneous materials properties (in 10 different cells). 257

D.1 Heron's proof of the shortest path. 307

D.2 The feasible region of design variables of a simple linear programming problem. 308

D.3 The plot of $\mathrm{sinc}(x) = \sin(x)/x$. 309

D.4 The plot of $f(x) = x^2 + 25\cos^2(x)$. 309

D.5 Quadratic penalty function $\Pi(x, \mu) = 100(x-1)^2 + \pi + \frac{\mu}{2}(x-a)^2$ and $\mu = 2000$. 311

D.6 A simple route to tour 4 cities. 312

PREFACE

Optimization is everywhere, from engineering design to computer sciences and from scheduling to economics. However, to realize that everything is optimization does not make the problem-solving easier. In fact, many seemingly simple problems are very difficult to solve. A well-known example is the so-called Traveling Salesman Problem in which the salesman intends to visit, say, 50 cities, exactly once so as to minimize the overall distance traveled or the overall traveling cost. No efficient algorithms exist for such hard problems. The latest developments over the last two decades tend to use metaheuristic algorithms. In fact, a vast majority of modern optimization techniques are usually heuristic and/or metaheuristic. Metaheuristic algorithms such as Simulated Annealing, Particle Swarm Optimization, Harmony Search, and Genetic Algorithms are becoming very powerful in solving hard optimization problems, and they have been applied in almost all major areas of science and engineering as well as industrial applications.

This book introduces all the major metaheuristic algorithms and their applications in optimization. This textbook consists of three parts: Part I: Introduction and fundamentals of optimization and algorithms; Part II: Metaheuristic algorithms; and Part III: applications of metaheuristics in engineering optimization. Part I provides a brief introduction to the nature of optimization and the common approaches to optimization problems, random number generation and Monte Carlo simulations. In Part II, we introduce all major/widely used metaheuristic algorithms in great detail, including Genetic Algorithms, Simulated Annealing, Ant Algorithms, Bee Algorithms, Particle Swarm Optimization, Firefly Algorithms, Harmony Search and others. In Part III, we briefly introduce multi-objective optimization. We also discuss a wide range of applications using metaheuristic algorithms in solving real-world optimization problems. In the appendices, we provide the implementation of some of the important/popular algorithms in Matlab® and/or Octave so that readers can use them for learning or solving other optimization problems. The files of the computer programs in the book are available at Wiley's FTP site

`ftp://ftp.wiley.com/public/sci_tech_med/engineering_optimization`

This unique book is self-contained with many step-by-step worked examples including various exercises. It can serve as an ideal textbook for both students and researchers to learn modern metaheuristic algorithms and engineering optimization.

XIN-SHE YANG

Cambridge, UK
April, 2010

ACKNOWLEDGMENTS

I would like to thank many of my mentors, friends, and colleagues for their help: J. Brindley, A. C. Fowler, A. B. Forbes, C. J. McDiarmid, A. C. McIntosh, G. T. Parks, S. Tsou, and L. Wright. Special thanks to my students at Cambridge University: E. Flower, M. Jordan, C. Pearson, J. Perry, P. De Souza, M. Stewart, and H. Scott Whittle.

I also would like to thank my Editor, Susanne Steitz-Filler, Editorial Program Coordinator, Jacqueline Palmieri, Production Editor, Melissa Yanuzzi, Copyeditor, Sharon Short, and staff at Wiley for their help and professionalism.

Last but not least, I thank my wife and son for their support and help.

X. S. Y.

INTRODUCTION

Optimization can mean many different things. However, mathematically speaking, it is possible to write an optimization problem in the generic form

$$\underset{\boldsymbol{x}\in\Re^n}{\text{minimize}} \quad f_i(\boldsymbol{x}), \qquad (i = 1, 2, ..., M), \tag{I.1}$$

$$\text{subject to } \phi_j(\boldsymbol{x}) = 0, \quad (j = 1, 2, ..., J), \tag{I.2}$$

$$\psi_k(\boldsymbol{x}) \leq 0, \quad (k = 1, 2, ..., K), \tag{I.3}$$

where $f_i(\boldsymbol{x}), \phi_j(\boldsymbol{x})$ and $\psi_k(\boldsymbol{x})$ are functions of the design vector

$$\boldsymbol{x} = (x_1, x_2, ..., x_n)^T, \tag{I.4}$$

where the components x_i of $\boldsymbol{x}$ are called design or decision variables, and they can be real continuous, discrete or a mixture of these two. The functions $f_i(\boldsymbol{x})$ where $i = 1, 2, ..., M$ are called the objective functions, and in the case of $M = 1$, there is only a single objective. The objective function is sometimes called the cost function or energy function in literature. The space spanned by the decision variables is called the search space $\Re^n$, while the space formed by the objective function values is called the solution space.

The objective functions can be either linear or nonlinear. The equalities for ϕ_j and inequalities for ψ_k are called constraints. It is worth pointing out that we can also write the inequalities in the other way ≥ 0, and we can also formulate the objectives as a maximization problem. This is because the maximization of $f(\boldsymbol{x})$ is equivalent to the minimization of $-f(\boldsymbol{x})$, and any inequality $g(\boldsymbol{x}) \leq 0$ is equivalent to $-g(\boldsymbol{x}) \geq 0$. For the constraints, the simplest case for a decision variable x_i is $x_{i,\min} \leq x_i \leq x_{i,\max}$, which is called bounds.

If the constraints ϕ_j and ψ_k are all linear, then it becomes a linearly constrained problem. If both the constraints and the objective functions are all linear, it becomes a linear programming problem. For linear programming problems, a significant progress was the development of the simplex method in 1947 by George B. Dantzig. However, generally speaking, since all f_i, ϕ_j and ψ_k are nonlinear, we have to deal with a nonlinear optimization problem. It is worth pointing out that all the functions (objective and constraints) are collectively called problem functions.

A special class of optimization is when there is no constraint at all (or $J = K = 0$), and the only task is to find the minimum or maximum of a single objective function $f(\boldsymbol{x})$. This usually makes things much easier, though not always. In this case, the optimization problem becomes an unconstrained one.

For example, we can find the minimum of the Rosenbrock banana function

$$f(x, y) = (1 - x)^2 + 100(y - x^2)^2. \tag{I.5}$$

In order to find its minimum, we can set its partial derivatives to zero, and we have

$$\frac{\partial f}{\partial x} = 2(1 - x) - 400(y - x^2)x = 0, \tag{I.6}$$

$$\frac{\partial f}{\partial y} = 200(y - x^2) = 0. \tag{I.7}$$

The second equation implies that $y = x^2$ can be substituted into the first one. We have

$$1 - x - 200(x^2 - x^2) = 1 - x = 0, \tag{I.8}$$

or $x = 1$. The minimum $f_{\min} = 0$ occurs at $x = y = 1$. This method uses important information from the objective function; that is, the gradient or first derivatives. Consequently, we can use gradient-based optimization methods such as Newton's method and conjugate gradient methods to find the minimum of this function.

A potential problem arises when we do not know the the gradient, or the first derivatives do not exist or are not defined. For example, we can design the following function

$$f(x, y) = (|x| + |y|) \exp[-\sin(x^2) - \sin(y^2)]. \tag{I.9}$$

The global minimum occurs at $(x, y) = (0, 0)$, but the derivatives at $(0, 0)$ are not well defined due to the factor $|x| + |y|$ and there is some discontinuity in the first derivatives. In this case, it is not possible to use gradient-based optimization methods. Obviously, we can use gradient-free method such as the Nelder-Mead downhill simplex method. But as the objective function is multimodal (because of the sine function), such optimization methods are very sensitive to the starting point. If the starting point is far from the the sought minimum, the algorithm will usually get stuck in a local minimum and/or simply fail.

Optimization can take other forms as well. Many mathematical and statistical methods are essentially a different form of optimization. For example, in data processing, the methods of least squares try to minimize the sum of the residuals or differences between the predicated values (by mathematical models) and the observed values. All major numerical methods such as finite difference methods intend to find some approximations that minimize the difference of the true solutions and the estimated solutions. In aircraft design, we try to design the shape in such a way so as to minimize the drag and maximize the lifting force. All these formulations could be converted or related to the generic form of the nonlinear optimization formulation discussed above. In some extreme cases, the objective functions do not have explicit form, or at least it cannot be easily linked with the design variables. For example, nowadays in product design and city planning, we have to optimize the energy efficiency and minimize the environmental impact. The study of such impact itself is a challenging topic and it is not always easy to characterize them; however, we still try to find some suboptimal or even optimal solutions in this context.

Nonlinearity and multimodality are the main problem, which renders most conventional methods such as the hill-climbing method inefficient and stuck in the wrong solutions. Another even more challenging problem arises when the number of decision variables increases or n is very large, say, $n = 50,000$. In addition, the nonlinearity coupled with the large scale complexity makes things even worse. For example, the well-known traveling salesman problem is to try to find the shortest route for a salesman to travel n cities once and only once. The number of possible combinations, without knowing the distribution of the cities, is $n!$. If $n = 100$, this number of combinations $n! \approx 9.3 \times 10^{157}$ is astronomical. The top supercomputers in the world such as IBM's Blue Gene can now do about 3 petraflops; there are about 3×10^{15} floating-point operations per second. In fact, with all the available computers in the world fully dedicated to the brutal force search of all the combinations of 100!, it would take much longer than the lifetime of the known universe. This clearly means that it is not practical to search all possible combinations. We have to use some alternative, yet efficient enough, methods.

Heuristic and metaheuristic algorithms are designed to deal with this type of problem. Most these algorithms are nature-inspired or bio-inspired as they have been developed based on the successful evolutionary behavior of natural systems – by learning from nature. Nature has been solving various tough problems over millions or even billions of years. Only the best and robust solutions remain – survival of the fittest. Similarly, heuristic algorithms use the trial-and-error, learning and adaptation to solve problems. We cannot expect them to find the best solution all the time, but expect them to find the good enough solutions or even the optimal solution most of the time, and more importantly, in a reasonably and practically short time. Modern metaheuristic algorithms are almost guaranteed to work well for a wide range of tough optimization problems. However, it is a well-known fact that there is 'no free lunch' in optimization. It has been proved by Wolpert and Macready in 1997 that if algorithm A is better than algorithm B for some problems, then B will outperform A for other problems. That is to say, a universally efficient algorithm does not exist. The main aim of research in optimization and algorithm development is to design and/or choose the most suitable and efficient algorithms for a given optimization task.

Loosely speaking, modern metaheuristic algorithms for engineering optimization include genetic algorithms (GA), simulated annealing (SA), particle swarm optimization (PSO), ant colony algorithm, bee algorithm, harmony search (HS), firefly algorithm (FA), and many others.

We will introduce all the major and widely used metaheuristics in Part II of this book, after a detailed introduction to the fundamentals of engineering optimization. In Part III, we will briefly outline other important algorithms and multiobjective optimization. We then focus on the applications of the algorithms introduced in the book to solve real-world optimization problems.

Each chapter will be self-contained or with minimal cross references to other chapters. We will include some exercises at the end of each chapter with detailed answers in the appendices. A further reading list is also provided at the end of each chapter. These make it ideal for the book to be used either as a textbook for relevant courses, or an additional reference as well as for self study. The self-contained nature of each chapter means that lecturers and students can use each individual chapter to suit their own purpose.

The main requirement for this book is the basic understanding of the calculus, particularly differentiation, and a good undestanding of algebraic manipulations. We will try to review these briefly in Part I.

In addition, the implementation of algorithms will inevitably use a programming language. However, we believe that the efficiency and performance of each algorithm, if properly implemented, should be independent of any programming. For this reason, we will explain each algorithm as detail as possible, but leave an actual implementation in the appendices where we will include some simple Matlab/Octave programs for demonstrating how the implemented algorithms work.

REFERENCES

1. G. B. Dantzig, *Linear Programming and Extensions*, Princeton University Press, 1963.

2. P. E. Gill, W. Murray, and M. H. Wright, *Practical Optimization*, Academic Press Inc., 1981.

3. S. Kirkpatrick, C. D. Gelatt, and M. P. Vecchi, "Optimization by simulated annealing", *Science*, **220** (4598), 671-680 (1983).

4. D. H. Wolpert and W. G. Macready, "No free lunch theorems for optimizaiton", *IEEE Transaction on Evolutionary Computation*, **1**, 67-82 (1997).

5. X. S. Yang, "Harmony search as a metaheuristic algorithm", in: *Music-Inspired Harmony Search Algorithm: Theory and Applications* (eds. Z. W. Geem), Springer, p. 1-14 (2009).

PART I

FOUNDATIONS OF OPTIMIZATION AND ALGORITHMS

CHAPTER 1

A BRIEF HISTORY OF OPTIMIZATION

Optimization is everywhere, from engineering design to financial markets, from our daily activity to planning our holidays, and computer sciences to industrial applications. We always intend to maximize or minimize something. An organization wants to maximize its profits, minimize costs, and maximize performance. Even when we plan our holidays, we want to maximize our enjoyment with least cost (or ideally free). In fact, we are constantly searching for the optimal solutions to every problem we meet, though we are not necessarily able to find such solutions.

It is no exaggeration to say that finding the solution to optimization problems, whether intentionally or subconsciously, is as old as human history itself. For example, the least effort principle can often explain many human behaviors. We know the shortest distance between any two different points on a plane is a straight line, though it often needs complex maths such as the calculus of variations to formally prove that a straight line segment between the two points is indeed the shortest.

In fact, many physical phenomena are governed by the so-called least action principle or its variants. For example, light travels and obeys Fermat's principle, that is to travel at the shortest time from one medium to another,

Engineering Optimization: An Introduction with Metaheuristic Applications.
By Xin-She Yang
Copyright © 2010 John Wiley & Sons, Inc.

thus resulting in Snell's law. The whole analytical mechanics is based on this least action principle.

1.1 BEFORE 1900

The study of optimization problems is also as old as science itself. It is known that the ancient Greek mathematicians solved many optimization problems. For example, Euclid in around 300BC proved that a square encloses the greatest area among all possible rectangles with the same total length of four sides. Later, Heron in around 100BC suggested that the distance between two points along the path reflected by a mirror is the shortest when light travels and reflects from a mirror obeying some symmetry, that is the angle of incidence is equal to the angle of reflection. It is a well-know optimization problem, called Heron's problem, as it was first described in Heron's *Catoptrica* (or *On Mirrors*).

The celebrated German astronomer, Johannes Kepler, is mainly famous for the discovery of his three laws of planetary motion; however, in 1613, he solved an optimal solution to the so-called marriage problem or secretary problem when he started to look for his second wife. He described his method in his personal letter dated October 23, 1613 to Baron Strahlendorf, including the balance of virtues and drawbacks of each candidate, her dowry, hesitation, and advice of friends. Among the eleven candidates interviewed, Kepler chose the fifth, though his friend suggested him to choose the fourth candidate. This may imply that Kepler was trying to optimize some utility function of some sort. This problem was formally introduced by Martin Gardner in 1960 in his *mathematical games* column in the February 1960 issue of *Scientific American.* Since then, it has developed into a field of probability optimization such as optimal stopping problems.

W. van Royen Snell discovered in 1621 the law of refraction, which remained unpublished; later, Christiaan Huygens mentioned Snell's results in his *Dioptrica* in 1703. This law was independently rediscovered by René Descartes and published in his treatise *Discours de la Methode* in 1637. About 20 years later, when Descartes' students contacted Pierre de Fermat collecting his correspondence with Descartes, Fermat looked again in 1657 at his argument with the unsatisfactory description of light refraction by Descartes, and derived Snell and Descartes' results from a more fundamental principle – light always travels in the shortest time in any medium, and this principle for light is now referred to as *Fermat's principle*, which laid the foundation of modern optics.

In his *Principia Mathematica* published in 1687, Sir Isaac Newton solved the problem of the body shape of minimal resistance that he posed earlier in 1685 as a pioneering problem in optimization, now a problem of the calculus of variations. The main aim was to find the shape of a symmetrical revolution body so as to minimize the resistance to motion in a fluid. Subsequently,

Newton derived the resistance law of the body. Interestingly, Galileo Galilei independently suggested a similar problem in 1638 in his *Discursi.*

In June 1696, J. Bernoulli made some significant progress in calculus. In an article in *Acta Eruditorum*, he challenged all the mathematicians in the world to find the shape or curve connecting two points at different heights so that a body will fall along the curve in the shortest time due to gravity – the line of quickest descent, though Bernoulli already knew the solution. On January 29, 1697 the challenge was received by Newton when he come home at four in the afternoon and he did not sleep until he had solved it by about four the next morning and on the same day he sent out his solution. Though Newton managed to solve it in less than 12 hours as he became the Warden of the Royal Mint on March 19, 1696, some suggested that he, as such a genius, should have been able to solve it in half an hour. Some said this was the first hint or evidence that too much administrative work will slow down one's progress. The solution as we now know is a part of a cycloid. This steepest descent is now called Brachistochrone problem, which inspired Euler and Lagrange to formulate the general theory of calculus of variations.

In 1746, the *principle of least action* was proposed by P. L. de Maupertuis to unify various laws of physical motion and its application to explain all phenomena. In modern terminology, it is a variational principle of stationary action in terms of an integral equation of a functional in the framework of calculus of variations, which plays a central role in the Lagrangian and Hamiltonian classical mechanics. It is also an important principle in mathematics and physics.

In 1781, Gaspard Monge, a French civil engineer, investigated the transportation problem for optimal transportation and allocation of resources, if the initial and final spatial distribution are known. In 1942, Leonid Kantorovich showed that this combinatorial optimization problem is in fact a case of a linear programming problem.

Around 1801, Frederich Gauss claimed that he used the *method of least-squares* to predict the orbital location of the asteroid Ceres, though his version of the least squares with more rigorous mathematical foundation was published later in 1809. In 1805, Adrien Legendre was the first to describe the method of least squares in an appendix of his book *Nouvelle meéthodes pour la determination des orbites des cometes*, and in 1806 he used the principle of least squares for curve fitting. Gauss later claimed that he had been using this method for more than 20 years, and laid the foundation for least-squares analysis in 1795. This led to some bitter disputes with Legendre. In 1808, Robert Adrain, unaware of Legendre's work, published the method of least squares studying the uncertainty and errors in making observations, not using the same terminology as those by Legendre.

In 1815, D. Ricardo proposed the *law of diminishing returns* for land cultivation, which can be applied in many activities. For example, the productivity of a piece of a land or a factory will only increase marginally with additional increase of inputs. This law is called law of increasing opportunity cost. It dictates that there is a fundamental relationship between opportunity and scarcity of resources, thus requiring that scarcely available resources be used efficiently.

In 1847 in a short note, L. A. Cauchy proposed a general method for solving systems of equations in an iterative way. This essentially leads to two iterative methods of minimization: now called the gradient method and steepest descent, for certain functions of more than one variable.

1.2 TWENTIETH CENTURY

In 1906, Danish mathematician J. Jensen introduced the concept of convexity and derived an inequality, now referred to as Jensen's inequality, which plays an important role in convex optimization and other areas such as economics. Convex optimization is a special but very important class of mathematical optimization as any optimality found is also guaranteed to be the global optimality. A wider range of optimization problems can be reformulated in terms of convex optimization. Consequently, it has many applications including control systems, data fitting and modelling, optimal design, signal processing, mathematical finance, and others.

As early as 1766, Leonhard Euler studied the Knight tour problem, and T. P. Kirkman published a research article on the way to find a circuit which passes through each vertex once and only once for a give graph of polyhedra. In 1856, Sir William Rowan Hamilton popularized his *Icosian Game*. Then, in February 1930, Karl Menger posed the Messenger's problem at a mathematical colloquium in Vienna, as this problem is often encountered by postal messengers and travelers. His work was published later in 1932. The task is to find the shortest path connecting a finite number of points/cities whose pairwise distances are known. Though the problem is solvable in a finite number of trials and permutations, there is no efficient algorithm for finding such solutions. In general, the simple rule of going to the nearest points does not result in the shortest path. This problem is now referred to as *Traveling Salesman Problem* which is closely related to many different applications such as network routing, resource allocation, scheduling and operations research in general. In fact, as early as 1832, the 1832 traveling salesman manual described a tour along 45 German cities with a shortest route of 1248 km, though the exact mathematical roots of this problem are quite obscure, and might be well around for some time before the 1930s.

Interestingly, H. Hancock published in 1917 the first book on optimization "*Theory of Minima and Maxima*".

In 1939, L. Kantorovich was the first to develop an algorithm for linear programming and use it in economics. He formulated the production problem of optimal planning and effective methods for finding solutions using linear programming. For this work, he shared the Noble prize with T. Koopmans in 1975. The next important step of progress is that George Dantzig invented in

1947 the *simplex method* for solving large-scale linear programming problems. Dorfman in an article published in 1984 wrote that linear programming was discovered three times, independently, between 1939 and 1947, but each time in a somewhat different form. The first discovery was by the Russian mathematician, L. Kantorovich, then by the Dutch economist, Koopmans, and the third in 1947 by the American mathematician George Dantzig. Dantzig's revolutionary simplex method is able to solve a wide range of optimal policy decision problems of great complexity. A classic example and one of the earliest of using linear programming as described in Dantzig's 1963 book was to find the solution to the special optimal diet problem involving 9 equations and 77 unknowns using hand-operated desk calculators.

In 1951, Harold Kuhn and A. W. Tucker studied the nonlinear optimiation problem and re-developed the optimality condition, as similar conditions were proposed by W. Karush in 1939 in his MSc dissertation. In fact, the optimality conditions are the generalization of Lagrange multipliers to nonlinear inequalities, and are now known as the Karush-Kuhn-Tucker conditions, or simply Kuhn-Tucker conditions, which are necessary conditions for a solution to be optimal in nonlinear programming.

Then, in 1957, Richard Bellman at Stanford University developed the dynamic programming and the optimality principle when studying multistage decision and planning processes while he spent some time at the RAND Corporation. He also coined the term *Dynamic Programming*. The idea of dynamic programming can date back to 1944 when John von Neumann and O. Morgenstern studied the sequential decision problems. John von Neumann also made important contribution to the development of operational research. As earlier as in 1840, Charles Babbage studied the cost of transportation and sorting mails; this could be the earliest research on the operational research. Significant progress was made during the Second World War, and ever since it expanded to find optimal or near optimal solutions in a wide range complex problems of interdisciplinary areas such as communication networks, project planning, scheduling, transport planning, and management.

After the 1960s, the literature on optimization exploded, and it would take a whole book to write even a brief history on optimization after the 1960s. As this book is mainly about the introduction to metaheuristic algorithms, we will then focus our attention on the development of heuristics and metaheursitics. In fact, quite a significant number of new algorithms in optimization are primarily metaheuristics.

1.3 HEURISTICS AND METAHEURISTICS

Heuristics is a solution strategy by trial-and-error to produce acceptable solutions to a complex problem in a reasonably practical time. The complexity of the problem of interest makes it impossible to search every possible solution or combination, the aim is to find good, feasible solutions in an acceptable timescale. There is no guarantee that the best solutions can be found, and we even do not know whether an algorithm will work and why if it does work. The idea is that an efficient but practical algorithm that will work most of the time and be able to produce good quality solutions. Among the found quality solutions, it is expected that some of them are nearly optimal, though there is no guarantee for such optimality.

Alan Turing was probably the first to use heuristic algorithms during the Second World War when he was breaking German Enigma ciphers at Bletchley Park where Turing, together with British mathematician Gordon Welchman, designed in 1940 a cryptanalytic electromechanical machine, the *Bombe*, to aid their code-breaking work. The bombe used a heuristic algorithm, as Turing called, to search, among about 10^{22} potential combinations, the possibly correct setting coded in an Enigma message. Turing called his search method *heuristic search*, as it could be expected it worked most of the time, but there was no guarantee to find the correct solution, but it was a tremendous success. In 1945, Turing was recruited to the National Physical Laboratory (NPL), UK where he set out his design for the Automatic Computing Engine (ACE). In an NPL report on "Intelligent machinery' in 1948, he outlined his innovative ideas of machine intelligence and learning, neural networks and evolutionary algorithms or an early version of genetic algorithms.

The next significant step is the development of evolutionary algorithms in the 1960s and 1970s. First, John Holland and his collaborators at the University of Michigan developed the genetic algorithms in the 1960s and 1970s. As early as 1962, Holland studied the adaptive system and was the first to use crossover and recombination manipulations for modeling such systems. His seminal book summarizing the development of genetic algorithms was published in 1975. In the same year, Kenneth De Jong finished his important dissertation showing the potential and power of genetic algorithms for a wide range of objective functions, either noisy, multimodal or even discontinuous.

Genetic algorithms (GA) is a search method based on the abstraction of Darwin's evolution and natural selection of biological systems and representing them in the mathematical operators: crossover or recombination, mutation, fitness, and selection of the fittest. Ever since, genetic algorithms become so successful in solving a wide range of optimization problems, several thousand research articles and hundreds of books have been written. Some statistics show that a vast majority of Fortune 500 companies are now using them routinely to solve tough combinatorial optimization problems such as planning, data-fitting, and scheduling.

During the same period, Ingo Rechenberg and Hans-Paul Schwefel both then at the Technical University of Berlin developed a search technique for solving optimization problem in aerospace engineering, called evolution strategy, in 1963. Later, Peter Bienert joined them and began to construct an automatic experimenter using simple rules of mutation and selection. There is no crossover in this technique; only mutation was used to produce an offspring and an improved solution was kept at each generation. This is essentially a simple trajectory-style hill-climbing algorithm with randomization. As early as 1960, Lawrence J. Fogel intended to use simulated evolution as a learning process as a tool to study artificial intelligence. Then, in 1966, L. J. Fogel, with A. J. Owen and M. J. Walsh, developed the evolutionary programming technique by representing solutions as finite-state machines and randomly mutating one of these machines. The above innovative ideas and methods have evolved into a much wider discipline, called evolutionary algorithms and evolutionary computation.

The decades of the 1980s and 1990s were the most exciting time for metaheuristic algorithms. The next big step is the development of simulated annealing (SA) in 1983, an optimization technique, pioneered by S. Kirkpatrick, C. D. Gellat and M. P. Vecchi, inspired by the annealing process of metals. It is a trajectory-based search algorithm starting with an initial guess solution at a high temperature, and gradually cooling down the system. A move or new solution is accepted if it is better; otherwise, it is accepted with a probability, which makes it possible for the system to escape any local optima. It is then expected that if the system is cooled slowly enough, the global optimal solution can be reached.

The actual first usage of metaheuristic is probably due to Fred Glover's Tabu search in 1986, though his seminal book on Tabu search was published later in 1997.

In 1992, Marco Dorigo finished his PhD thesis on optimization and natural algorithms, in which he described his innovative work on ant colony optimization (ACO). This search technique was inspired by the swarm intelligence of social ants using pheromone as a chemical messenger. Then, in 1992, John R. Koza of Stanford University published a treatise on genetic programming which laid the foundation of a whole new area of machine learning, revolutionizing computer programming. As early as in 1988, Koza applied his first patent on genetic programming. The basic idea is to use the genetic principle to breed computer programs so as to produce the best programs for a given type of problem.

Slightly later in 1995, another significant step of progress is the development of the particle swarm optimization (PSO) by American social psychologist James Kennedy, and engineer Russell C. Eberhart. Loosely speaking, PSO is an optimization algorithm inspired by the swarm intelligence of fish and birds and even by human behavior. The multiple agents, called particles, swarm around the search space starting from some initial random guess. The swarm communicates the current best and shares the global best so as to focus on the quality solutions. Since its development, there have been about 20 different variants of particle swarm, and have been applied to almost all areas of tough optimization problems. There is some strong evidence that PSO is better than traditional search algorithms and even better than genetic algorithms for most type of problems, though this is far from conclusive.

In 1997, the publication of the 'no free lunch theorems for optimization' by D. H. Wolpert and W. G. Macready sent out a shock wave to the optimization community. Researchers have always been trying to find better algorithms, or even universally robust algorithms, for optimization, especially for tough NP-hard optimization problems. However, these theorems state that if algorithm A performs better than algorithm B for some optimization functions, then B will outperform A for other functions. That is to say, if averaged over all possible function space, both algorithms A and B will perform on average equally well. Alternatively, there is no universally better algorithms exist. That is disappointing, right? Then, people realized that we do not need the average over all possible functions as for a given optimization problem. What we want is to find the best solutions; this has nothing to do with average over all the whole function space. In addition, we can accept the fact that there is no universal or magical tool, but we do know from our experience that some algorithms indeed outperform others for given types of optimization problems. So the research now focuses on finding the best and most efficient algorithm(s) for a given problem. The task is to design better algorithms for most types of problems, not for all the problems. Therefore, the search is still on.

At the turn of the twenty-first century, things became even more exciting. First, Zong Woo Geem *et al.* in 2001 developed the Harmony Search (HS) algorithm, which has been widely applied in solving various optimization problems such as water distribution, transport modelling and scheduling. In 2004, S. Nakrani and C. Tovey proposed the Honey Bee algorithm and its application for optimizing Internet hosting centers, which followed by the development of a novel bee algorithm by D. T. Pham *et al.* in 2005 and the Artificial Bee Colony (ABC) by D. Karaboga in 2005. In 2008, the author of this book developed the Firefly Algorithm (FA). Quite a few research articles on the Firefly Algorithm then followed, and this algorithm has attracted a wide range of interests.

As we can see, more and more metaheuristic algorithms are being developed. Such a diverse range of algorithms necessitates a system summary of various metaheuristic algorithms, and this book is such an attempt to introduce all the latest and major metaheuristics with applications.

EXERCISES

1.1 Find the minimum value of $f(x) = x^2 - x - 6$ in $[-\infty, \infty]$.

1.2 For the previous problem, use simple differentiation to obtain the same results.

1.3 In about 300BC, Euclid proved that a square encloses the greatest area among all rectangles, assuming the total length of the four edges is fixed. Provide your own version of such proof.

1.4 Design a cylindrical water tank which uses the minimal materials and holds the largest volume of water. What is the relationship between the radius r of the base and the height h?

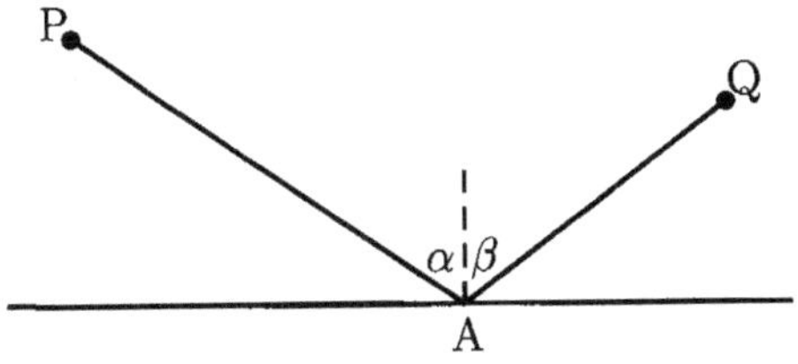

Figure 1.1: Reflection of light at a mirror.

1.5 In about 100BC, Heron proved that a path PAQ is the shortest when reflecting at a mirror with the angle of incidence α is equal to angle of reflectance β (see Figure 1.1). Show that $\alpha = \beta$ leads to the shortest distance PAQ.

REFERENCES

1. M. S. Bazaraa, H. D. Serali, and C. M. Shetty, *Nonlinear Programming: Theory and Algorithms*, John Wiley & Sons, 1993.
2. L. A. Cauchy, "Méthode générale pour la résolution des systeèmes d'équations simulatanées, *Comptes Rendus de l'Académie de Sciences de Paris*, **25**, 536-538 (1847).
3. B. J. Copeland, *The Essential Turing*, Oxford University Press, 2004.
4. B. J. Copeland, *Alan Turing's Automatic Computing Engine*, Oxford University Press, 2005.
5. K. De Jong, *Analysis of the Behaviour of a Class of Genetic Adaptive Systems*, PhD thesis, University of Michigan, Ann Anbor, 1975.
6. R Dorfman, "The discovery of linear programming", *Ann. Hist. Comput.*, **6** (3), 283-295 (1984).
7. M. Dorigo, *Optimization, Learning and Natural Algorithms*, PhD thesis, Politecnico di Milano, Italy, 1992.
8. S. Dreyfus, "Richard Bellman on the birth of dynamic programming", *Operations Research*, **50** (1), 48-51 (2002).
9. T. S. Ferguson, "Who solved the secretary problem?", *Statist. Sci.*, **4** (3), 282-296 (1989).
10. R. P. Feynman, R. B. Leighton, and M. Sands, *The Feynman Lectures on Physics*, vol. 2, Addison-Wesley, Reading, Mass., 1963.
11. L. J. Fogel, A. J. Owens, and M. J. Walsh, *Artificial Intelligence Through Simulated Evolution*, John Wiley & Sons, 1966.
12. Z. W. Geem, J. H. Kim and G. V. Loganathan, "A new heuristic optimization: Harmony search", *Simulation*, **76**(2), 60-68 (2001).
13. F. Glover, "Future paths for integer programming and links to artificial intelligence", *Comput. Operational Res.*, **13** (5), 533-549 (1986).

14. F. Glover and M. Laguna, *Tabu Search*, Kluwer Academic Publishers, Boston, 1997.

15. H. H. Holstine, *A History of the Calculus of Variations from the 17th through the 19th Century*, Springer-Verlag, Heidelberg, 1980.

16. A. A. Goldstein, "Cauchy's method of minimization", *Numerische Mathematik*, **4**(1), 146-150 (1962).

17. J. Holland, *Adaptation in Natural and Artficial systems*, University of Michigan Press, Ann Anbor, 1975.

18. J. L. W. V. Jensen, "Sur les fonctions convexes et les inégalités entre les valeurs moyenners", *Acta Mathematica*, **30**, 175-193 (1906).

19. P. Judea, *Heuristics*, Addison-Wesley, 1984.

20. D. Karaboga, "An idea based on honey bee swarm for numerical optimization", Technical Report, Erciyes University, 2005.

21. W. Karush, *Minima of Functions of Several Variables with Inequalities as Side Constraints*, MSc Dissertation, Department of Mathematics, University of Chicago, Illinois, 1939.

22. J. Kennedy and R. Eberhart, "Particle swarm optimization", in: *Proc. of the IEEE Int. Conf. on Neural Networks*, Piscataway, NJ, p. 1942-1948 (1995).

23. S. Kirkpatrick, C. D. Gellat, and M. P. Vecchi, "Optimization by simulated annealing", *Science*, **220**, 671-680 (1983).

24. J. R. Koza, *Genetic Programming: One the Programming of Computers by Means of Natural Selection*, MIT Press, 1992.

25. H. W. Kuhn and A. W. Tucker, "Nonlinear programming", *Proc. 2nd Berkeley Symposium*, University of California Press, Berkeley, p. 481-492 (1951).

26. K. Menger, "Dass botenproblem", in: *Ergebnisse eines Mathematischen Kolloquiums* **2**, Ed. K. Menger), Teubner, Leipzig, p. 11-12 (1932).

27. L. T. Moore, *Isaac Newton: A Biography*, Dover, New York, 1934.

28. S. Nakrani and C. Tovey, "On honey bees and dynamic server allocation in Internet hostubg centers", *Adaptive Behavior*, **12**, 223-240 (2004).

29. D. T. Pham, A. Ghanbarzadeh, E. Koc, S. Otri, S. Rahim and M. Zaidi, "The bees algorithm", Technical Note, Manufacturing Engineering Center, Cardiff University, 2005.

30. O. Sheynin, "On the history of the principle of least squares", *Archive for History of Exact Sciences*, **46** (1), 39-54 (1993).

31. A. Schrijver, "On the history of combinatorial optimization (till 1960)", in: *Handbook of Discrete Optimization* (Eds K. Aardal, G. L. Nemhauser, R. Weismantel), Elsevier, Amsterdam, p.1-68 (2005).

32. V. M. Tikhomirov, *Stories about Mixima and Minima*, American Mathematical Society, 1990.

33. A. M. Turing, *Intelligent Machinery*, National Physical Laboratory, 1948.

34. W. T. Tutte, *Graph Theory as I Have Known It*, Oxford Science Publications, 1998.

35. D. Wells, *The Penguin Book of Curious and Interesting Geometry*, Penguin, 1991.

36. D. H. Wolpert and W. G. Macready, "No free lunch theorems for optimization", *IEEE Transaction on Evolutionary Computation*, **1**, 67-82 (1997).

37. X. S. Yang, *Nature-Inspired Metaheuristic Algorithms*, Luniver Press, (2008).

38. X. S. Yang, "Firefly algorithms for multimodal optimization", *Proc. 5th Symposium on Stochastic Algorithms, Foundations and Applications, SAGA 2009*, Eds. O. Watanabe and T. Zeugmann, Lecture Notes in Computer Science, **5792**, 169-178 Japan, 2009.

39. History of optimization, http://hse-econ.fi/kitti/opthist.html

40. The Turing Archive for the History of Computing, http://www.alanturing.net/

41. History of Mathematics and Mathematicians, http://turnbull.dcs.st-and.ac.uk /history/Mathematicians

CHAPTER 2

ENGINEERING OPTIMIZATION

Optimization can include a wide range of problems with the aim of searching for certain optimality. Subsequently, there are many different ways of naming and classifying optimization problems, and typically the optimization techniques can also vary significantly from problem to problem. A unified approach is not possible, and the complexity of an optimization problem largely depends on the function forms of its objective functions and constraints.

2.1 OPTIMIZATION

As we discussed in our introduction, it is possible to write most optimization problems in the generic form mathematically

$$\underset{\boldsymbol{x} \in \Re^n}{\text{minimize}} \quad f_i(\boldsymbol{x}), \qquad (i = 1, 2, ..., M), \tag{2.1}$$

$$\text{subject to} \quad \phi_j(\boldsymbol{x}) = 0, \quad (j = 1, 2, ..., J), \tag{2.2}$$

$$\psi_k(\boldsymbol{x}) \leq 0, \quad (k = 1, 2, ..., K), \tag{2.3}$$

where $f_i(\boldsymbol{x}), \phi_j(\boldsymbol{x})$ and $\psi_k(\boldsymbol{x})$ are functions of the design vector

$$\boldsymbol{x} = (x_1, x_2, ..., x_n)^T. \tag{2.4}$$

Engineering Optimization: An Introduction with Metaheuristic Applications.
By Xin-She Yang
Copyright © 2010 John Wiley & Sons, Inc.

Here the components x_i of $\boldsymbol{x}$ are called design or decision variables, and they can be real continuous, discrete or the mixed of these two. The functions $f_i(\boldsymbol{x})$ where $i = 1, 2, ..., M$ are called the objective functions, and in the case of $M = 1$, there is only a single objective. The objective function is sometimes called the cost function or energy function in literature. The space spanned by the decision variables is called the design space or search space $\Re^n$, while the space formed by the objective function values is called the solution space or response space. The equalities for ϕ_j and inequalities for ψ_k are called constraints. It is worth pointing out that we can also write the inequalities in the other way ≥ 0, and we can also formulate the objectives as a maximization problem.

In a rare but extreme case where there is no objective at all, there are only constraints. Such a problem is called a feasibility problem because any feasible solution is an optimal solution.

However, sometimes it is not always possible to write the objective functions in the explicit form. For example, we may want to design a car engine with the highest possible fuel efficiency and lowest carbon dioxide emission. These objectives will depend on many factors, including the geometry of the engine, the type of fuel, ignition system, air-fuel mixing and injection, working temperature, and many other factors. Such dependence is often investigated by using the simulation tool such as computational fluid dynamics (CFD), and the relationship between the efficiency and many design variables is so complex, and no explicit form is possible. In this case, we are dealing with black-box type of optimization problems, which is very difficult to solve. In this book, we will mainly focus on the optimization problems with explicit objectives as in (2.1). We will also briefly come back to the black-box optimization in Part III when discussing applications.

In other cases, the objective function is not easily measurable, but we still want to maximize it. For example, when we go on holiday, we want to maximize our enjoyment while minimizing our spending. It is difficult to measure the level of enjoyment in terms of a simple mathematical function, and each person may have a slightly different definition of his or her enjoyment. Mathematical speaking, we can only maximize something if we can write it in some form of mathematical equation.

In principle, all the functions f_i, ϕ_j, ψ_k can also include some integrals, which make things extremely complicated. In this case, we are dealing with optimization of functionals, and in most cases we have to use calculus of variations. For example, to determine the shape of a hanging rope anchored with two points A and B, we have to find a curve or shape $y(x)$ such that the total potential energy E_p is minimal

$$\underset{y(x)}{\text{minimize}} \ E_p = \rho g \int_A^B y\sqrt{1 + y'^2}\, dx, \tag{2.5}$$

where ρ is the mass per unit length of the rope, and g is the acceleration due to gravity. In addition, the length L of the rope is also fixed, that is, the constraint is

$$\int_A^B \sqrt{1+y'^2}\, dx = L. \tag{2.6}$$

This type of optimization often requires the solution of the Euler-Lagrange equation. Furthermore, when the objective is a functional or integral, and the constraints are expressed in terms of differential equations, not simple functions, the optimization problem becomes an optimal control problem. We will only briefly discuss these later in this book.

Therefore, most of this book will focus on the algorithms for solving optimization problems with explicit objective functions. Before we proceed, let us discuss various types of optimization and relevant terminologies.

2.2 TYPE OF OPTIMIZATION

The classification of optimization is not well established and there is some confusion in literature, especially about the use of some terminologies. Here we will use the most widely used terminologies. However, we do not intend to be rigorous in classifications; rather we would like to introduce all relevant concepts in a concise manner. Loosely speaking, classification can be carried out in terms of the number of objectives, number of constraints, function forms, landscape of the objective functions, type of design variables, uncertainty in values, and computational effort (see Figure 2.1).

If we try to classify optimization problems according to the number of objectives, then there are two categories: single objective $M = 1$ and multiobjective $M > 1$. Multiobjective optimization is also referred to as multicriteria or even multi-attributes optimization in literature. Most real-world optimization problems are multiobjective. For example, when designing a car engine, we want to maximize its fuel efficiency, to minimize the carbon-dioxide emission, and to lower its noise level. Here we have three objectives. Sometimes, these objectives are often conflicting, and some compromise is needed. For example, when we rent or buy a house, we want it in the best possible location with a large space while we intend to spend as little as possible. Though the algorithms we will discuss in this book are equally applicable to multiobjective optimization with some modifications, we will mainly focus on single objective optimization problems.

Similarly, we can also classify optimization in terms of number of constraints $J + K$. If there is no constraint at all $J = K = 0$, then it is called an unconstrained optimization problem. If $K = 0$ and $J \geq 1$, it is called an equality-constrained problem, while $J = 0$ and $K \geq 1$ becomes an inequality-constrained problem. It is worth pointing out that in some formulations in optimization literature, equalities are not explicitly included, and only inequalities are included. This is because an equality can be written as two inequalities. For example $\phi(\boldsymbol{x}) = 0$ is equivalent to $\phi(\boldsymbol{x}) \leq 0$ and $\phi(\boldsymbol{x}) \geq 0$.

$$
\text{optimization} \left\{
\begin{array}{l}
\text{objective} \left\{ \begin{array}{l} \text{single objective} \\ \text{multiobjective} \end{array} \right. \\[2ex]
\text{constraint} \left\{ \begin{array}{l} \text{unconstrained} \\ \text{constrained} \end{array} \right. \\[2ex]
\text{landscape} \left\{ \begin{array}{l} \text{unimodal (convex)} \\ \text{mulimodal} \end{array} \right. \\
\text{function form} \left\{ \begin{array}{l} \text{linear} \\ \text{nonlinear} \left\{ \begin{array}{l} \text{quadratic} \\ \ldots \end{array} \right. \end{array} \right. \\
\text{variables/response} \left\{ \begin{array}{l} \text{discrete} \left\{ \begin{array}{l} \text{integer} \\ \ldots \end{array} \right. \\ \text{continuous} \\ \text{mixed} \end{array} \right. \\
\text{determinacy} \left\{ \begin{array}{l} \text{deterministic (all the above)} \\ \\ \text{stochastic} \end{array} \right.
\end{array}
\right.
$$

Figure 2.1: Classification of optimization problems.

We can also use the actual function forms for classification. The objective functions can be either linear or nonlinear. If the constraints ϕ_j and ψ_k are all linear, then it becomes a linearly constrained problem. If both the constraints and the objective functions are all linear, it becomes a linear programming problem. Here 'programming' has nothing to do with computing programming; it means planning and/or optimization. However, generally speaking, all f_i, ϕ_j and ψ_k are nonlinear, we have to deal with a nonlinear optimization problem.

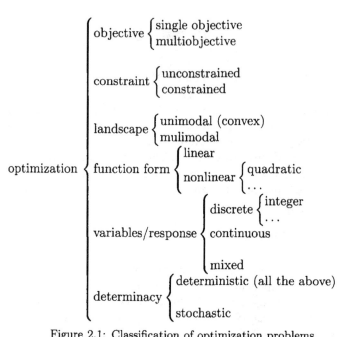

From a slightly different angle, we can classify optimization in terms of the landscape of the objective functions. For a single objective function, the shape or the landscape may vary significantly for the nonlinear case. If there is only a single valley or peak with a unique global optimum, then the optimization problem is said to be unimodal. In this case, the local optimum is the only optimum, which is also its global optimum. For example, $f(x, y) = x^2 + y^2$ is unimodal, as the local minimum at $(0, 0)$ is the global minimum (also see exercises at the end of this chapter). A special and very important class of unimodal optimization is the convex optimization, when the objective functions have certain convexity and the global optimal solution is guaranteed. In engineering optimization, we intend to design or reformulate the problem in terms of some quadratic matrix forms which are often convex, and thus their optimal solutions are the global best. However, most objective functions have more than one mode, and such multimodal functions are much more difficult to solve. For example, $f(x, y) = \sin(x)\sin(y)$ is multimodal.

The type of value of the design variables can also be used for classification. If the values of all design variables are discrete, then the optimization is called discrete optimization. If all these values are integers, then it becomes an integer programming problem. Loosely speaking, discrete optimization is also referred to as the combinatorial optimization, though some literature prefers to state that the discrete optimization consists of integer programming and combinatorial optimization. Combinatorial optimization is by far the most popular type of optimization and it is often linked with graph theory and routing, such as the traveling salesman problem, minimum spanning tree problem, vehicle routing, airline scheduling and knapsack problem. Furthermore, if all the design variables are continuous or taking real values in some interval, the optimization is referred to as a continuous optimization problem. However, the design variables in many problems can be both discrete and continuous, we then call such optimization the mixed type.

So far, we only consider the case when the values in both design variables and objective/constraints are exact, and there is no uncertainty or noise in their values. All the optimization problems are deterministic in the sense that for any given set of design variables, the values of both objective functions and the constraint functions are determined exactly. In the real world, we can only know some parameters to a certain extent with uncertainty. For example, when we design an engineering product, the material properties such as the Young's modulus can only be measured to a certain accuracy, and the inhomogeneity in material properties of real materials is often intrinsic. Thus, if there is any uncertainty and noise in the design variables and the objective functions and/or constraints, then the optimization becomes a stochastic optimization problem, or a robust optimization problem with noise. For most stochastic problems, we have to redefine or reformulate them in a way so that they becomes meaningful when using standard optimization techniques discussed in this book. This often involves the averaging over some space and the objective should be evaluated in terms of mean and/or in combination with their related uncertainties.

2.3 OPTIMIZATION ALGORITHMS

For different types of optimization problems, we often have to use different optimization techniques because some algorithms such as the Newton-Raphson method are more suitable for certain types of optimization than others.

Figuratively speaking, searching for the optimal solution is like treasure hunting. Imagine we are trying to hunt for a hidden treasure in a hilly landscape within a time limit. In one extreme, suppose we are blind-folded without any guidance, so the search process is essentially a pure random search, which is usually not efficient as we can expect. In another extreme, if we are told the treasure is placed at the highest peak of a known region, we will then directly climb up to the steepest cliff and try to reach to the highest peak, and this scenario corresponds to the classical hill-climbing techniques. In most cases, our search is between these extremes. We are not blind-folded, and we do not know where to look. It is a silly idea to search every single square inch of an extremely large hilly region so as to find the treasure. In most cases, we will do some random walking while looking for some hints; we look at some places almost randomly, then move to another plausible place, then another and so on. Such random walking is a main characteristic of modern search algorithms. Obviously, we can either do the treasure-hunting alone, so the whole path is a trajectory-based search, and simulated annealing is such a kind. Alternatively, we can ask a group of people to do the hunting and share the information (and any treasure found), and this scenario uses the so-called swarm intelligence and corresponds to the particle swarm optimization to be discussed in detail in Part II. If the treasure is really important and if the area is really extremely large, the search process will take a very long time. If there is no time limit and if any region is accessible (for example, no islands in a lake), it is theoretically possible to find the ultimate treasure (the global optimal solution).

Obviously, we can refine our search strategy a little bit further. Some hunters are better than others. We can only keep the better hunters and recruit new ones; this is similar to the genetic algorithms or evolutionary algorithms where the search agents are improving. In fact, as we will see in almost all modern metaheuristic algorithms, we try to use the best solutions or agents, and randomize (or replace) the not-so-good ones, while evaluating each individual's competence (fitness) in combination with the system history (use of memory). In such a balanced way, we intend to design better and more efficient optimization algorithms.

In general, optimization algorithms can be divided into two categories: deterministic algorithms, and stochastic algorithms (see Figure 2.2). Deterministic algorithms follow a rigorous procedure and its path and values of both design variables and the functions are repeatable. For example, hill-climbing is a deterministic algorithm, and for the same starting point, they will follow the same path whether you run the program today or tomorrow. On the other hand, the stochastic algorithms always have some randomness. Genetic algorithms are a good example, the strings or solutions in the population will be different each time you run a program since the algorithms use some pseudo-random numbers, though the final results may be no big difference, but the paths of each individual are not exactly repeatable. Furthermore, there is a third type of algorithm which is a mixture or hybrid of deterministic and stochastic algorithms. For example, hill-climbing with a random restart is a good example. The basic idea is to use the deterministic algorithm, but start with different initial points. This has certain advantages over a simple hill-climbing technique which may be stuck in a local peak. However, since

- algorithms
  - deterministic
    - linear programming
    - nonlinear programming
    - ...
    - gradient-based
    - gradient-free
  - stochastic
    - heuristic
    - metaheuristic
      - population-based
      - trajectory-based

Figure 2.2: Classification of algorithms.

there is a random component in this hybrid algorithm, we often classify it as a type of stochastic algorithm in the optimization literature.

Most conventional or classic algorithms are deterministic. For example, the Simplex method in linear programming is deterministic. Some deterministic optimization algorithms used the gradient information, they are called gradient-based algorithms. For example, the well-known Newton-Raphson algorithm is gradient-based as it uses the function values and their derivatives, and it works extremely well for smooth unimodal problems. However, if there is some discontinuity in the objective function, it does not work well. In this case, a non-gradient algorithm is preferred. Non-gradient-based or gradient-free algorithms do not use any derivative, but only the function values. Hooke-Jeeves pattern search and Nelder-Mead downhill simplex are examples of gradient-free algorithms.

For stochastic algorithms, we have in general two types: heuristic and metaheuristic, though their difference is small. Loosely speaking, *heuristic* means 'to find' or 'to discover by trial and error'. Quality solutions to a tough optimization problem can be found in a reasonable amount of time, but there is no guarantee that optimal solutions are reached. It is expected that these algorithms work most of the time, but not all the time. This is usually good enough when we do not necessarily want the best solutions but rather good solutions which are easily reachable.

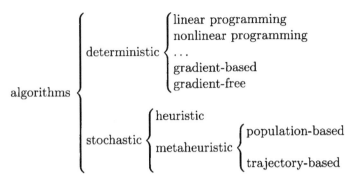

Further development over the heuristic algorithms is the so-called metaheuristic algorithms. Here *meta-* means 'beyond' or 'higher level', and they generally perform better than simple heuristics. In addition, all metaheuristic algorithms use certain tradeoff of randomization and local search. It is worth pointing out that no agreed definitions of heuristics and metaheuristics exist in literature, some use 'heuristics' and 'metaheuristics' interchangeably. However, recent trends tend to name all stochastic algorithms with randomization and local search as metaheuristic. Here we will also use this convention. Randomization provides a good way to move away from local search to the search on the global scale. Therefore, almost all metaheuristic algorithms intend to be suitable for global optimization.

2.4 METAHEURISTICS

Most metaheuristic algorithms are nature-inspired as they have been developed based on some abstraction of nature. Nature has evolved over millions of years and has found perfect solutions to almost all the problems she met. We can thus learn the success of problem-solving from nature and develop nature-inspired heuristic and/or metaheuristic algorithms. More specifically, some nature-inspired algorithms are inspired by Darwin's evolutionary theory. Consequently, they are said to be biology-inspired or simply bio-inspired.

Two major components of any metaheuristic algorithms are: selection of the best solutions and randomization. The selection of the best ensures that the solutions will converge to the optimality, while the randomization avoids the solutions being trapped at local optima and, at the same, increase the diversity of the solutions. The good combination of these two components will usually ensure that the global optimality is achievable.

Metaheuristic algorithms can be classified in many ways. One way is to classify them as: population-based and trajectory-based. For example, genetic algorithms are population-based as they use a set of strings, so is the particle swarm optimization (PSO) which uses multiple agents or particles. PSO is also referred to as agent-based algorithms.

On the other hand, simulated annealing uses a single agent or solution which moves through the design space or search space in a piecewise style. A better move or solution is always accepted, while a not-so-good move can be accepted with certain probability. The steps or moves trace a trajectory in the search space, with a non-zero probability that this trajectory can reach the global optimum.

We will discuss all major modern metaheuristic methods in Part II of this book, including genetic algorithms (GA), ant colony optimization (ACO), bee algorithms (BA), particle swarm optimization (PSO), simulated annealing (SA), harmony Search (HS), the firefly algorithm (FA) and others.

The efficiency of an algorithm is largely determined by the complexity of the algorithm. The algorithm complexity is often denoted by the order notation, which will be introduced below.

2.5 ORDER NOTATION

The efficiency of an algorithm is often measured by the algorithmic complexity or computational complexity. In literature, this complexity is also called Kolmogorov complexity. For a given problem size n, the complexity is denoted using Big-O notations such as $O(n^2)$ or $O(n \log n)$.

Loosely speaking, for two functions $f(x)$ and $g(x)$, if

$$\lim_{x \to x_0} \frac{f(x)}{g(x)} \to K, \tag{2.7}$$

where K is a finite, non-zero limit, we write

$$f = O(g). \tag{2.8}$$

The big O notation means that f is asymptotically equivalent to the order of $g(x)$. If the limit is unity or $K = 1$, we say $f(x)$ is order of $g(x)$. In this special case, we write

$$f \sim g, \tag{2.9}$$

which is equivalent to $f/g \to 1$ and $g/f \to 1$ as $x \to x_0$. Obviously, x_0 can be any value, including 0 and ∞. The notation $\sim$ does not necessarily mean $\approx$ in general, though they might give the same results, especially in the case when $x \to 0$ [for example, $\sin x \sim x$ and $\sin x \approx x$ if $x \to 0$]. When we say f is order of 100 (or $f \sim 100$), this does not mean $f \approx 100$, but it can mean that f is between about 50 to 150.

The small o notation is often used if the limit tends to 0. That is

$$\lim_{x \to x_0} \frac{f}{g} \to 0, \tag{2.10}$$

or

$$f = o(g). \tag{2.11}$$

If $g > 0$, $f = o(g)$ is equivalent to $f \ll g$. For example, for $\forall x \in \mathcal{R}$, we have $e^x \approx 1 + x + O(x^2) \approx 1 + x + \frac{x^2}{2} + o(x)$.

■ **EXAMPLE 2.1**

From the order definition, we can easily conclude that the addition of an $O(1)$ number does not change the order of an expression. That is, $n+1, n+5$ and $n+15$ are all $O(n)$. Similarly, an $O(1)$ factor has not much influence at all. That is, $0.7n, 1.5n, 2n+1$ and $3n+7$ are all $O(n)$.

As in computation, we always deal with integers $n \gg 1$, this means that $n \ll n^2$ as $n/n^2 \ll 1$. The term with the highest power of n dominates. Therefore, n^2+2n+3 is equivalent to $O(n^2)$, and $n^3+2n^2+5n+20$ is $O(n^3)$.

■ **EXAMPLE 2.2**

A classical example is Stirling's asymptotic series for factorials

$$n! \sim \sqrt{2\pi n}\,(\frac{n}{e})^n(1 + \frac{1}{12n} + \frac{1}{288n^2} - \frac{139}{51480n^3} - ...). \tag{2.12}$$

This is a good example of asymptotic series. For standard power expansions, the error $R_k(h^k) \to 0$, but for an asymptotic series, the error of the truncated series R_k decreases and gets smaller compared with the leading term [here $\sqrt{2\pi n}(n/e)^n$]. However, R_n does not necessarily tend to zero. In fact,

$$R_2 = \frac{1}{12n} \cdot \sqrt{2\pi n}(n/e)^n,$$

is still very large as $R_2 \to \infty$ if $n \gg 1$. For example, for $n = 100$, we have

$$n! = 9.3326 \times 10^{157},$$

while the leading approximation is $\sqrt{2\pi n}(n/e)^n = 9.3248 \times 10^{157}$. The difference between these two values is 7.7740×10^{154}, which is still very large, though three orders smaller than the leading approximation.

2.6 ALGORITHM COMPLEXITY

Let us come back to the computational complexity of an algorithm. For the sorting algorithm for a given number of n data entries, sorting these numbers into either ascending or descending order will take the computational time as a function of the problem size n. $O(n)$ means a linear complexity, while $O(n^2)$ has a quadratic complexity. That is, if n is doubled, then the time will double for linear complexity, but it will quadruple for quadratic complexity.

■ **EXAMPLE 2.3**

For example, the bubble sorting algorithm starts at the beginning of the data set by comparing the first two elements. If the first is smaller than the second, then swap them. This comparison and swap process continues for each possible pair of adjacent elements. There are $n \times n$ pairs as we need two loops over the whole data set, then the algorithm complexity is $O(n^2)$. On the other hand, the quicksort algorithm uses a divide-and-conquer approach via partition. By first choosing a pivot element, we then put all the elements into two sublists with all the smaller elements before the pivot and all the greater elements after it. Then, the sublists are recursively sorted in a similar manner. This algorithm will result in a complexity of $O(n \log n)$. The quicksort is much more efficient than the bubble algorithm. For $n = 1000$, then the bubble algorithm will need about $O(n^2) \approx O(10^6)$ calculations, while the quicksort only requires $O(n \log n) \approx O(3 \times 10^3)$ calculations (at least two orders less).

From the viewpoint of computational effort associated with the optimization problem, we can classify them as NP and polynomial complexity. A solution to an optimization can be verifiable in polynomial time in a deterministic manner, often in terms of a determinist Turing machine, then we say the optimization is solvable in computational time or polynomial (P) time. This kind of problem has P-complexity. For example, Agrawal *et al.* showed in 2002 whether a number n is prime or not can be determined in a polynomial time. In this case, we also say that such a procedure or algorithm has P-complexity. If an algorithm has P-complexity, then we say it is efficient. On the other hand, a problem is not solvable in polynomial time, we use the 'non-deterministic polynomial (NP) time' to refer to this case. Subsequently, the problem has the NP-complexity. However, a solution can still be verified in a polynomial time. The main difficulty is to find such a solution in the first place in an efficient way. Finding such a solution in the first place may require more than NP time.

In mathematical programming, an easy or tractable problem is a problem whose solution can be obtained by computer algorithms with a solution time (or number of steps) as a polynomial function of problem size n. Algorithms with polynomial-time are considered efficient. A problem is called the P-problem or polynomial-time problem if the number of steps needed to find the solution is bounded by a polynomial in n and it has at least one algorithm to solve it.

On the other hand, a hard or intractable problem requires solution time that is an exponential function of n, and thus exponential-time algorithms are considered inefficient. A problem is called nondeterministic polynomial (NP) if its solution can only be guessed and evaluated in polynomial time, and there is no known rule to make such guess (hence, nondeterministic). Consequently, guessed solutions cannot guarantee to be optimal or even near optimal. In fact, no known efficient algorithms exist to solve NP-hard problems, and only approximate solutions or heuristic solutions are possible. Thus, heuristic and metaheuristic methods are very promising in obtaining approximate solutions or nearly optimal/suboptimal solutions.

A problem is called NP-complete if it is an NP-hard problem and all other problems in NP are reducible to it via certain reduction algorithms. The reduction algorithm has a polynomial time. An example of NP-hard problem is the Traveling Salesman Problem, and its objective is to find the shortest route or minimum traveling cost to visit all given n cities exactly once and then return to the starting city.

The solvability of NP-complete problems (whether by polynomial time or not) is still an unsolved problem for which Clay Mathematical Institute is offering a million dollars reward for a formal proof. Most real-world problem are NP-hard, and thus any advances in dealing with NP problems will have potential impact on many applications.

2.7 NO FREE LUNCH THEOREMS

The methods used to solve a particular problem depend largely on the type and characteristics of the optimization problem itself. There is no universal method that works for all problems, and there is generally no guarantee to finding the optimal solution in global optimization problems. In fact, there are several so-called Wolpert and Macready's 'No Free Lunch Theorems' (NLF theorems) which state that if any algorithm A outperforms another algorithm B in the search for an extremum of a cost function, then algorithm B will outperform A over other cost functions. NFL theorems apply to the scenario (either deterministic or stochastic) where a set of continuous (or discrete or mixed) parameter θ maps the cost functions into a finite set.

Let n_θ be the number of values of θ (either due to discrete values or the finite machine precisions), and n_f be the number of values of the cost function. Then, the number of all possible combinations of cost functions is $N = n_f^{n_\theta}$ which is finite, but usually huge. The NFL theorems prove that the average performance over all possible cost functions is the same for all search algorithms. Mathematically, if $P(s_m^y|f, m, A)$ denotes the performance, based on the probability theory, of an algorithm A iterated m times on a cost function f over the sample s_m, then we have the averaged performance for two algorithms

$$\sum_f P(s_m^y|f, m, A) = \sum_f P(s_m^y|f, m, B), \tag{2.13}$$

where $s_m = \{(s_m^x(1), s_m^y(1)), ..., (s_m^x(m), s_m^y(m))\}$ is a time-ordered set of m distinct visited points with a sample of size m. The interesting thing is that the performance is independent of algorithm A itself. That is to say, all algorithms for optimization will give the same performance when averaged over *all possible* functions. This means that the universally best method does not exist.

Well, you might say, there is no need to formulate new algorithms because all algorithms will perform equally well. The truth is that the performance is measured in the statistical sense and over *all possible* functions. This does not mean all algorithms perform equally well over some *specific* functions. The reality is that no optimization problems require averaged performance over all possible functions.

Even though, the NFL theorems are valid mathematically, their influence on parameter search and optimization is limited. For any specific set of functions, some algorithms do perform much better than others. In fact, for any specific problem with specific functions, there usually exist some algorithms that are more efficient than others if we do not need to measure their *average* performance. The main problem is probably how to find these better algorithms for a given particular type of problem.

On the other hand, we have to emphasize on the best estimate or sub-optimal solutions under the given conditions if the best optimality is not achievable. The knowledge about the particular problem of interest is always helpful for the appropriate choice of the best or most efficient methods for the optimization procedure.

It is worth pointing out that the NFL theorems have been proved for the single-objective optimization problems. For multiobjective optimization problems, the theorems still remain unproved.

EXERCISES

2.1 Find the global minimum of $f(x, y) = x^2 + y^2$. What is the type of optimization?

2.2 State the type of the following optimization problems:

a) $\max \exp(-x^2)$ subject to $-2 \le x \le 5$;
b) $\min x^2 + y^2, (x-2)^2, (y-3)^3$;
c) $\max 2x + 3y$ subject to $x > 0, 0 < y \le 10, y \ge x$;
d) $\min x^2 y^6$ subject to $x^2 + y^2/4 = 1$;
e) Design a quieter, more efficient, low cost car engine;
f) Design a software package to predict the stock market.

2.3 Work out the algorithm complexity of the following algorithms:

a) Addition of n real numbers;
b) Multiplication of two $n \times n$ square matrices;
c) Find the maximum value among n random numbers;
d) Evaluations of a polynomial of degree n;
e) Sorting n numbers in increasing/decreasing order by simple pair-wise comparison;

2.4 Simplify the following order expression to the closest orders.

a) $0.9n + 15$;
b) $5n^2 - 5$;
c) $1.5n^3 + 2n^2 + 10n$;
d) $n^5 + 100n^3 + 200$;
e) $n \log(n) + \log(2n)$;

2.5 Find the global maximum for linear optimization $f(x, y) = x + y$ subject to $0 \le x \le 5$, $0 \le y \le 2$ and $2x + y = 8$.

REFERENCES

1. S. Arora and B. Barak, *Computational Complexity: A Modern Approach*, Cambridge University Press, 2009.
2. M. Agrawal, N. Kayal, and N. Saxena, "Primes is in P", *Ann. Mathematics*, **160** (2), 781-793 (2002).
3. D. H. Wolpert and W. G. Macready, "No free lunch theorems for optimization", *IEEE Transaction on Evolutionary Computation*, **1**, 67-82 (1997).
4. X. S. Yang, *Nature-Inspired Metaheuristic Algorithms*, Luniver Press, 2008.

5. X. S. Yang, *Introduction to Computational Mathematics*, World Scientific, 2008.

CHAPTER 3

MATHEMATICAL FOUNDATIONS

As this book is mainly about the introduction of engineering optimization and metaheuristic algorithms, we will assume that the readers have some basic knowledge of basic calculus and vectors. In this chapter, we will briefly review some relevant concepts that are frequently used in optimization.

3.1 UPPER AND LOWER BOUNDS

For a given non-empty set $\mathcal{S} \in \Re$ of real numbers, we now introduce some important concepts such as the supremum and infimum. A number U is called an upper bound for $\mathcal{S}$ if $x \leq U$ for all $x \in \mathcal{S}$. An upper bound β is said to be the least (or smallest) upper bound for $\mathcal{S}$, or the supremum, if $\beta \leq U$ for any upper bound U. This is often written as

$$\beta \equiv \sup_{x \in \mathcal{S}} x \equiv \sup \mathcal{S} \equiv \sup(\mathcal{S}). \tag{3.1}$$

All such notations are widely used in the literature of mathematical analysis.

On the other hand, a number L is called a lower bound for $\mathcal{S}$ if $x \geq L$ for all x in $\mathcal{S}$ (that is $\forall x \in \mathcal{S}$). A lower bound α is referred to as the greatest (or

Engineering Optimization: An Introduction with Metaheuristic Applications.
By Xin-She Yang

Copyright © 2010 John Wiley & Sons, Inc.

largest) lower bound if $\alpha \geq L$ for any lower bound L, which is written as

$$\alpha \equiv \sup_{x \in \mathcal{S}} x \equiv \inf \mathcal{S} \equiv \inf(\mathcal{S}). \tag{3.2}$$

In general, both the supremum β and the infimum α, if they exist, may or may not belong to $\mathcal{S}$.

For example, any numbers greater than 5, say, 7.2 and 500 are an upper bound for the interval $-2 \leq x \leq 5$ or $[-2, 5]$. However, its smallest upper bound (or sup) is 5. Similarly, numbers such as -10 and -10^5 are lower bound of the interval, but -2 is the greatest lower bound (or inf). In addition, the interval $\mathcal{S} = [15, \infty)$ has an infimum of 15 but it has no upper bound. That is to say, its supremum does not exist, or $\sup \mathcal{S} \to \infty$.

There is an important completeness axiom which says that if a non-empty set $\mathcal{S} \in \Re$ of real numbers is bounded above, then it has a supremum. Similarly, if a non-empty set of real numbers is bounded below, then it has an infimum.

Both suprema and infima have the following properties:

$$\inf(\mathcal{Q}) = -\sup(-\mathcal{Q}), \tag{3.3}$$

$$\sup_{p \in \mathcal{P}, q \in \mathcal{Q}} (p + q) = \sup(\mathcal{P}) + \sup(\mathcal{Q}), \tag{3.4}$$

$$\sup_{x \in \mathcal{S}}(f(x) + g(x)) \leq \sup_{x \in \mathcal{S}} f(x) + \sup_{x \in \mathcal{S}} g(x). \tag{3.5}$$

Since the suprema or infima may not exist in the unbounded cases, it is necessary to extend the real numbers $\Re$ or

$$-\infty < r < \infty, \tag{3.6}$$

to the affinely extended real numbers $\bar{\Re}$

$$\bar{\Re} = \Re \cup \{\pm\infty\}. \tag{3.7}$$

If we further define the $\sup(\emptyset) = -\infty$, for any sets of the affinely extended real numbers, the suprema and infima always exist if we let $\sup \Re = \infty$ and $\inf \Re = -\infty$.

Furthermore, the maximum for $\mathcal{S}$ is the largest value of all elements $s \in \mathcal{S}$, and often written as $\max(\mathcal{S})$ or $\max \mathcal{S}$, while the minimum, $\min(\mathcal{S})$ or $\min \mathcal{S}$, is the smallest value among all $s \in \mathcal{S}$. For the same interval $[-2, 5]$, the maximum of this interval is 5 which is equal to its supremum, while its minimum 5 is also equal to its infimum. Though the supremum and infimum are not necessarily part of the set $\mathcal{S}$; however, the maximum and minimum (if they exist) always belong to the set.

However, the concepts of supremum (or infimum) and maximum (or minimum) are not the same, and maximum/minimum may not always exist.

■ **EXAMPLE 3.1**

For example, the interval $\mathcal{S} = [-2, 7)$ or $-2 \leq x < 7$ has the supremum of $\sup \mathcal{S} = 7$, but $\mathcal{S}$ has no maximum. If any value $a \in \mathcal{S}$ is a maximum, then

$$a < b = \frac{a+7}{2} \in \mathcal{S}, \tag{3.8}$$

because b is the midpoint of a and 7. However, 7 is not part of $\mathcal{S}$. Therefore, the set $\mathcal{S}$ does not have a maximum. In addition, the set

$$\sup\{x \in \Re : x^7 < \pi\} = \sqrt[7]{\pi}.$$

Similarly, the interval $(-10, 15]$ does not have a minimum, though its infimum is -10. Furthermore, the open interval $(-2, 7)$ has no maximum or minimum; however, its supremum is 7, and infimum is -2.

The maximum or minimum may not exist at all; however, we will assume they will always exist in many applications, especially in the discussion of optimization problems in this chapter. Now the objective becomes how to find the maximum or minimum in various optimization problems.

3.2 BASIC CALCULUS

For a univariate function $f(x)$, its gradient or first derivative at any point P is defined as

$$\frac{df(x)}{dx} \equiv f'(x) = \lim_{\Delta x \to 0} \frac{f(x + \Delta x) - f(x)}{\Delta x}, \tag{3.9}$$

on the condition that such a limit exists. If the limit does not exist, we say that the function is not differentiable at this point P. For example, the simple absolute function $|x|$ is not differentiable at $x = 0$. Similarly, the differentiation of the gradient is the second derivative

$$\frac{d^2 f(x)}{dx^2} \equiv f''(x). \tag{3.10}$$

Conventionally, the standard notation $\frac{df}{dx}$ is called Leibnitz's notation, while the prime notation is called Lagrange's notation. Newton's dot notation $\dot{f} = df/dt$ is now exclusively used for time derivative. We will use the prime notation $'$ and the standard notation df/dx interchangeably for convenience. Higher derivatives can be defined in a similar manner.

For differentiation, there are three important rules: the product rule, quotient rule, and chain rule. For a combined function $f(x) = u(x)v(x)$, we have the product rule

$$f'(x) = [u(x)v(x)]' = u'(x)v(x) + u(x)v'(x). \tag{3.11}$$

Similarly, if we replace $v(x)$ by $1/v(x)$ and using $(1/v)' = -v'/v^2$, we have the quotient rule

$$(\frac{u}{v})' = \frac{u'v - uv'}{v^2}. \tag{3.12}$$

■ EXAMPLE 3.2

We can now use the above rules to carry out the following differentiation:

$$[\sin(x)\exp(x)]' = \sin'(x)\exp(x) + \sin(x)[\exp(x)]'$$

$$= \cos(x)\exp(x) + \sin(x)\exp(x).$$

$$\left[\frac{\sin(x)}{x}\right]' = \frac{\sin'(x) \cdot x - \sin(x) \cdot (x)'}{x^2} = \frac{x\cos x - \sin(x)}{x^2}.$$

In case of a function $f(x)$ can be written as a function of a simpler function $f[u(x)]$, we have the chain rule

$$\frac{df(x)}{dx} = \frac{df}{du} \cdot \frac{du}{dx}. \tag{3.13}$$

■ EXAMPLE 3.3

Let us look at some simple examples.

$$[\sin(x^2)]' = \cos(x^2) \cdot 2x = 2x\cos(x^2).$$

$$\{\exp[-\sin(x^2)]\}' = \exp[-\sin(x^2)] \cdot [-\cos(x^2)] \cdot 2x$$

$$= -2x\cos(x^2)\exp[-\sin(x^2)].$$

$$[\cos(x)\exp(-x^2)]' = -\sin(x)\exp(-x^2) + \cos(x)\exp(-x^2) \cdot (-2x)$$

$$= -\sin(x)\exp(-x^2) - 2x\cos(x)\exp(-x^2).$$

For functions with more than one independent variable, we have to use partial derivatives. For a bivariate function $f(x, y)$, the partial derivatives are defined as

$$\frac{\partial f(x,y)}{\partial x} \equiv f_x = \lim_{\Delta x \to 0, y=\text{fixed}} \frac{f(x + \Delta x, y) - f(x, y)}{\Delta x}, \tag{3.14}$$

and

$$\frac{\partial f(x,y)}{\partial y} \equiv f_y = \lim_{\Delta y \to 0, x=\text{fixed}} \frac{f(x, y + \Delta y) - f(x, y)}{\Delta y}, \tag{3.15}$$

providing that these limits exist. This is essentially equivalent to carrying out the standard differentiation for a univariate function while assuming other variables remaining constant. The total infinitesimal change df of a function due to the infinitesimal changes dx and dy can be written as

$$df = \frac{\partial f}{\partial x}dx + \frac{\partial f}{\partial y}dy. \tag{3.16}$$

Higher partial derivatives can be defined in a similar manner, and we have

$$\frac{\partial^2 f}{\partial x^2} = \frac{\partial}{\partial x}(\frac{\partial f}{\partial x}), \quad \frac{\partial^2 f}{\partial y^2} = \frac{\partial}{\partial y}(\frac{\partial f}{\partial y}), \quad \frac{\partial^2 f}{\partial x \partial y} = \frac{\partial}{\partial x}(\frac{\partial f}{\partial y}). \tag{3.17}$$

Since $\Delta x \Delta y = \Delta y \Delta x$, from the basic definition, we have

$$\frac{\partial^2 f}{\partial x \partial y} = \frac{\partial^2 f}{\partial y \partial x}. \tag{3.18}$$

The differentiation rules still apply for partial derivatives.

For multivariate functions with more than two independent variables, partial derivatives can be defined similarly.

■ EXAMPLE 3.4

The partial derivatives of $f(x, y) = x^2 + xy + \sin(y^2)$ are

$$\frac{\partial f}{\partial x} = 2x + y, \quad \frac{\partial f}{\partial y} = x + \cos(y^2) \cdot 2y = x + 2y\cos(y^2),$$

$$\frac{\partial^2 f}{\partial x^2} = 2, \quad \frac{\partial^2 f}{\partial x \partial y} = \frac{\partial}{\partial x}[x + 2y\cos(y^2)] = 1,$$

$$\frac{\partial^2 f}{\partial y \partial x} = \frac{\partial}{\partial y}(\frac{\partial f}{\partial x}) = \frac{\partial}{\partial y}(2x + y) = 1,$$

$$\frac{\partial^2 f}{\partial y^2} = 2\cos(y^2) + 2y \cdot [-\sin(y^2)] \cdot (2y) = 2\cos(y^2) - 4y^2\sin(y^2).$$

We can see that $\frac{\partial^2 f}{\partial x \partial y} = \frac{\partial^2 f}{\partial y \partial x}$ is true indeed.

Integration is basically a reverse process of differentiation, though it is much trickier. For example, from $(x^n)' = nx^{n-1}$, we have

$$\int x^n dx = \frac{1}{n+1}x^{n+1} + C, \tag{3.19}$$

where C is an integration constant. This comes from the fact that the gradient of a function remains the same if a constant is added. In the special case $n = -1$ for the integrand x^n, this has to change to

$$\int x^{-1} dx = \ln x + C. \tag{3.20}$$

When carrying out the differentiation of a definite integral with variable limits and the bivariate integrand, care should be taken. For example, we have

$$\frac{d}{dx}\int_{a(x)}^{b(x)} f(x,y)dy = \Big[f(x,b)\frac{db}{dx} - f(x,a)\frac{da}{dx}\Big] + \int_{a(x)}^{b(x)} \frac{\partial f(x,y)}{\partial x}dy. \quad (3.21)$$

The Jacobian is an important concept concerning the change of variables and transformation. For a simple integral, the change of variables from x to a new variables u so that $x = x(u)$, we have $dx = (dx/du)du$, which leads to

$$\int_{x_a}^{x_b} f(x)dx = \int_a^b f[x(u)]\frac{dx}{du}du. \quad (3.22)$$

This factor dx/du is in fact the Jacobian. For the change of variables in bivariate cases $x = x(\xi,\eta)$ and $y = y(\xi,\eta)$, we have the Jacobian J defined by

$$J \equiv \frac{\partial(x,y)}{\partial(\xi,\eta)} = \begin{vmatrix} \frac{\partial x}{\partial \xi} & \frac{\partial y}{\partial \xi} \\ \frac{\partial x}{\partial \eta} & \frac{\partial y}{\partial \eta} \end{vmatrix}, \quad (3.23)$$

which is always a scalar or a simple value. The notation $\partial(x,y)/\partial(\xi,\eta)$ is just a useful shorthand commonly used in literature.

■ EXAMPLE 3.5

As an example, let us calculate the derivative of $\int_x^{x^2}\sin(xy)dy$ with respect to x. We have

$$\frac{d}{dx}\int_x^{x^2}\sin(xy)dy = \Big[\sin(x\cdot x^2)\cdot(2x) - \sin(x\cdot x)\cdot 1\Big] + \int_x^{x^2}\cos(xy)\cdot y dy$$

$$= [2x\sin(x^3) - \sin(x^2)] + \int_x^{x^2} y\cos(xy)dy.$$

To transform from a 2D Cartesian system (x, y) to polar coordinates (r, θ), we have

$$x = r\cos\theta, \qquad y = r\sin\theta.$$

Therefore, the Jacobian is

$$J = \frac{\partial(x,y)}{\partial(r,\theta)} = \begin{vmatrix} \frac{\partial x}{\partial r} & \frac{\partial y}{\partial r} \\ \frac{\partial x}{\partial \theta} & \frac{\partial y}{\partial \theta} \end{vmatrix} = \frac{\partial x}{\partial r}\frac{\partial y}{\partial \theta} - \frac{\partial x}{\partial \theta}\frac{\partial y}{\partial r}$$

$$= \cos\theta\cdot r\cos\theta - (-r\sin\theta)\cdot\sin\theta = r(\cos^2\theta + \sin^2\theta) = r.$$

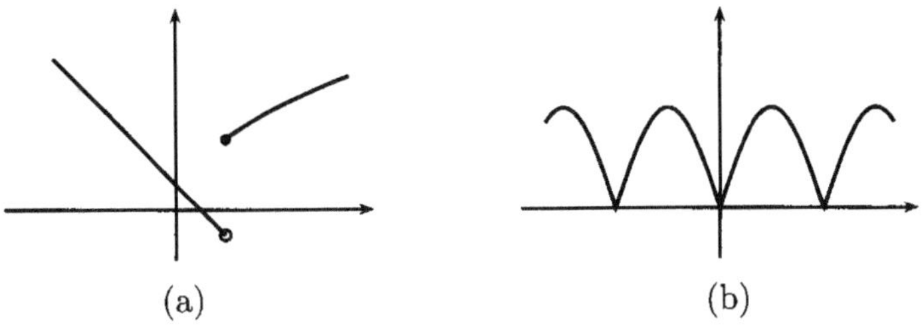

Figure 3.1: (a) A jump discontinuity at x_0, but piecewise continuous where the solid point means the point is included, while a circle is excluded. (b) $|\sin(x)|$ is piecewise smooth (but not differentiable at $x = 0, \pm\pi, \pm 2\pi, ...$).

3.3 OPTIMALITY

3.3.1 Continuity and Smoothness

In mathematical programming, there are many important concepts, and we will only review the most relevant ones in this book. Let us first introduce the concepts of smoothness, singularity, stationary condition, inflection point and critical points.

A function $f(x)$ is said to be continuous if a small change in x will result in a small change in $f(x)$. Alternatively, the value of the function is always finite with the continuity condition for any $\epsilon > 0$ at $x = a$

$$\lim_{\epsilon \to 0} |f(a+\epsilon) - f(a-\epsilon)| \to 0. \tag{3.24}$$

If we use the notation $f_\pm$ for $f(a \pm \epsilon)$, we have $f_+ = f_-$ at point $x = a$. In mathematics, we often use C^0 to denote the set of all continuous functions. A function that does not satisfy the above continuous condition is called discontinuous. When $|f_+ - f_-| \to k > 0$ is a constant, then the discontinuity is called a jump discontinuity (see Figure 3.1). If the function in an interval $[a, b]$ can be divided into a finite number smaller intervals, and the function is continuous in each of these intervals, even with a finite number of jump discontinuity at the end points of the subintervals, we call it piecewise continuous.

A function f is said to be of class C^n if its derivatives up to n-th derivatives $(f', f'', ..., f^{(n)})$ all exist and continuous. It is also called n-continuously differentiably. So C^1 is the set of all differentiable functions whose first derivatives are continuous. In other words, a class C^1 function has a derivative of C^0.

A function is said to be smooth if it has derivatives of all orders. That is to say, a smooth function belongs to class C^∞. For example, $\exp(x)$ is a smooth function because its derivatives of all orders exist. Similarly, x^2 is also smooth as it has derivatives of all orders, though the third and higher derivatives are all zero. However, not all functions have derivatives of all orders. So in opti-

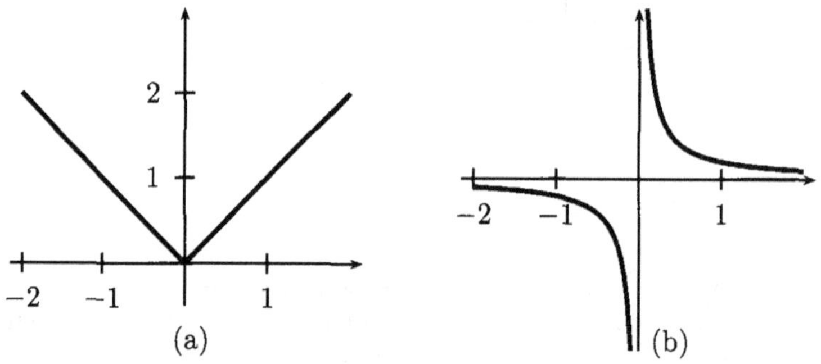

Figure 3.2: (a) $|x|$ is not differentiable at $x = 0$, (b) $1/x$ has a singular point at $x = 0$.

mization, we often define a smooth function that has continuous derivatives up to some desired order over certain domains of interest. For this reason, we can loosely refer C^n functions as n-order smooth, though this is not a rigorous mathematical term. A curve $f(x)$ in an interval $[a, b]$ is called piecewise smooth if it can be partitioned into a finite number of smaller 'pieces' or intervals, continuity holds cross the joins of the pieces and smoothness holds in each interval. In each interval, the function has a bounded first derivative that is continuous except at a finite number of points at the joints.

A singularity occurs if a function $f(x)$ blows up or diverges or becomes unbounded. For example $1/x$ is singular at the point $x = 0$. We also call this singularity as a singular point (see Figure 3.2). In addition, as in each subdomain $(-\infty, 0)$ and $(0, \infty)$, $|x|$ is differentiable, and mathematically we call it piecewise differentiable.

3.3.2 Stationary Points

A stationary point of $f(x)$ is a point x at which the stationary condition holds

$$f'(x) = 0. \tag{3.25}$$

We know that the first derivative is the gradient or the rate of change of the function. When the rate of change $f'(x)$ is zero, we call it stationary. Consequently, such points are called stationary points. A smooth function can only achieve its maxima or minima at stationary points (and/or end points of an interval), so the stationary condition is a necessary condition for maxima or minima. However, being stationary is not sufficient to be maximal or minimal. In this case, a further condition is needed. That is, if $f''(x) > 0$, the function $f(x)$ achieves a local minimum at the stationary point. If $f''(x) < 0$, the stationary point corresponds to a local maximum.

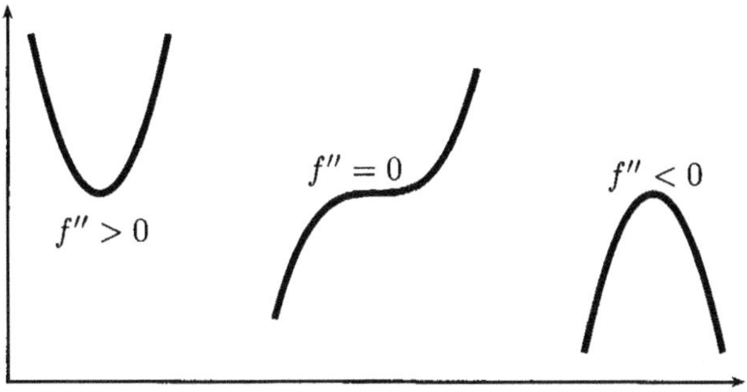

Figure 3.3: The sign of the second derivative at a stationary point. (a) $f''(x) > 0$, (b) $f''(x) = 0$, and (c) $f''(x) < 0$.

■ **EXAMPLE 3.6**

For a simple function $f(x) = x^2$, the stationary condition $f'(x) = 2x = 0$ leads to $x = 0$ which is a stationary point. Since $f''(x) = 2 > 0$, $f(0) = 0$ is a local minimum. In fact, it is the only minimum in the whole domain $(-\infty, \infty)$, and thus it is also the global minimum.

For $f(x) = xe^{-x}$ in the domain $[0, \infty)$, its stationary condition becomes

$$f'(x) = e^{-x} - xe^{-x} = (1 - x)e^{-x} = 0.$$

Since $\exp(-x) > 0$, we have $x = 1$ as the stationary point. In addition, from $f''(x) = (x - 2)e^{-x}$, we know that $f''(1) = -e^{-1} < 0$. So $f_{\max} = e^{-1}$ at $x = 1$ reaches the maximum.

In the special case of $f''(x) = 0$, we have to use other methods such as the sign changes on both sides. The condition $f''(x) = 0$ determines a point of inflection, however, $f'(x)$ is not necessarily zero. For example, $f(x) = x^3$, the point $x = 0$ is an point of inflection, not an extremum (see Figure 3.3).

A point of a real function $f(x)$ is called a critical point if its first derivative is zero or if it is not differentiable (see Figure 3.2). For example, $|x|$ at $x = 0$ is not differentiable, and thus it is a critical point. So a stationary point is a critical point, but a critical point may not be stationary. That is, critical points consists of stationary points and points at which the function is not differentiable (including singular points). In addition, a maximum or minimum can only occur at a critical point. For example, from Figure 3.2, we know that the global minimum of $f(x) = |x|$ is at $x = 0$, but this point is not stationary. However, $x = 0$ is a critical point as $|x|$ is not differentiable.

The stationary point of $f(x) = \sin(x)$ is simply determined by

$$f'(x) = \cos(x) = 0,$$

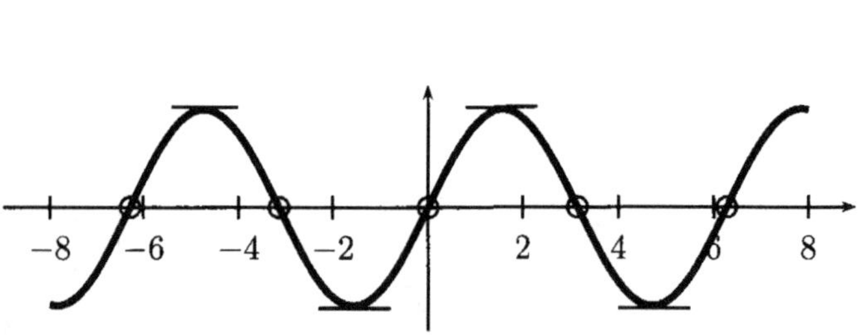

Figure 3.4: Sine function $\sin(x)$ and its stationary points (marked with – and points of inflection (marked with $\circ$).

which leads to $x = \pm\pi/2, \pm 3\pi/2, ...$ which are marked with '–' in Figure 3.4. The points of inflection are determined by

$$f''(x) = -\sin(x) = 0.$$

That is $x = 0, \pm\pi, \pm 2\pi, ...$ which are marked with $\circ$ in the same figure.

3.3.3 Optimality Criteria

Now we can introduce more related concepts: feasible solutions, the strong local maximum and weak local maximum.

A point $\boldsymbol{x}$ that satisfies all the constraints is called a feasible point and thus is a feasible solution to the problem. The set of all feasible points is called the feasible region. A point $\boldsymbol{x}_*$ is called a strong local maximum of the nonlinearly constrained optimization problem if $f(\boldsymbol{x})$ is defined in a δ-neigbourhood $N(\boldsymbol{x}_*, \delta)$ and satisfies $f(\boldsymbol{x}_*) > f(\boldsymbol{u})$ for $\forall \boldsymbol{u} \in N(\boldsymbol{x}_*, \delta)$ where $\delta > 0$ and $\boldsymbol{u} \neq \boldsymbol{x}_*$. If $\boldsymbol{x}_*$ is not a strong local maximum, the inclusion of equality in the condition $f(\boldsymbol{x}_*) \geq f(\boldsymbol{u})$ for $\forall \boldsymbol{u} \in N(\boldsymbol{x}_*, \delta)$ defines the point $\boldsymbol{x}_*$ as a weak local maximum (see Figure 3.5). The local minima can be defined in the similar manner when $>$ and $\geq$ are replaced by $<$ and $\leq$, respectively.

Figure 3.5 shows various local maxima and minima. Segment A is a weak local maximum, while point B is a local maximum with a jump discontinuity. C and D are the local minimum and maximum, respectively. Region E has a weak local minimum because there are many (well, infinite) different values of x which will lead to the same value of $f(\boldsymbol{x}_*)$. Point F is the global maximum which is also a strong maximum. Finally, G is the strong global minimum. However, at point B, there exists a discontinuity, and $f'(x)$ is not defined unless we are only concerned with the special case of approaching from the right. We will not deal with this type of minima or maxima in detail. In our present discussion, we will assume that both $f(\boldsymbol{x})$ and $f'(\boldsymbol{x})$ are always continuous or $f(\boldsymbol{x})$ is everywhere twice-continuously differentiable. In a more general case, we will assume that all objective functions belong to class C^0.

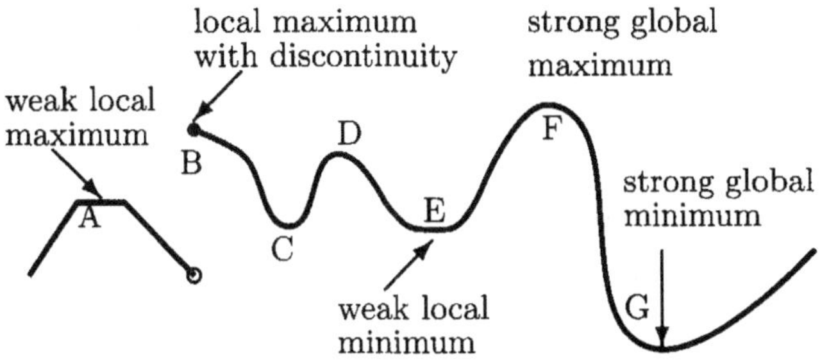

Figure 3.5: Strong and weak maxima and minima. A is a weak local maximum; point B is a local maximum with discontinuity; C and D are the minimum and maximum, respectively. E is a weak local minimum. Point F corresponds to a strong maximum and also the global maximum, while point G is the strong global minimum.

■ EXAMPLE 3.7

The minimum of $f(x) = x^2$ at $x = 0$ is a strong local minimum. The minimum of $g(x, y) = (x - y)^2$ at $x = y = 0$ is a weak local minimum because $g(x, y) = 0$ along the line $x = y$ so that $g(x, y = x) = 0 = g(0, 0)$.

Almost all optimization problems can be written as a generic form discussed in the introduction and earlier chapters. The main notation is often

$$\underset{\text{(design variables/space)}}{\text{maximize or minimize}} \quad \text{objective functions.} \tag{3.26}$$

However, there are other notations such as

$$\min_{x \in \Re} x^4 + (x-2)^2. \tag{3.27}$$

It is worth pointing out that other notation using argmin or argmax may also be used in literature. Mathematically, the argmin or arg min such as

$$\text{argmin }_x f(x), \tag{3.28}$$

means to find the value of x for which $f(x)$ is the minimum. For example,

$$\text{argmin}_{x \in \Re} \; x^2 = 0. \tag{3.29}$$

The argmax or arg max can be defined in a similar manner.

The notation argmin or argmax focuses on the values of independent or design variables. Obviously, the aim of optimization is to find the optimal solution. Once we know that values of the design variables, it is straightforward to calculate the objectives. Therefore, essentially two types of notation are the same. Conventionally, the optimization notation using 'min' or 'max' is just a shorthand, and it emphasize the fact that we intend to minimize or maximize certain objectives or targets.

In general, the maximum or minimum of a function can only occur at the following places:

- A stationary point determined by $f'(x) = 0$;
- A critical point when the function is not differentiable;
- At the boundaries.

3.4 VECTOR AND MATRIX NORMS

For a vector with n components (or simply an n-vector)

$$\boldsymbol{v} = \begin{pmatrix} v_1 & v_2 & \dots & v_n \end{pmatrix}^T = \begin{pmatrix} v_1 \\ v_2 \\ \vdots \\ v_n \end{pmatrix}, \tag{3.30}$$

its p-norm or ℓ_p-norm is denoted by $\|\boldsymbol{v}\|_p$ and defined as

$$\|\boldsymbol{v}\|_p = \Big(\sum_{i=1}^{n} |v_i|^p\Big)^{1/p}, \tag{3.31}$$

where p is a positive integer. From this definition, it is straightforward to show that the p-norm satisfies the following conditions: $\|\boldsymbol{v}\| \geq 0$ for all $\boldsymbol{v}$, and $\|\boldsymbol{v}\| = 0$ if and only if $\boldsymbol{v} = \mathbf{0}$. This is the non-negativeness condition. In addition, for any real number α, we have the scaling condition: $\|\alpha\boldsymbol{v}\| = \alpha\|\boldsymbol{v}\|$.

The three most common norms are one-, two- and infinity-norms (or ℓ_1, ℓ_2, ℓ_∞) when $p = 1, 2$, and ∞, respectively. For $p = 1$, the one-norm is just the simple sum of the absolute value of each component $|v_i|$, while the ℓ_2-norm $\|\boldsymbol{v}\|_2$ for $p = 2$ is the standard Euclidean norm because $\|\boldsymbol{v}\|_2$ is the length of the vector $\boldsymbol{v}$

$$\|\boldsymbol{v}\|_2 = \sqrt{\boldsymbol{v} \cdot \boldsymbol{v}} = \sqrt{v_1^2 + v_2^2 + \dots + v_n^2}, \tag{3.32}$$

where the notation $\boldsymbol{u} \cdot \boldsymbol{v}$ is the inner product of two vectors $\boldsymbol{u}$ and $\boldsymbol{v}$.

The inner product of two vectors is defined as

$$\boldsymbol{u} \cdot \boldsymbol{v} \equiv <\boldsymbol{u}, \boldsymbol{v}> \equiv \boldsymbol{u}^T\boldsymbol{v} = \sum_{i=1}^{n} u_i v_i. \tag{3.33}$$

Therefore, the Euclidean norm or ℓ_2-norm of $\boldsymbol{v}$ is the square root of its inner product. That is

$$\|\boldsymbol{v}\|_2 = \sqrt{\boldsymbol{v}^T\boldsymbol{v}} = (v_1^2 + \dots + v_n^2)^{1/2}. \tag{3.34}$$

The norms and inner product satisfy the following Cauchy-Schwartz inequality

$$|\boldsymbol{u}^T \boldsymbol{v}| \leq \|\boldsymbol{u}\|_2 \, \|\boldsymbol{v}\|_2. \tag{3.35}$$

The ℓ_1-norm is in fact the sum-absolute value and is given by

$$\|\boldsymbol{v}\|_1 = |v_1| + |v_2| + ... + |v_n|. \tag{3.36}$$

For the special case $p = \infty$, the ℓ_∞-norm is also called the Chebyshev norm. We denote $v_{\max}$ as the maximum absolute value of all the components v_i, or

$$\|\boldsymbol{v}\|_\infty = v_{\max} \equiv \max |v_i| = \max(|v_1|, |v_2|, ..., |v_n|). \tag{3.37}$$

$$\|\boldsymbol{v}\|_\infty = \lim_{p\to\infty} (\sum_{i=1}^n |v_i|^p)^{1/p} = \lim_{p\to\infty} \Big(v_{\max}^p \sum_{i=1}^n |\frac{v_i}{v_{\max}}|^p\Big)^{1/p}$$

$$= \lim_{p\to\infty} (v_{\max}^p)^{\frac{1}{p}} (\sum_{i=1}^n |\frac{v_i}{v_{\max}}|^p)^{\frac{1}{p}} = v_{\max} \lim_{p\to\infty} (\sum_{i=1}^n |\frac{v_i}{v_{\max}}|^p)^{\frac{1}{p}}. \tag{3.38}$$

Since $|v_i/v_{\max}| \leq 1$ and for all terms $|v_i/v_{\max}| < 1$, we have

$$|v_i/v_{\max}|^p \to 0, \text{ for } p \to \infty.$$

Thus, the only non-zero term in the sum is the one with $|v_i/v_{\max}| = 1$, which means that

$$\lim_{p\to\infty} \sum_{i=1}^n |v_i/v_{\max}|^p = 1. \tag{3.39}$$

Therefore, we finally have

$$\|\boldsymbol{v}\|_\infty = v_{\max} = \max |v_i|. \tag{3.40}$$

For the uniqueness and consistency of norms, it is necessary for the norms to satisfy the triangle inequality

$$\|\boldsymbol{u} + \boldsymbol{v}\| \leq \|\boldsymbol{u}\| + \|\boldsymbol{v}\|. \tag{3.41}$$

It is straightforward to check that this is true for $p = 1, 2,$ and ∞ from their definitions. The equality occurs when $\boldsymbol{u} = \boldsymbol{v}$. It remains as an exercise to verify this inequality for any $p > 0$.

■ EXAMPLE 3.8

For two 4-dimensional vectors

$$\boldsymbol{u} = \begin{pmatrix} 5 & 2 & 3 & -2 \end{pmatrix}^T, \qquad \boldsymbol{v} = \begin{pmatrix} -2 & 0 & 1 & 2 \end{pmatrix}^T,$$

then the p-norms of $\boldsymbol{u}$ are

$$\|\boldsymbol{u}\|_1 = |5| + |2| + |3| + |-2| = 12,$$

$$\|\boldsymbol{u}\|_2 = \sqrt{5^2 + 2^2 + 3^2 + (-2)^2} = \sqrt{42},$$

and

$$\|\boldsymbol{u}\|_\infty = \max(5, 2, 3, -2) = 5.$$

Similarly, $\|\boldsymbol{v}\|_1 = 5$, $\|\boldsymbol{v}\|_2 = 3$ and $\|\boldsymbol{v}\|_\infty = 2$. We know that

$$\boldsymbol{u} + \boldsymbol{v} = \begin{pmatrix} 5 + (-2) \\ 2 + 0 \\ 3 + 1 \\ -2 + 2 \end{pmatrix} = \begin{pmatrix} 3 \\ 2 \\ 4 \\ 0 \end{pmatrix},$$

and its corresponding norms are $\|\boldsymbol{u} + \boldsymbol{v}\|_1 = 9$, $\|\boldsymbol{u} + \boldsymbol{v}\|_2 = \sqrt{29}$ and $\|\boldsymbol{u} + \boldsymbol{v}\|_\infty = 4$. It is straightforward to check that

$$\|\boldsymbol{u} + \boldsymbol{v}\|_1 = 9 < 12 + 5 = \|\boldsymbol{u}\|_1 + \|\boldsymbol{v}\|_1,$$

$$\|\boldsymbol{u} + \boldsymbol{v}\|_2 = \sqrt{29} < \sqrt{42} + 3 = \|\boldsymbol{u}\|_2 + \|\boldsymbol{v}\|_2,$$

and

$$\|\boldsymbol{u} + \boldsymbol{v}\|_\infty = 4 < 5 + 4 = \|\boldsymbol{u}\|_\infty + \|\boldsymbol{v}\|_\infty.$$

Matrices are the extension of vectors, so we can define the corresponding norms. For an $m \times n$ matrix $\boldsymbol{A} = [a_{ij}]$, a simple way to extend the norms is to use the fact that $\boldsymbol{A}\boldsymbol{u}$ is a vector for any vector $\|\boldsymbol{u}\| = 1$. So the p-norm is defined as

$$\|\boldsymbol{A}\|_p = \Big(\sum_{i=1}^{m} \sum_{j=1}^{n} |a_{ij}|^p\Big)^{1/p}. \tag{3.42}$$

Alternatively, we can consider that all the elements or entries a_{ij} form a vector. A popular norm, called Frobenius form (also called the Hilbert-Schmidt norm), is defined as

$$\|\boldsymbol{A}\|_F = \Big(\sum_{i=1}^{m} \sum_{j=1}^{n} a_{ij}^2\Big)^{1/2}. \tag{3.43}$$

In fact, Frobenius norm is a 2-norm.

Other popular norms are based on the absolute column sum or row sum. For example,

$$\|\boldsymbol{A}\|_1 = \max_{1 \le j \le n} \Big(\sum_{i=1}^{m} |a_{ij}|\Big), \tag{3.44}$$

which is the maximum of absolute column sum, while

$$\|\boldsymbol{A}\|_\infty = \max_{1 \le i \le m} \Big(\sum_{j=1}^{n} |a_{ij}|\Big), \tag{3.45}$$

is the maximum of the absolute row sum. The max norm is defined as

$$\|A\|_{\max} = \max\{|a_{ij}|\}. \tag{3.46}$$

From the definitions of these norms, we know that they satisfy the non-negativeness condition $\|A\| \geq 0$, the scaling condition $\|\alpha A\| = |\alpha|\|A\|$, and the triangle inequality $\|A + B\| \leq \|A\| + \|B\|$.

■ EXAMPLE 3.9

For the matrix $A = \begin{pmatrix} 2 & 3 \\ 4 & -5 \end{pmatrix}$, we know that

$$\|A\|_F = \|A\|_2 = \sqrt{2^2 + 3^2 + 4^2 + (-5)^2} = \sqrt{54},$$

$$\|A\|_\infty = \max \begin{bmatrix} |2| + |3| \\ \\ |4| + |-5| \end{bmatrix} = 9,$$

and $\|A\|_{\max} = 5$.

3.5 EIGENVALUES AND DEFINITENESS

3.5.1 Eigenvalues

The eigenvalues λ of an $n \times n$ square matrix A are determined by

$$Au = \lambda u, \tag{3.47}$$

or

$$(A - \lambda I)u = 0. \tag{3.48}$$

Any non-trivial solution requires that

$$\det|A - \lambda I| = 0, \tag{3.49}$$

which is often referred to as the characteristic equation. We can also write it as

$$\begin{vmatrix} a_{11} - \lambda & a_{12} & \dots & a_{1n} \\ a_{21} & a_{22} - \lambda & \dots & a_{2n} \\ \vdots & & \ddots & \\ a_{n1} & a_{n2} & \dots & a_{nn} - \lambda \end{vmatrix} = 0, \tag{3.50}$$

which again can be written as a characteristic polynomial

$$\lambda^n + \alpha_{n-1}\lambda^{n-1} + \dots + \alpha_0 = (\lambda - \lambda_1)^{k_1} \dots (\lambda - \lambda_p)^{k_p} = 0, \tag{3.51}$$

where λ_i are the eigenvalues that could be complex numbers. k_p is the multiplicity of the eigenvalue λ_p, and $k_1 + k_2 + ... + k_p = n$. For each eigenvalue λ, there is a corresponding eigenvector $\boldsymbol{u}$ whose direction can be determined uniquely. However, the length of the eigenvector is not unique because any non-zero multiple of $\boldsymbol{u}$ will also satisfy equation (3.47), and thus can be considered as an eigenvector. For this reason, it is usually necessary to apply an additional condition by setting the length as unity, and subsequently the eigenvector becomes a unit eigenvector.

In general, a real $n \times n$ matrix $\boldsymbol{A}$ has n eigenvalues $\lambda_i (i = 1, 2, ..., n)$; however, these eigenvalues are not necessarily distinct. If the real matrix is symmetric, that is to say $\boldsymbol{A}^T = \boldsymbol{A}$, then the matrix has n distinct eigenvectors, and all the eigenvalues are real numbers. The eigenvalues of an inverse $\boldsymbol{A}^{-1}$ are the reciprocals of the eigenvalues of $\boldsymbol{A}$.

Eigenvalues have the interesting connections with the matrix,

$$\mathrm{tr}(\boldsymbol{A}) = \sum_{i=1}^{n} a_{ii} = \lambda_1 + \lambda_2 + ... + \lambda_n. \tag{3.52}$$

In addition, we also have

$$\det(\boldsymbol{A}) = \prod_{i=1}^{n} \lambda_i. \tag{3.53}$$

For a symmetric square matrix, the two eigenvectors for two distinct eigenvalues λ_i and λ_j are orthogonal $\boldsymbol{u}_i^T \boldsymbol{u}_j = 0$.

■ EXAMPLE 3.10

The eigenvalues of the square matrix

$$\boldsymbol{A} = \begin{pmatrix} 4 & 9 \\ 2 & -3 \end{pmatrix},$$

can be obtained by solving

$$\begin{vmatrix} 4-\lambda & 9 \\ 2 & -3-\lambda \end{vmatrix} = 0.$$

We have

$$(4-\lambda)(-3-\lambda) - 18 = (\lambda - 6)(\lambda + 5) = 0.$$

Thus, the eigenvalues are $\lambda = 6$ and $\lambda = -5$. Let $\boldsymbol{v} = (v_1 \ v_2)^T$ be the eigenvector, we have for $\lambda = 6$

$$|\boldsymbol{A} - \lambda \boldsymbol{I}| = \begin{pmatrix} -2 & 9 \\ 2 & -9 \end{pmatrix} \begin{pmatrix} v_1 \\ v_2 \end{pmatrix} = 0,$$

which means that

$$-2v_1 + 9v_2 = 0, \qquad 2v_1 - 9v_2 = 0.$$

These two equations are virtually the same (not linearly independent), so the solution is

$$v_1 = \frac{9}{2} v_2.$$

Any vector parallel to $\boldsymbol{v}$ is also an eigenvector. In order to get a unique eigenvector, we have to impose an extra requirement, that is, the length of the vector is unity. We now have

$$v_1^2 + v_2^2 = 1,$$

or

$$(\frac{9v_2}{2})^2 + v_2^2 = 1,$$

which gives $v_2 = \pm 2/\sqrt{85}$, and $v_1 = \pm 9/\sqrt{85}$. As these two vectors are in opposite directions, we can choose either of the two directions. So the eigenvector for the eigenvalue $\lambda = 6$ is

$$\boldsymbol{v} = \begin{pmatrix} 9/\sqrt{85} \\ \\ 2/\sqrt{85} \end{pmatrix}.$$

Similarly, the corresponding eigenvector for the eigenvalue $\lambda = -5$ is $\boldsymbol{u} = (-\sqrt{2}/2 \;\; \sqrt{2}/2)^T$.

Furthermore, we know the trace of $\boldsymbol{A}$ is $\mathrm{tr}(\boldsymbol{A}) = 4 + (-3) = 1$, while its determinant is $\det \boldsymbol{A} = 4 \times (-3) - 2 \times 9 = -30$. From its eigenvalues 6 and -5, we have indeed

$$\mathrm{tr}(\boldsymbol{A}) = 1 = \sum_{i=1}^{2} \lambda_i = 6 + (-5),$$

and

$$\det \boldsymbol{A} = \prod_{i=1}^{2} \lambda_i = 6 \times (-5) = -30.$$

We know that the inverse of $\boldsymbol{A}$ is

$$\boldsymbol{A}^{-1} = \frac{1}{4 \times (-3) - 2 \times 9} \begin{pmatrix} -3 & -9 \\ -2 & 4 \end{pmatrix} = \begin{pmatrix} 1/10 & 3/10 \\ 1/15 & -4/30 \end{pmatrix}.$$

The eigenvalues of $\boldsymbol{A}^{-1}$ are

$$\Lambda = 1/6, -1/5,$$

which are exactly the reciprocals of the eigenvalues $\lambda = 6$ and -5 of the original matrix $\boldsymbol{A}$. That is $\Lambda = 1/\lambda$.

3.5.2 Definiteness

A square symmetric matrix $\boldsymbol{A}$ (i.e., $\boldsymbol{A}^T = \boldsymbol{A}$) is said to be positive definite if all its eigenvalues are strictly positive ($\lambda_i > 0$ where $i = 1, 2, ..., n$). By multiplying (3.47) by $\boldsymbol{u}^T$, we have

$$\boldsymbol{u}^T \boldsymbol{A} \boldsymbol{u} = \boldsymbol{u}^T \lambda \boldsymbol{u} = \lambda \boldsymbol{u}^T \boldsymbol{u}, \tag{3.54}$$

which leads to

$$\lambda = \frac{\boldsymbol{u}^T \boldsymbol{A} \boldsymbol{u}}{\boldsymbol{u}^T \boldsymbol{u}}. \tag{3.55}$$

This means that

$$\boldsymbol{u}^T \boldsymbol{A} \boldsymbol{u} > 0, \qquad \text{if } \lambda > 0. \tag{3.56}$$

In fact, for any vector $\boldsymbol{v}$, the following relationship holds

$$\boldsymbol{v}^T \boldsymbol{A} \boldsymbol{v} > 0. \tag{3.57}$$

For $\boldsymbol{v}$ can be a unit vector; thus all the diagonal elements of $\boldsymbol{A}$ should be strictly positive as well. If the equal sign is included in the definition, we have semi-definiteness. That is, $\boldsymbol{A}$ is called positive semidefinite if $\boldsymbol{u}^T \boldsymbol{A} \boldsymbol{u} \geq 0$, and negative semidefinite if $\boldsymbol{u}^T \boldsymbol{A} \boldsymbol{u} \leq 0$ for all $\boldsymbol{u}$.

If all the eigenvalues are non-negative or $\lambda_i \geq 0$, then the matrix is called positive semi-definite. If all the eigenvalues are non-positive or $\lambda_i \leq 0$, then the matrix is called negative semi-definite. In general, an indefinite matrix can have both positive and negative eigenvalues. Furthermore, the inverse of a positive definite matrix is also positive definite. For a linear system $\boldsymbol{A}\boldsymbol{u} = \boldsymbol{f}$, if $\boldsymbol{A}$ is positive definite, the system can be solved more efficiently by matrix decomposition methods.

■ EXAMPLE 3.11

In order to determine the definiteness of a 2×2 symmetric matrix $\boldsymbol{A}$

$$\boldsymbol{A} = \begin{pmatrix} \alpha & \beta \\ \beta & \alpha \end{pmatrix},$$

we first have to determine its eigenvalues. From $|\boldsymbol{A} - \lambda \boldsymbol{I}| = 0$, we have

$$\det \begin{pmatrix} \alpha - \lambda & \beta \\ \beta & \alpha - \lambda \end{pmatrix} = (\alpha - \lambda)^2 - \beta^2 = 0,$$

or

$$\lambda = \alpha \pm \beta.$$

Their corresponding eigenvectors are

$$\boldsymbol{v}_1 = \frac{1}{\sqrt{2}} \begin{pmatrix} 1 \\ 1 \end{pmatrix}, \qquad \boldsymbol{v}_2 = \frac{1}{\sqrt{2}} \begin{pmatrix} 1 \\ -1 \end{pmatrix}.$$

Eigenvectors associated with distinct eigenvalues of a symmetric square matrix are orthogonal. Indeed, they are orthogonal since we have

$$v_1^T v_2 = \begin{pmatrix} 1/\sqrt{2} & 1/\sqrt{2} \end{pmatrix} \begin{pmatrix} 1/\sqrt{2} \\ -1/\sqrt{2} \end{pmatrix} = \frac{1}{\sqrt{2}} \times \frac{1}{\sqrt{2}} + \frac{1}{\sqrt{2}} \times (\frac{-1}{\sqrt{2}}) = 0.$$

The matrix $\boldsymbol{A}$ will be positive definite if

$$\alpha \pm \beta > 0,$$

which means $\alpha > 0$ and $\alpha > \max(+\beta, -\beta)$. The inverse

$$\boldsymbol{A}^{-1} = \frac{1}{\alpha^2 - \beta^2} \begin{pmatrix} \alpha & -\beta \\ -\beta & \alpha \end{pmatrix},$$

will also be positive definite. For example,

$$\boldsymbol{A} = \begin{pmatrix} 3 & -2 \\ -2 & 3 \end{pmatrix}, \quad \boldsymbol{B} = \begin{pmatrix} 10 & 9 \\ 9 & 10 \end{pmatrix},$$

are positive definite, while

$$\boldsymbol{D} = \begin{pmatrix} 2 & 3 \\ 3 & 2 \end{pmatrix},$$

is indefinite.

For a symmetric matrix, it may be easier to test its definiteness using determinants of principal minor matrices, rather than calculating the eigenvalues. The ith principal minor $\boldsymbol{A}^{(1)}$ of an $n \times n$ square matrix $\boldsymbol{A} = [a_{ij}]$ is formed by the first i rows and i columns of $\boldsymbol{A}$. That is

$$\boldsymbol{A}^{(1)} = (a_{11}), \;\; \boldsymbol{A}^{(2)} = \begin{pmatrix} a_{11} & a_{12} \\ a_{21} & a_{22} \end{pmatrix}, \;\; \boldsymbol{A}^{(3)} = \begin{pmatrix} a_{11} & a_{12} & a_{13} \\ a_{21} & a_{22} & a_{23} \\ a_{31} & a_{32} & a_{33} \end{pmatrix}, \;\; ..., \qquad (3.58)$$

whose determinants can be used to verify the definiteness. If all the determinants of its principal minors $A^{(1)}, ..., A^{(n)}$ are strictly positive, then $\boldsymbol{A}$ is positive definite. If all the determinants are all nonnegative, then $\boldsymbol{A}$ is positive semidefinite. On the other hand, if the determinants alternate in signs with $\det(A^{(1)}) < 0$ (or ≤ 0), then $\boldsymbol{A}$ is negative (semi)definite.

■ EXAMPLE 3.12

For example, for a symmetric real matrix

$$\boldsymbol{A} = \begin{pmatrix} \alpha & \beta \\ \beta & \gamma \end{pmatrix},$$

we have its principal minors

$$\boldsymbol{A}^{(1)} = (\alpha), \qquad \boldsymbol{A}^{(2)} = \begin{pmatrix} \alpha & \beta \\ \beta & \gamma \end{pmatrix}.$$

For $\boldsymbol{A}$ to be positive definite, it requires that

$$\det(\boldsymbol{A}^{(1)}) = \alpha > 0, \qquad \det(\boldsymbol{A}^{(2)}) = \alpha\gamma - \beta^2 > 0.$$

For example, if

$$\boldsymbol{A} = \begin{pmatrix} 802 & -400 \\ -400 & 200 \end{pmatrix},$$

we have $\alpha = 802, \beta = -400$ and $\gamma = 200$. Since $802 > 0$ and $\alpha\gamma - \beta^2 = 400 > 0$. So it is positive definite. We will use this result later in this chapter.

3.6 LINEAR AND AFFINE FUNCTIONS

3.6.1 Linear Functions

Generally speaking, a function is a mapping from the independent variables or inputs to a dependent variable or variables/outputs. For example, $f(x, y) = x^2 + y^2 + xy$ is a function which depends on two independent variables. This function maps the domain $\Re^2$ (for $-\infty < x < \infty$ and $-\infty < y < \infty$) to f on the real axis as its range. So we use the notation $f : \Re^2 \to \Re$ to denote this. In general, a function $f(x, y, z, ...)$ will map n independent variables to m dependent variables, and we use the notation $f : \Re^n \to \Re^m$ to mean that the domain of the function is a subset of $\Re^n$ while its range is a subset of $\Re^m$. The domain of a function is sometimes denoted by dom(f) or dom f.

The inputs or independent variables can often be written as a vector. For simplicity, we often use a vector $\boldsymbol{x} = (x, y, z, ...)^T$ for multiple variables. Therefore, $f(\boldsymbol{x})$ is often used to mean $f(x, y, z, ...)$ or $f(x_1, x_2, ..., x_n)$.

A function $\mathcal{L}(\boldsymbol{x})$ is called linear if it satisfies

$$\mathcal{L}(\boldsymbol{x} + \boldsymbol{y}) = \mathcal{L}(\boldsymbol{x}) + \mathcal{L}(\boldsymbol{y}), \qquad \mathcal{L}(\alpha\boldsymbol{x}) = \alpha\mathcal{L}(\boldsymbol{x}), \tag{3.59}$$

for any vectors $\boldsymbol{x}$ and $\boldsymbol{y}$, and any scalar $\alpha \in \Re$.

■ EXAMPLE 3.13

To see if $f(\boldsymbol{x}) = f(x_1, x_2) = 2x_1 + 3x_2$ is linear, we use

$$f(x_1 + y_1, x_2 + y_2) = 2(x_1 + y_1) + 3(x_2 + y_2) = 2x_1 + 2y_1 + 3x_2 + 3y_2$$
$$= [2x_1 + 3x_2] + [2y_1 + 3y_2] = f(x_1, x_2) + f(y_1, y_2).$$

In addition, for any scalar α, we have

$$f(\alpha x_1, \alpha x_2) = 2\alpha x_1 + 3\alpha x_2 = \alpha[2x_1 + 3x_2] = \alpha f(x_1, x_2).$$

Therefore, this function is indeed linear. This function can also be written as a vector form

$$f(\boldsymbol{x}) = (2 \quad 3)\begin{pmatrix} x_1 \\ x_2 \end{pmatrix} = \boldsymbol{a} \cdot \boldsymbol{x} = \boldsymbol{a}^T \boldsymbol{u},$$

where $\boldsymbol{a} = (2 \ \ 3)^T$ and $\boldsymbol{x} = (x_1 \ \ x_2)^T$.

3.6.2 Affine Functions

A function $\boldsymbol{F}$ is called affine if there exists a linear function $\mathcal{L} : \Re^n \to \Re^m$ and a vector constant $\mathbf{c}$ such that

$$\boldsymbol{F} = \mathcal{L}(\boldsymbol{x}) + \mathbf{c}. \tag{3.60}$$

In general, an affine function is a linear function with a translation, which can be written as a matrix form

$$\boldsymbol{F} = \boldsymbol{A}\boldsymbol{x} + \boldsymbol{c}, \tag{3.61}$$

where $\boldsymbol{A}$ is an $m \times n$ matrix, and $\mathbf{c}$ is a column vector in $\Re^n$.

■ EXAMPLE 3.14

The function

$$\boldsymbol{F} = \begin{pmatrix} 5x + 5 \\ -2x + 3y + 2 \end{pmatrix},$$

is affine because it can be written as

$$\boldsymbol{F} = \begin{pmatrix} 5 & 0 \\ -2 & 3 \end{pmatrix}\begin{pmatrix} x \\ y \end{pmatrix} + \begin{pmatrix} 5 \\ 2 \end{pmatrix}.$$

Indeed, it is a linear function.

3.6.3 Quadratic Form

Quadratic forms are widely used in optimization, especially in convex optimization and engineering optimization. Loosely speaking, a quadratic form is homogenous polynomial of degree 2 of n variables. For example, $3x^2 + 10xy + 7y^2$ is a binary quadratic form, while $x^2 + 2xy + y^2 - y$ is not.

For a real $n \times n$ symmetric matrix $\boldsymbol{A}$ and an n-vector $\boldsymbol{u}$, their combination

$$Q = \boldsymbol{u}^T \boldsymbol{A} \boldsymbol{u}, \tag{3.62}$$

is called a quadratic form. Since $\boldsymbol{A} = [a_{ij}]$, we have

$$Q = \boldsymbol{u}^T \boldsymbol{A} \boldsymbol{u} = \sum_{i=1}^{n} \sum_{j=1}^{n} u_i a_{ij} u_j = \sum_{i=1}^{n} \sum_{j=1}^{n} a_{ij} u_i u_j$$

$$= \sum_{i=1}^{n} a_{ii} u_i^2 + 2 \sum_{i=2}^{n} \sum_{j=1}^{i-1} a_{ij} u_i u_j. \tag{3.63}$$

■ EXAMPLE 3.15

For the symmetric matrix $\boldsymbol{A} = \begin{pmatrix} 1 & 2 \\ 2 & 5 \end{pmatrix}$ and $\boldsymbol{u} = \begin{pmatrix} u_1 & u_2 \end{pmatrix}^T$, we have

$$\boldsymbol{u}^T \boldsymbol{A} \boldsymbol{u} = \begin{pmatrix} u_1 & u_2 \end{pmatrix} \begin{pmatrix} 1 & 2 \\ 2 & 5 \end{pmatrix} \begin{pmatrix} u_1 \\ u_2 \end{pmatrix}$$

$$= \begin{pmatrix} u_1 & u_2 \end{pmatrix} \begin{pmatrix} u_1 + 2u_2 \\ 2u_1 + 5u_2 \end{pmatrix} = 3u_1^2 + 10u_1u_2 + 7u_2^2.$$

In fact, for a binary quadratic form $Q(x, y) = ax^2 + bxy + cy^2$, we have

$$\begin{pmatrix} x & y \end{pmatrix} \begin{pmatrix} \alpha & \beta \\ \beta & \gamma \end{pmatrix} \begin{pmatrix} x \\ y \end{pmatrix} = (\alpha + \beta)x^2 + (\alpha + 2\beta + \gamma)xy + (\beta + \gamma)y^2.$$

If this is equivalent to $Q(x, y)$, it requires that

$$\alpha + \beta = a, \qquad \alpha + 2\beta + \gamma = b, \qquad \beta + \gamma = c,$$

which leads to $a + c = b$. This means that not all arbitrary quadratic functions $Q(x, y)$ are quadratic form.

If $\boldsymbol{A}$ is real symmetric, its eigenvalues λ_i are real and its eigenvectors $\boldsymbol{v}_i$ are orthogonal. Therefore, we can write $\boldsymbol{u}$ using eigenvector basis and we have

$$\boldsymbol{u} = \sum_{i=1}^{n} \alpha_i \boldsymbol{v}_i. \tag{3.64}$$

In addition, $\boldsymbol{A}$ becomes diagonal in this basis. That is

$$\boldsymbol{A} = \begin{pmatrix} \lambda_1 & & \\ & \ddots & \\ & & \lambda_n \end{pmatrix}. \tag{3.65}$$

Subsequently, we have

$$Au = \sum_{i=1}^{n} \alpha_i A v_i = \sum_{i=1}^{n} \lambda_i \alpha_i v_i, \tag{3.66}$$

which means that

$$u^T A u = \sum_{j=1}^{n} \sum_{i=1}^{n} \lambda_i \alpha_j \alpha_i v_i^T v_i = \sum_{i=1}^{n} \lambda_i \alpha_i^2, \tag{3.67}$$

where we have used the fact that $v_i^T v_i = 1$.

3.7 GRADIENT AND HESSIAN MATRICES

3.7.1 Gradient

The gradient vector of a multivariate function $f(x)$ is defined as a column vector

$$G(x) \equiv \nabla f(x) \equiv \left(\frac{\partial f}{\partial x_1}, \frac{\partial f}{\partial x_2}, \ldots, \frac{\partial f}{\partial x_n} \right)^T, \tag{3.68}$$

where $x = (x_1, x_2, ..., x_n)^T$ is a vector. As the gradient $\nabla f(x)$ of a linear function $f(x)$ is always a constant vector k, then any linear function can be written as

$$f(x) = k^T x + b, \tag{3.69}$$

where b is a vector constant.

3.7.2 Hessian

The second derivatives of a generic function $f(x)$ form an $n \times n$ matrix, called Hessian matrix, given by

$$H(x) \equiv \nabla^2 f(x) \equiv \begin{pmatrix} \frac{\partial^2 f}{\partial x_1^2} & \cdots & \frac{\partial^2 f}{\partial x_1 \partial x_n} \\ \vdots & & \vdots \\ \frac{\partial^2 f}{\partial x_n \partial x_1} & \cdots & \frac{\partial^2 f}{\partial {x_n}^2} \end{pmatrix}, \tag{3.70}$$

which is symmetric due to the fact that $\frac{\partial^2 f}{\partial x_i \partial x_j} = \frac{\partial^2 f}{\partial x_j \partial x_i}$. When the Hessian matrix $H(x) = A$ is a constant matrix (the values of its entries are independent of x), the function $f(x)$ is called a quadratic function, and can subsequently be written as the following generic form

$$f(x) = \frac{1}{2} x^T A x + k^T x + b. \tag{3.71}$$

The use of the factor $1/2$ in the expression is to avoid the appearance of a factor 2 everywhere in the derivatives, and this choice is purely for convenience.

EXAMPLE 3.16

The gradient of $f(x, y, z) = xy - y\exp(-z) + z\cos(x)$ is simply

$$G = \left(y - z\sin x \quad x - e^{-z} \quad ye^{-z} + \cos x\right)^T.$$

The Hessian matrix is

$$H = \begin{pmatrix} \frac{\partial^2 f}{\partial x^2} & \frac{\partial^2 f}{\partial x \partial y} & \frac{\partial^2 f}{\partial x \partial z} \\ \frac{\partial^2 f}{\partial y \partial x} & \frac{\partial^2 f}{\partial y^2} & \frac{\partial^2 f}{\partial y \partial z} \\ \frac{\partial^2 f}{\partial z \partial x} & \frac{\partial^2 f}{\partial z \partial y} & \frac{\partial^2 f}{\partial z^2} \end{pmatrix} = \begin{pmatrix} -z\cos x & 1 & -\sin x \\ 1 & 0 & e^{-z} \\ -\sin x & e^{-z} & -ye^{-z} \end{pmatrix}.$$

3.7.3 Function approximations

In many optimization problems, we have to use some form of estimations to approximate the objective function, and this is often based on the Taylor expansion

$$f(x) = f(x_0) + f'(x_0)(x - x_0) + \frac{f''(x_0)}{2!}(x - x_0)^2 + ... + \frac{f^{(n)}}{n!}(x - x_0)^n + ..., \quad (3.72)$$

where x_0 is a known point around which the approximation is made. This univariate form can be extended to functions of multiple variables using vectors, and we have

$$f(\boldsymbol{x}) = f(\boldsymbol{x}_0) + \nabla f(\boldsymbol{x}_0)^T(\boldsymbol{x} - \boldsymbol{x}_0) + \frac{1}{2!}(\boldsymbol{x} - \boldsymbol{x}_0)^T(\nabla^2 f)(\boldsymbol{x} - \boldsymbol{x}_0) + ..., \quad (3.73)$$

where we can see that the second term involving the gradient ∇f and the third term involving the Hessian matrix $\nabla^2 f$. Here we have used a short-hand notation ∇f^T to mean $(\nabla f(\boldsymbol{x}_0))^T$, and this notation is very commonly used in optimization. As $f(\boldsymbol{x}_0)$ is known and ∇f is small near an optimal point, the third term is a quadratic form which dominates the whole expansions.

3.7.4 Optimality of multivariate functions

Similar to the univariate case, the optima of a multivarate function can only occur at three places: a stationary point where the gradient vector is zero, a non-differentiable point, and the boundary. For both non-differentiable point and boundary, we have to deal with them separately, often depending on each individual case.

For the case of stationary points, the maximum or minimum of a multivariate function $f(\boldsymbol{x}) = f(x_1, ..., x_n)$ requires

$$G = \nabla f(\boldsymbol{x}) = 0. \quad (3.74)$$

That is

$$\frac{\partial f}{\partial x_i} = 0, \qquad (i = 1, 2, ..., n), \tag{3.75}$$

which leads to a stationary point $\boldsymbol{x}_*$. Whether the point $\boldsymbol{x}_*$ is maximum or minimum is determined by the definiteness of its Hessian matrix. It is a local minimum if $\boldsymbol{H}(\boldsymbol{x}_*)$ is positive definite, and a local maximum if $\boldsymbol{H}(\boldsymbol{x}_*)$ is negative definite. This extension is similar to the case for univariate functions. However, if $\boldsymbol{H}(\boldsymbol{x}_*)$ is indefinite, it may be a saddle point. For example, $f(x, y) = xy$ has a saddle point at $(0, 0)$.

EXAMPLE 3.17

For the banana function $f(x, y) = (x - 1)^2 + 100(y - x^2)^2$, its gradient and Hessian can be written as

$$\boldsymbol{G} = \nabla f = \begin{pmatrix} 2(x-1) - 400x(y - x^2) \\ 200(y - x^2) \end{pmatrix} = 0,$$

$$\boldsymbol{H} = \begin{pmatrix} 2 + 1200x^2 - 400y & -400u \\ -400u & 200 \end{pmatrix}.$$

From $\boldsymbol{G} = \boldsymbol{0}$, we have

$$2(x - 1) - 400x(y - x^2) = 0, \qquad 200(y - x^2) = 0.$$

This means that $x_* = 1$ and $y_* = 1$. The Hessian matrix at this point becomes

$$\boldsymbol{H}(1, 1) = \begin{pmatrix} 802 & -400 \\ -400 & 200 \end{pmatrix}.$$

We know from an earlier example (see Example 3.12), it is positive definite. Therefore, the point $(1, 1)$ is a minimum.

3.8 CONVEXITY

3.8.1 Convex Set

Knowing the properties of a function can be useful for finding the maximum or minimum of the function. In fact, in mathematical optimization, nonlinear problems are often classified according to the convexity of the defining function(s). Geometrically speaking, an object is convex if for any two points within the object, every point on the straight line segment joining them is also within the object. Examples are a solid ball, a cube or a pyramid. Obviously, a hollow object is not convex. Four examples are given in Figure 3.6.

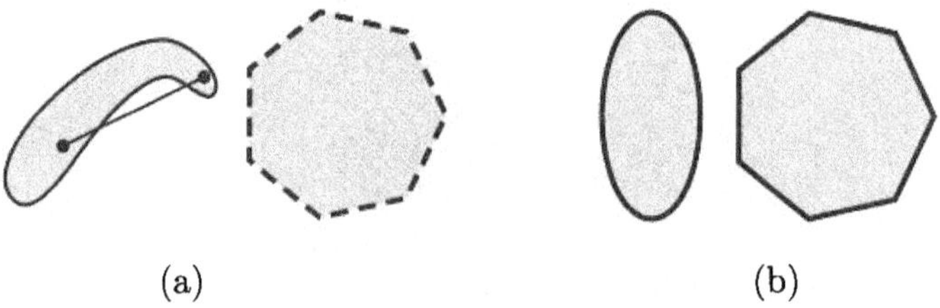

Figure 3.6: Convexity: (a) non-convex, and (b) convex.

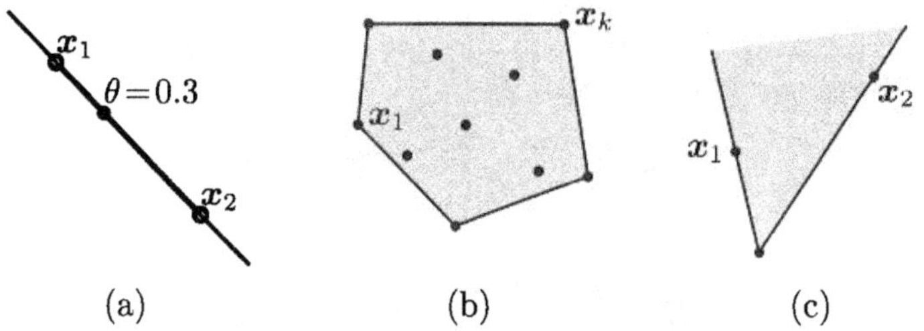

Figure 3.7: Convexity: (a) affine set $\boldsymbol{x} = \theta\boldsymbol{x}_1 + (1-\theta)\boldsymbol{x}_2$ where $\theta \in \Re$, (b) convex hull $\boldsymbol{x} = \sum_{i=1}^{k} \theta_i \boldsymbol{x}_i$ with $\sum_{i=1}^{k} \theta_i = 1$ and $\theta_i \geq 0$, and (c) convex cone $\boldsymbol{x} = \theta_1\boldsymbol{x}_1 + \theta_2\boldsymbol{x}_2$ with $\theta_1 \geq 0$ and $\theta_2 \geq 0$.

A set is called an affine set if the set contains the line through any two distinct points $\boldsymbol{x}_1$ and $\boldsymbol{x}_2$ in the set S as shown in Figure 3.7. That is, the whole line as a linear combination

$$\boldsymbol{x} = \theta\boldsymbol{x}_1 + (1-\theta)\boldsymbol{x}_2, \qquad (\theta \in \Re), \tag{3.76}$$

is contained in S. All the affine functions form an affine set. For example, all the solutions to a linear system $\boldsymbol{Au} = \boldsymbol{b}$ form an affine set $\{\boldsymbol{u}|\boldsymbol{Au} = \boldsymbol{b}\}$.

Mathematically speaking, a set $S \in \Re^n$ in a real vector space is called a convex set if

$$\theta\boldsymbol{x} + (1-\theta)\boldsymbol{y} \in S, \qquad \forall(\boldsymbol{x}, \boldsymbol{y}) \in S, \;\; \theta \in [0, 1]. \tag{3.77}$$

Thus, an affine set is always convex, but a convex set is not necessarily affine.

A convex hull, also called convex envelope, is a set $\boldsymbol{S}_{\text{convex}}$ formed by the convex combination

$$\boldsymbol{x} = \sum_{i=1}^{k} \theta_i \boldsymbol{x}_i = \theta_1\boldsymbol{x}_1 + ... + \theta_k\boldsymbol{x}_k, \qquad \sum_{i=1}^{k} \theta_i = 1, \tag{3.78}$$

where $\theta_i \geq 0$ are non-negative for all $i = 1, 2, ..., k$. It can be considered as the minimal set containing all the points.

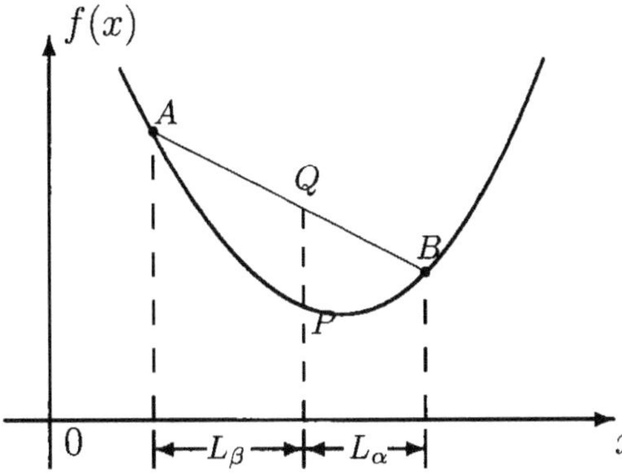

Figure 3.8: Convexity of a function $f(x)$. Chord AB lies above the curve segment joining A and B. For any point P, we have $L_\alpha = \alpha L$, $L_\beta = \beta L$ and $L = |x_B - x_A|$.

On the other hand, a convex cone is a set containing all the conic combinations of the points in the form

$$\boldsymbol{x} = \theta_1 \boldsymbol{x}_1 + \theta_2 \boldsymbol{x}_2, \qquad (\theta \geq 0, \theta_2 \geq), \tag{3.79}$$

for any two points $\boldsymbol{x}_1, \boldsymbol{x}_2$ in the set.

A componentwise inequality for vectors and matrices is defined

$$\boldsymbol{u} \preceq \boldsymbol{v} \Longleftrightarrow u_i \leq v_i \;\; i = 1, 2, ..., n. \tag{3.80}$$

It is a short-hand notation for n inequalities for all n components of the vectors $\boldsymbol{u}$ and $\boldsymbol{v}$. Here $\Longleftrightarrow$ means 'if and only if'. It is worth pointing out in some literature the notation $\boldsymbol{u} \leq \boldsymbol{v}$ is used to mean the same thing, though $\preceq$ provides a specific reminder to mean a componentwise inequality. If there is no confusion caused, we can use either. If the inequality is strict, we have

$$\boldsymbol{u} \prec \boldsymbol{v} \Longleftrightarrow u_i < v_i \quad \forall i = 1, ..., n. \tag{3.81}$$

Using this notation, we can now define hyperplanes and halfspaces. A hyperplane is a set satisfying $\{\boldsymbol{x} | \boldsymbol{A}\boldsymbol{x} = \boldsymbol{b}\}$ with a non-zero normal vector $\boldsymbol{A} \neq 0$, while a halfspace is a set formed by $\{\boldsymbol{x} | \boldsymbol{A}\boldsymbol{x} \preceq \boldsymbol{b}\}$ with $\boldsymbol{A} \neq 0$. It is straightforward to verify that a hyperplane is affine and convex, while a halfspace is convex.

3.8.2 Convex Functions

A function $f(x)$ defined on a convex set Ω is called convex if and only if it satisfies

$$f(\alpha x + \beta y) \leq \alpha f(x) + \beta f(y), \qquad \forall x, y \in \Omega, \tag{3.82}$$

and

$$\alpha \geq 0, \quad \beta \geq 0, \quad \alpha + \beta = 1. \tag{3.83}$$

Geometrically speaking, the chord AB lies above the curve segment APB joining A and B (see Figure 3.8). For example, for any point P between A and B, we have $x_P = \alpha x_A + \beta x_B$ with

$$\alpha = \frac{L_\alpha}{L} = \frac{x_B - x_P}{x_B - x_A} \geq 0,$$

$$\beta = \frac{L_\beta}{L} = \frac{x_P - x_A}{x_B - x_A} \geq 0, \tag{3.84}$$

which indeed gives $\alpha + \beta = 1$. In addition, we know that

$$\alpha x_A + \beta x_B = \frac{x_A(x_B - x_P)}{x_B - x_A} + \frac{x_B(x_P - x_A)}{x_B - x_A} = x_P. \tag{3.85}$$

The value of the function $f(x_P)$ at P should be less than or equal to the weighted combination $\alpha f(x_A) + \beta f(x_B)$ (or the value at point Q). That is

$$f(x_P) = f(\alpha x_A + \beta x_B) \leq \alpha f(x_A) + \beta f(x_B). \tag{3.86}$$

■ EXAMPLE 3.18

For example, the convexity of $f(x) = x^2 - 1$ requires

$$(\alpha x + \beta y)^2 - 1 \leq \alpha(x^2 - 1) + \beta(y^2 - 1), \quad \forall x, y \in \Re,$$

where $\alpha, \beta \geq 0$ and $\alpha + \beta = 1$. This is equivalent to

$$\alpha x^2 + \beta y^2 - (\alpha x + \beta y)^2 \geq 0,$$

where we have used $\alpha + \beta = 1$. We now have

$$\alpha x^2 + \beta y^2 - \alpha^2 x^2 - 2\alpha\beta xy - \beta^2 y^2$$

$$= \alpha(1 - \alpha)(x - y)^2 = \alpha\beta(x - y)^2 \geq 0,$$

which is always true because $\alpha, \beta \geq 0$ and $(x - y)^2 \geq 0$. Therefore, $f(x) = x^2 - 1$ is convex for $\forall x \in \Re$.

A function $f(x)$ on Ω is concave if and only if $g(x) = -f(x)$ is convex. An interesting property of a convex function f is that the vanishing of the gradient $df/dx|_{x_*} = 0$ guarantees that the point x_* is the global minimum of f. Similarly, for a concave function, any local maximum is also the global maximum. If a function is not convex or concave, then it is much more difficult to find its global minima or maxima.

The test of convexity using the above definition is tedious. A more quick and efficient way is to use the definiteness of the Hessian matrix. A function is convex if its Hessian matrix is positive semidefinite for every point in the domain. Conversely, a function becomes concave if its Hessian is negative semidefinite for every point in its domain. This is a very powerful method, and let us revisit the above example $f(x) = x^2 - 1$. We know its second derivative or Hessian is simply $f''(x) = 2$, which is always positive for every point in the domain $\Re$. We can quickly draw the conclusion that $f(x) = x^2 - 1$ is convex in the domain.

■ EXAMPLE 3.19

To see if $f(x, y) = x^2 + 2y^2$ is convex, let us calculate its Hessian first, and we have

$$\boldsymbol{H} = \begin{pmatrix} 2 & 0 \\ 0 & 4 \end{pmatrix},$$

which is obviously positive definite because it is a diagonal matrix and its eigenvalues 2 and 4 are both positive. Therefore, $f(x, y) = x^2 + 2y^2$ is convex on the whole domain $(x, y) = \Re^2$.

Examples of convex functions define on $\Re$ are $\alpha x + \beta$ for $\alpha, \beta \in \Re$, $\exp[\alpha x]$ for $\alpha \in \Re$, and $|x|^\alpha$ for $\alpha \geq 1$. Examples of concave functions are $\alpha x + \beta$ for $\alpha, \beta \in \Re$, x^α for $x > 0$ and $0 \leq p \leq 1$, and $\log x$ for $x > 0$. It is worth pointing out that a linear function is both convex and concave.

For any convex function $f(x)$, we know

$$f(\theta \boldsymbol{x} + (1 - \theta)\boldsymbol{y}) \leq \theta f(\boldsymbol{x}) + (1 - \theta) f(\boldsymbol{y}), \tag{3.87}$$

which can be extended to the probabilistic case when $\boldsymbol{x}$ is a random variable. Now we have the well-known Jensen's inequality

$$f(E[w]) \leq E[f(w)], \tag{3.88}$$

where $E[w]$ is the expectation of the random variables w.

There are some important mathematical operations that still preserve the convexity such as non-negative weighted sum, composition using affine functions, and maximization or minimization. For example, if f is convex, then βf is also convex for $\beta \geq 0$. The non-negative sum $\alpha f_1 + \beta f_2$ is convex if f_1, f_2 are convex and $\alpha, \beta \geq 0$.

The composition using an affine function also holds. For example, $f(\boldsymbol{A}\boldsymbol{x} + \boldsymbol{b})$ is convex if f is convex. In addition, if $f_1, f_2, ..., f_n$ are convex, then $\max\{f_1, f_2, ..., f_n\}$ is also convex. Similarly, the piecewise-linear function $\max_{i=1}^{n}(\boldsymbol{A}_i\boldsymbol{x} + \boldsymbol{b}_i)$ is also convex.

If both f and g are convex, then $\psi(\boldsymbol{x}) = f(g(\boldsymbol{x}))$ can also be convex under certain non-decreasing conditions. For example, $\exp[f(\boldsymbol{x})]$ is convex if $f(\boldsymbol{x})$ is convex. This can be extended to the vector composition, and most interestingly, the log-sum-exp function

$$f(\boldsymbol{x}) = \log[\sum_{k=1}^{n} e^{x_k}], \tag{3.89}$$

is convex.

For the applications discussed later in this book, two important examples are quadratic functions and least-squares functions. If $\boldsymbol{A}$ is positive definite, then the quadratic function

$$f(\boldsymbol{x}) = \frac{1}{2}\boldsymbol{u}^T \boldsymbol{A}\boldsymbol{u} + \boldsymbol{b}^T \boldsymbol{u} + \boldsymbol{c}, \tag{3.90}$$

is convex. In addition, the least-squares minimization

$$f(\boldsymbol{x}) = ||\boldsymbol{A}\boldsymbol{u} - \boldsymbol{b}||_2^2, \tag{3.91}$$

is convex for any $\boldsymbol{A}$. This is because $\nabla f = 2\boldsymbol{A}^T(\boldsymbol{A}\boldsymbol{u} - \boldsymbol{b})$ and $\nabla^2 f(\boldsymbol{x}) = 2\boldsymbol{A}^T\boldsymbol{A}$ is positive definite for any $\boldsymbol{A}$.

EXERCISES

3.1 Find the stationary points and inflection points of $\cos(x)$.

3.2 Find the stationary and inflection points of $f(x) = \exp(-x^2)$.

3.3 Find the global maximum of $f(x) = \mathrm{sinc}(x) = \sin(x)/x$.

3.4 Find the maximum of $f(x) = \exp[-x^x]$ in $(0, \infty)$.

3.5 Find the minima of $f(x) = x^2 + k\cos^2(x)$ with $k > 0$. How many minima? Will the number of minima depend on k?

3.6 Find the minimum of $f(x) = 2\cos(x) + x^2 + y^2$?

3.7 Find the optima of $f(x) = 1/(1 + |\tan x|)$ in $\pi \le x \le \pi$. What is the potential difficulty?

3.8 Find if the following matrices are positive definite ?

a) $\boldsymbol{A} = \begin{pmatrix} 2 & 3 \\ 3 & 5 \end{pmatrix}$;

b) $\boldsymbol{B} = \begin{pmatrix} -2 & 2 \\ 2 & 3 \end{pmatrix}$;

c) $\boldsymbol{C} = \begin{pmatrix} -5 & 2 \\ 2 & -2 \end{pmatrix}$.

3.9 Find if the following functions are convex or concave or neither?

a) $x^2 + \exp(y)$;

b) $1 - x^2 - y^2$;

c) xy;

d) x^2/y for $y > 0$.

REFERENCES

1. S. P. Boyd and L. Vandenberghe, *Convex Optimization*, Cambridge University Press, 2004.
2. D. B. Bridges, *Computability*, Springer, New York, 1994.
3. P. E. Gill, W. Murray, M. H. Wright, *Practical Optimization*, Academic Press, 1981.
4. A. Jeffrey, *Advanced Engineering Mathematics*, Academic Press, 2002.
5. X. S. Yang, *Introduction to Computational Mathematics*, World Scientific, 2008.

CHAPTER 4

CLASSIC OPTIMIZATION METHODS I

4.1 UNCONSTRAINED OPTIMIZATION

The simplest optimization without any constraints is probably the search of the maxima or minima of a function $f(x)$. This requires finding the root of the first derivatives or the stationary condition

$$f'(x) = 0. \tag{4.1}$$

However, the stationary condition $f'(x) = 0$ is just a necessary condition, but it is not a sufficient condition. If $f'(x_*) = 0$ and $f''(x_*) > 0$, it is a local minimum. Conversely, if $f'(x_*) = 0$ and $f''(x_*) < 0$, then it is a local maximum. However, if $f'(x_*) = 0$ but $f''(x)$ is indefinite when $x \to x_*$, then x_* may correspond to a saddle point.

For example, in order to find the maximum or minimum of a univariate function $f(x)$

$$f(x) = xe^{-x^2}, \qquad -\infty < x < \infty, \tag{4.2}$$

Engineering Optimization: An Introduction with Metaheuristic Applications.
By Xin-She Yang
Copyright © 2010 John Wiley & Sons, Inc.

we have to find first the stationary point x_* when the first derivative $f'(x)$ is zero. That is

$$\frac{df(x_*)}{dx_*} = e^{-x_*^2} - 2x_*^2 e^{-x_*^2} = 0. \tag{4.3}$$

Since $\exp(-x_*^2) \neq 0$, we have

$$x_* = \pm\frac{\sqrt{2}}{2}. \tag{4.4}$$

From the basic calculus we know that the maximum requires $f''(x_*) \leq 0$ while minimum requires $f''(x_*) \geq 0$. At $x_* = \sqrt{2}/2$, we have

$$f''(x_*) = (4x_*^2 - 6)x_* e^{-x_*^2} = -2\sqrt{2}e^{-1/2} < 0, \tag{4.5}$$

so it corresponds to a maximum $f(x_*) = \frac{1}{2}e^{-1/2}$. Similarly, at $x_* = -\sqrt{2}/2$, $f''(x_*) = 2\sqrt{2}e^{-1/2} > 0$, we have a minimum $f(x_*) = -\frac{1}{2}e^{-1/2}$.

4.2 GRADIENT-BASED METHODS

4.2.1 Newton's Method

Newton's method is a root-finding algorithm, but it can be modified for solving optimization problems. This is because optimization is equivalent to finding the root of the first derivative $f'(\boldsymbol{x})$ based on the stationary conditions once the objective function $f(\boldsymbol{x})$ is given. For a continuously differentiable function $f(\boldsymbol{x})$, we have the Taylor expansion in terms of $\Delta x = \boldsymbol{x} - \boldsymbol{x}_n$ about a fixed point $\boldsymbol{x}_n$

$$f(\boldsymbol{x}) = f(\boldsymbol{x}_n) + (\nabla f(\boldsymbol{x}_n))^T \Delta\boldsymbol{x} + \frac{1}{2}\Delta\boldsymbol{x}^T \nabla^2 f(\boldsymbol{x}_n)\Delta\boldsymbol{x} + ...,$$

whose third term is a quadratic form. Hence $f(\boldsymbol{x})$ is minimized if $\Delta\boldsymbol{x}$ is the solution of the following linear equation

$$\nabla f(\boldsymbol{x}_n) + \nabla^2 f(\boldsymbol{x}_n)\Delta\boldsymbol{x} = 0. \tag{4.6}$$

This leads to

$$\boldsymbol{x} = \boldsymbol{x}_n - \boldsymbol{H}^{-1}\nabla f(\boldsymbol{x}_n), \tag{4.7}$$

where $\boldsymbol{H}^{-1}(\boldsymbol{x}^{(n)})$ is the inverse of the Hessian matrix $\boldsymbol{H} = \nabla^2 f(\boldsymbol{x}_n)$, which is defined as

$$\boldsymbol{H}(\boldsymbol{x}) \equiv \nabla^2 f(\boldsymbol{x}) \equiv \begin{pmatrix} \frac{\partial^2 f}{\partial x_1^2} & \cdots & \frac{\partial^2 f}{\partial x_1 \partial x_n} \\ \vdots & & \vdots \\ \frac{\partial^2 f}{\partial x_n \partial x_1} & \cdots & \frac{\partial^2 f}{\partial x_n{}^2} \end{pmatrix}. \tag{4.8}$$

This matrix is symmetric due to the fact that

$$\frac{\partial^2 f}{\partial x_i \partial x_j} = \frac{\partial^2 f}{\partial x_j \partial x_i}. \tag{4.9}$$

If the iteration procedure starts from the initial vector $\boldsymbol{x}^{(0)}$, usually a guessed point in the feasible region, then Newton's formula for the nth iteration becomes

$$\boldsymbol{x}^{(n+1)} = \boldsymbol{x}^{(n)} - \boldsymbol{H}^{-1}(\boldsymbol{x}^{(n)}) f(\boldsymbol{x}^{(n)}). \tag{4.10}$$

It is worth pointing out that if $f(\boldsymbol{x})$ is quadratic, then the solution can be found exactly in a single step. However, this method is not efficient for non-quadratic functions.

In order to speed up the convergence, we can use a smaller step size $\alpha \in (0, 1]$ and we have the modified Newton's method

$$\boldsymbol{x}^{(n+1)} = \boldsymbol{x}^{(n)} - \alpha \boldsymbol{H}^{-1}(\boldsymbol{x}^{(n)}) f(\boldsymbol{x}^{(n)}). \tag{4.11}$$

Sometimes, it might be time-consuming to calculate the Hessian matrix for second derivatives. A good alternative is to use an identity matrix $\boldsymbol{I}$ to approximate $\boldsymbol{H}$ so that $\boldsymbol{H}^{-1} = \boldsymbol{I}$, and we have the quasi-Newton method

$$\boldsymbol{x}^{(n+1)} = \boldsymbol{x}^{(n)} - \alpha \boldsymbol{I} \nabla f(\boldsymbol{x}^{(n)}), \tag{4.12}$$

which is essentially the steepest descent method.

4.2.2 Steepest Descent Method

The essence of the steepest descent method is to find the lowest possible value of the objective function $f(\boldsymbol{x})$ from the current point $\boldsymbol{x}^{(n)}$. From the Taylor expansion of $f(\boldsymbol{x})$ about $\boldsymbol{x}^{(n)}$, we have

$$f(\boldsymbol{x}^{(n+1)}) = f(\boldsymbol{x}^{(n)} + \Delta \boldsymbol{s}) \approx f(\boldsymbol{x}^{(n)} + (\nabla f(\boldsymbol{x}^{(n)}))^T \Delta \boldsymbol{s}, \tag{4.13}$$

where $\Delta \boldsymbol{s} = \boldsymbol{x}^{(n+1)} - \boldsymbol{x}^{(n)}$ is the increment vector. Since we are trying to find a better approximation to the objective function, it requires that the second term on the right hand is negative. So

$$f(\boldsymbol{x}^{(n)} + \Delta \boldsymbol{s}) - f(\boldsymbol{x}^{(n)}) = (\nabla f)^T \Delta \boldsymbol{s} < 0. \tag{4.14}$$

From vector analysis, we know that the inner product $\boldsymbol{u}^T \boldsymbol{v}$ of two vectors $\boldsymbol{u}$ and $\boldsymbol{v}$ is the largest when they are parallel but in opposite directions. Therefore, we have

$$\Delta \boldsymbol{s} = -\alpha \nabla f(\boldsymbol{x}^{(n)}), \tag{4.15}$$

where $\alpha > 0$ is the step size. This is the case when the direction $\Delta \boldsymbol{s}$ is along the steepest descent in the negative gradient direction. In the case of finding maxima, this method is often referred to as hill-climbing.

The choice of the step size α is very important. A very small step size means slow movement towards the local minimum, while a large step may overshoot and subsequently makes it move far away from the local minimum. Therefore, the step size $\alpha = \alpha^{(n)}$ should be different at each iteration and should be chosen so that it minimizes the objective function $f(\boldsymbol{x}^{(n+1)}) = f(\boldsymbol{x}^{(n)}, \alpha^{(n)})$. Therefore, the steepest descent method can be written as

$$f(\boldsymbol{x}^{(n+1)}) = f(\boldsymbol{x}^{(n)}) - \alpha^{(n)}(\nabla f(\boldsymbol{x}^{(n)}))^T \nabla f(\boldsymbol{x}^{(n)}). \tag{4.16}$$

In each iteration, the gradient and step size will be calculated. Again, a good initial guess of both the starting point and the step size is useful.

■ EXAMPLE 4.1

Let us minimize the function

$$f(x_1, x_2) = 10x_1^2 + 5x_1x_2 + 10(x_2 - 3)^2,$$

where $(x_1, x_2) \in [-10, 10] \times [-15, 15]$. Using the steepest descent method, we can start from a corner point as the initial guess $\boldsymbol{x}^{(0)} = (10, 15)^T$. We know that the gradient is

$$\nabla f = (20x_1 + 5x_2, 5x_1 + 20x_2 - 60)^T,$$

therefore $\nabla f(\boldsymbol{x}^{(0)}) = (275, 290)^T$. In the first iteration, we have

$$\boldsymbol{x}^{(1)} = \boldsymbol{x}^{(0)} - \alpha_0 \begin{pmatrix} 275 \\ 290 \end{pmatrix}.$$

The step size α_0 should be chosen such that $f(\boldsymbol{x}^{(1)})$ is at the minimum, which means that

$$f(\alpha_0) = 10(10 - 275\alpha_0)^2$$
$$+5(10 - 275\alpha_0)(15 - 290\alpha_0) + 10(12 - 290\alpha_0)^2,$$

should be minimized. This becomes an optimization problem for a single independent variable α_0. All the techniques for univariate optimization problems such as Newton's method can be used to find α_0. We can also obtain the solution by setting

$$\frac{df}{d\alpha_0} = -159725 + 3992000\alpha_0 = 0,$$

whose solution is $\alpha_0 \approx 0.04001$.

At the second step, we have

$$\nabla f(\boldsymbol{x}^{(1)}) = (-3.078, 2.919)^T, \ \boldsymbol{x}^{(2)} = \boldsymbol{x}^{(1)} - \alpha_1 \begin{pmatrix} -3.078 \\ 2.919 \end{pmatrix}.$$

The minimization of $f(\alpha_1)$ gives $\alpha_1 \approx 0.066$, and the new location is

$$\boldsymbol{x}^{(2)} \approx (-0.797, 3.202)^T.$$

At the third iteration, we have

$$\nabla f(\boldsymbol{x}^{(2)}) = (0.060, 0.064)^T, \quad \boldsymbol{x}^{(3)} = \boldsymbol{x}^{(2)} - \alpha_2 \begin{pmatrix} 0.060 \\ 0.064 \end{pmatrix}.$$

The minimization of $f(\alpha_2)$ leads to $\alpha_2 \approx 0.040$, and thus

$$\boldsymbol{x}^{(3)} \approx (-0.8000299, 3.20029)^T.$$

Then, the iterations continue until a prescribed tolerance is met.

From the basic calculus, we know that first partial derivatives are equal to zero

$$\frac{\partial f}{\partial x_1} = 20x_1 + 5x_2 = 0, \quad \frac{\partial f}{\partial x_2} = 5x_1 + 20x_2 - 60 = 0.$$

This means that the minimum occurs exactly at

$$\boldsymbol{x}_* = (-4/5, 16/5)^T = (-0.8, 3.2)^T.$$

We see that the steepest descent method gives almost the exact solution after only 3 iterations.

In finding the step size α_n in the above steepest descent method, we have used $df(\alpha_n)/d\alpha_n = 0$. Well, you may say that if we can use this stationary condition for $f(\alpha_0)$, why not use the same method to get the minimum point of $f(\boldsymbol{x})$ in the first place. There are two reasons here. The first reason is that this is a simple example for demonstrating how the steepest descent method works. The second reason is that even for complicated functions of multiple variables $f(x_1, ..., x_p)$ (say $p = 500$), $f(\alpha_n)$ at any step n is still a univariate function, and the optimization of such $f(\alpha_n)$ is much simpler compared with the original multivariate problem.

It is worth pointing out that in our example, the convergence from the second iteration to the third iteration is slow. In fact, the steepest descent is typically slow once the local minimization is near. This is because the gradient is nearly zero in the neighborhood of any local minimum, and thus the rate of descent is also slow. If high accuracy is needed near the local minimum, other local search methods should be used.

4.2.3 Line Search

In the steepest descent method, there are two important parts: the descent direction and the step size (or how far to descent). The calculations of the exact step size may be very time consuming. In reality, we intend to find the

Line Search Method

Initial guess $\boldsymbol{x}_0$ at $k = 0$
while ($\|\nabla f(\boldsymbol{x}_k)\| >$ accuracy)
Find the search direction $\boldsymbol{s}_k = -\nabla f(\boldsymbol{x}_k)$
Solve for α_k by decreasing $f(\boldsymbol{x}_k + \alpha \boldsymbol{s}_k)$ significantly
while satisfying the Wolfe's conditions
Update the result $\boldsymbol{x}_{k+1} = \boldsymbol{x}_k + \alpha_k \boldsymbol{s}_k$
$k = k + 1$
end

Figure 4.1: The basic steps of a line search method.

right descent direction. Then a reasonable amount of descent, not necessarily the exact amount, during each iteration will usually be sufficient. For this, we essentially use a line search method.

To find the local minimum of the objective function $f(\boldsymbol{x})$, we try to search along a descent direction $\boldsymbol{s}_k$ with an adjustable step size α_k so that

$$\psi(\alpha_k) = f(\boldsymbol{x}_k + \alpha_k \boldsymbol{s}_k), \tag{4.17}$$

decreases as much as possible, depending on the value of α_k. Loosely speaking, the reasonably right step size should satisfy the Wolfe's conditions:

$$f(\boldsymbol{x}_k + \alpha_k \boldsymbol{s}_k) \leq f(\boldsymbol{x}_k) + \gamma_1 \alpha_k \boldsymbol{s}_k^T \nabla f(\boldsymbol{x}_k), \tag{4.18}$$

and

$$\boldsymbol{s}_k^T \nabla f(\boldsymbol{x}_k + \alpha_k \boldsymbol{s}_k) \geq \gamma_2 \boldsymbol{s}_k^T \nabla f(\boldsymbol{x}_k), \tag{4.19}$$

where $0 < \gamma_1 < \gamma_2 < 1$ are algorithm-dependent parameters. The first condition is a sufficient decrease condition for α_k, often called the Armijo condition or rule, while the second inequality is often referred to as the curvature condition. For most functions, we can use $\gamma_1 = 10^{-4}$ to 10^{-2}, and $\gamma_2 = 0.1$ to 0.9. These conditions are usually sufficient to ensure the algorithm converge in most cases; however, more strong conditions may be needed for some tough functions. The basic steps of the line search method can be summarized in Figure 4.1.

4.2.4 Conjugate Gradient Method

The method of conjugate gradient belongs to a wider class of the so-called Krylov subspace iteration methods. The conjugate gradient method was pioneered by Magnus Hestenes, Eduard Stiefel and Cornelius Lanczos in the 1950s. It was named as one of the top 10 algorithms of the 20th century.

The conjugate gradient method can be used to solve the following linear system

$$\boldsymbol{A}\boldsymbol{u} = \boldsymbol{b}, \tag{4.20}$$

where $\boldsymbol{A}$ is often a symmetric positive definite matrix. The above system is equivalent to minimizing the following function $f(\boldsymbol{u})$

$$f(\boldsymbol{u}) = \frac{1}{2}\boldsymbol{u}^T \boldsymbol{A}\boldsymbol{u} - \boldsymbol{b}^T \boldsymbol{u} + \boldsymbol{v}, \tag{4.21}$$

where $\boldsymbol{v}$ is a vector constant and can be taken to be zero. We can easily see that $\nabla f(\boldsymbol{u}) = 0$ leads to $\boldsymbol{A}\boldsymbol{u} = \boldsymbol{b}$.

In general, the size of $\boldsymbol{A}$ can be very large and sparse with $n > 100,000$, but it is not required that $\boldsymbol{A}$ is strictly symmetric positive definite. In fact, the main condition is that $\boldsymbol{A}$ should be a normal matrix. A square matrix $\boldsymbol{A}$ is called normal if $\boldsymbol{A}^T\boldsymbol{A} = \boldsymbol{A}\boldsymbol{A}^T$. Therefore, a symmetric matrix is a normal matrix, so is an orthogonal matrix because an orthogonal matrix $\boldsymbol{Q}$ satisfying $\boldsymbol{Q}\boldsymbol{Q}^T = \boldsymbol{Q}^T\boldsymbol{Q} = \boldsymbol{I}$.

The theory behind these iterative methods is closely related to the Krylov subspace $\mathcal{K}_n$ spanned by $\boldsymbol{A}$ and $\boldsymbol{b}$ as defined by

$$\mathcal{K}_n(\boldsymbol{A}, \boldsymbol{b}) = \{\boldsymbol{I}\boldsymbol{b}, \boldsymbol{A}\boldsymbol{b}, \boldsymbol{A}^2\boldsymbol{b}, ..., \boldsymbol{A}^{n-1}\boldsymbol{b}\}, \tag{4.22}$$

where $\boldsymbol{A}^0 = \boldsymbol{I}$.

If we use an iterative procedure to obtain the approximate solution $\boldsymbol{u}_n$ to $\boldsymbol{A}\boldsymbol{u} = \boldsymbol{b}$ at nth iteration, the residual is given by

$$\boldsymbol{r}_n = \boldsymbol{b} - \boldsymbol{A}\boldsymbol{u}_n, \tag{4.23}$$

which is essentially the negative gradient $\nabla f(\boldsymbol{u}_n)$. The search direction vector in the conjugate gradient method is subsequently determined by

$$\boldsymbol{d}_{n+1} = \boldsymbol{r}_n - \frac{\boldsymbol{d}_n^T \boldsymbol{A}\boldsymbol{r}_n}{\boldsymbol{d}_n^T \boldsymbol{A}\boldsymbol{d}_n}\boldsymbol{d}_n. \tag{4.24}$$

The solution often starts with an initial guess $\boldsymbol{u}_0$ at $n = 0$, and proceeds iteratively. The above steps can compactly be written as

$$\boldsymbol{u}_{n+1} = \boldsymbol{u}_n + \alpha_n \boldsymbol{d}_n, \quad \boldsymbol{r}_{n+1} = \boldsymbol{r}_n - \alpha_n \boldsymbol{A}\boldsymbol{d}_n, \tag{4.25}$$

and

$$\boldsymbol{d}_{n+1} = \boldsymbol{r}_{n+1} + \beta_n \boldsymbol{d}_n, \tag{4.26}$$

where

$$\alpha_n = \frac{\boldsymbol{r}_n^T \boldsymbol{r}_n}{\boldsymbol{d}_n^T \boldsymbol{A}\boldsymbol{d}_n}, \qquad \beta_n = \frac{\boldsymbol{r}_{n+1}^T \boldsymbol{r}_{n+1}}{\boldsymbol{r}_n^T \boldsymbol{r}_n}. \tag{4.27}$$

Iterations stop when a prescribed accuracy is reached. This can easily be programmed in any programming language, especially Matlab (see Appendix B).

It is worth pointing that the initial guess $\boldsymbol{r}_0$ can be any educated guess; however, $\boldsymbol{d}_0$ should be taken as $\boldsymbol{d}_0 = \boldsymbol{r}_0$, otherwise, the algorithm may not converge. In the case when $\boldsymbol{A}$ is not symmetric, we can use the generalized minimal residual (GMRES) algorithm developed by Y. Saad and M. H. Schultz in 1986.

4.3 CONSTRAINED OPTIMIZATION

Whatever the real world problem is, it is usually possible to formulate any optimization problem in a generic form. All optimization problems with explicit objectives can in general be expressed as nonlinearly constrained optimization problems in the following generic form

$$\underset{\boldsymbol{x}\in\Re^n}{\text{maximize/minimize}}\, f(\boldsymbol{x}), \quad \boldsymbol{x}=(x_1,x_2,...,x_n)^T\in\Re^n,$$
$$\text{subject to } \phi_j(\boldsymbol{x})=0, \quad (j=1,2,...,M),$$
$$\psi_k(\boldsymbol{x})\le 0, \quad (k=1,...,N), \tag{4.28}$$

where $f(\boldsymbol{x})$, $\phi_i(\boldsymbol{x})$ and $\psi_j(\boldsymbol{x})$ are scalar functions of the real column vector $\boldsymbol{x}$. We will discuss this in great detail later in this chapter and next chapter. Before we proceed, let us briefly review the fundamentals of linear programming.

4.4 LINEAR PROGRAMMING

Linear programming is a powerful mathematical programming technique which is widely used in business planning, engineering design, oil industry, and many other optimization applications. The basic idea in linear programming is to find the maximum or minimum of a linear objective under linear constraints.

For example, an Internet service provider (ISP) can provide two different services x_1 and x_2. The first service is, say, the fixed monthly rate with limited download limits and bandwidth, while the second service is the higher rate with no download limit. The profit of the first service is αx_1 while the second is βx_2, though the profit of the second product is higher $\beta > \alpha > 0$, so the total profit is

$$P(\boldsymbol{x}) = \alpha x_1 + \beta x_2, \qquad \beta/\alpha > 1, \tag{4.29}$$

which is the objective function because the aim of the ISP company is to increase the profit as much as possible. Suppose the provided service is limited by the total bandwidth of the ISP company, thus at most $n_1 = 16$ (in 1000 units) of the first and at most $n_2 = 10$ (in 1000 units) of the second can be provided per unit time, say, each day. Therefore, we have

$$x_1 \le n_1, \qquad x_2 \le n_2. \tag{4.30}$$

If the management of each of the two service packages take the same staff time, so that a maximum of $n = 20$ (in 1000 units) can be maintained, which means

$$x_1 + x_2 \le n. \tag{4.31}$$

The additional constraints are that both x_1 and x_2 must be non-negative since negative numbers are unrealistic. We now have the following constraints

$$0 \le x_1 \le n_1, \qquad 0 \le x_2 \le n_2. \tag{4.32}$$

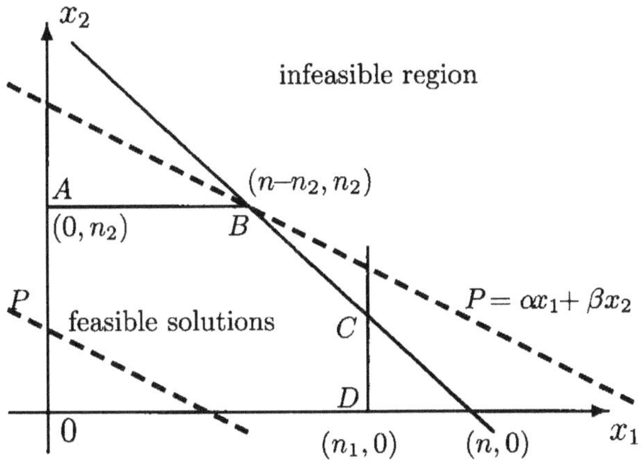

Figure 4.2: Schematic representation of linear programming. If $\alpha = 2$, $\beta = 3$, $n_1 = 16$, $n_2 = 10$ and $n = 20$, then the optimal solution is at $B(10, 10)$.

The problem now is to find the best x_1 and x_2 so that the profit P is a maximum. Mathematically, we have

$$\underset{(x_1,x_2)\in\mathcal{N}^2}{\text{maximize}} \quad P(x_1, x_2) = \alpha x_1 + \beta x_2,$$

$$\text{subject to} \quad x_1 + x_2 \le n,$$

$$0 \le x_1 \le n_1, \; 0 \le x_2 \le n_2. \tag{4.33}$$

The feasible solutions to the problem (4.33) can graphically be represented as the inside region of the polygon $OABCD$ as shown in Figure 4.2. As the aim is to maximize the profit P, thus the optimal solution is at the extreme point B with $(n - n_2, n_2)$ and $P = \alpha(n - n_2) + \beta n_2$.

For example, if $\alpha = 2$, $\beta = 3$, $n_1 = 16$, $n_2 = 10$, and $n = 20$, then the optimal solution occurs at $x_1 = n - n_2 = 10$ and $x_2 = n_2 = 10$ with the total profit $P = 2 \times (20 - 10) + 3 \times 10 = 50$ thousand pounds.

Since the solution (x_1 and x_2) must be integers, an interesting thing is that the solution is independent of β/α if and only if $\beta/\alpha > 1$. However, the profit P does depend on the parameters α and β.

The number of feasible solutions is infinite if x_1 and x_2 are real numbers. Even for integers $x_1, x_2 \in \mathcal{N}$, the number of feasible solutions is quite large. Therefore, there is a need to use a systematic method to find the optimal solution. In order to find the best solution, we first plot out all the constraints as straight lines, and all the feasible solutions satisfying all the constraints form the inside region of the polygon $OABCD$. The vertices of the polygon form the set of the extreme points. Then, we plot the objective function P as a family of parallel lines (shown as dashed lines) so as to find the maximum value of P. Obviously, the highest value of P corresponds to the case when the objective line goes through the extreme point B. Therefore, $x_1 = n - n_2$ and $x_2 = n_2$ at the point B are the best solutions. The current example is relatively simple because it has only two decision variables and three constraints, which can be solved easily using a graphic approach. For more complicated problems, we need a formal approach. One of the most widely used methods is the simplex method.

4.5 SIMPLEX METHOD

The simplex method was introduced by George Dantzig in 1947. The simplex method essentially works in the following way: for a given linear optimization problem such as the example of the ISP service we discussed earlier, it assumes that all the extreme points are known. If the extreme points are not known, the first step is to determine these extreme points or to check whether there are any feasible solutions. With known extreme points, it is easy to test whether or not an extreme point is optimal using the algebraic relationship and the objective function. If the test for optimality is not passed, then move to an adjacent extreme point to do the same test. This process stops until an optimal extreme point is found or the unbounded case occurs.

4.5.1 Basic Procedure

Mathematically, the simplex method first transforms the constraint inequalities into equalities by using slack variables.

To convert an inequality such as

$$5x_1 + 6x_2 \leq 20, \tag{4.34}$$

we can use a new variable x_3 or $s_1 = 20 - 5x_1 - 6x_2$ so that the original inequality becomes an equality

$$5x_1 + 6x_2 + s_1 = 20, \tag{4.35}$$

with an auxiliary non-negativeness condition

$$s_1 \geq 0. \tag{4.36}$$

Such a variable is referred to as a slack variable.

Thus, the inequalities in our example

$$x_1 + x_2 \leq n, \qquad 0 \leq x_1 \leq n_1, \qquad 0 \leq x_2 \leq n_2, \tag{4.37}$$

can be written, using three slack variables s_1, s_2, s_3, as the following equalities

$$x_1 + x_2 + s_1 = n, \tag{4.38}$$

$$x_1 + s_2 = n_1, \qquad x_2 + s_3 = n_2, \tag{4.39}$$

and

$$x_i \geq 0 \ (i = 1, 2), \ \ s_j \geq 0 (j = 1, 2, 3). \tag{4.40}$$

The original problem (4.33) becomes

$$\underset{\boldsymbol{x} \in \Re^5}{\text{maximize}} \, P(\boldsymbol{x}) = \alpha x_1 + \beta x_2 + 0 s_1 + 0 s_2 + 0 s_3,$$

$$\text{subject to} \begin{pmatrix} 1 & 1 & 1 & 0 & 0 \\ 1 & 0 & 0 & 1 & 0 \\ 0 & 1 & 0 & 0 & 1 \end{pmatrix} \begin{pmatrix} x_1 \\ x_2 \\ s_1 \\ s_2 \\ s_3 \end{pmatrix} = \begin{pmatrix} n \\ n_1 \\ n_2 \end{pmatrix},$$

$$x_i \geq 0, \qquad (i = 1, 2, ..., 5), \tag{4.41}$$

which has two control variables (x_1, x_2) and three slack variables $x_3 = s_1$, $x_4 = s_2$, $x_5 = s_3$.

In general, a linear programming problem can be written in the following standard form

$$\underset{\boldsymbol{x} \in \Re^n}{\text{maximize}} \, f(\boldsymbol{x}) = Z = \sum_{i=1}^{p} \alpha_i x_i = \boldsymbol{\alpha}^T \boldsymbol{x},$$

$$\text{subject to } \boldsymbol{A}\boldsymbol{x} = \boldsymbol{b}, \ x_i \geq 0 \ (i = 1, ..., p), \tag{4.42}$$

where $\boldsymbol{A}$ is a $q \times p$ matrix, $\boldsymbol{b} = (b_1, ..., b_q)^T$, and

$$\boldsymbol{x} = [\boldsymbol{x}_p \ \ \boldsymbol{x}_s]^T = (x_1, ..., x_m, s_1, ..., s_{p-m})^T. \tag{4.43}$$

This problem has p variables, and q equalities and all p variables are non-negative. In the standard form, all constraints are expressed as equalities and all variables including slack variables are non-negative.

A basic solution to the linear system $\boldsymbol{A}\boldsymbol{x} = \boldsymbol{b}$ of q linear equations in p variables in the standard form is usually obtained by setting $p - q$ variables equal to zero, and subsequently solving the resulting $q \times q$ linear system to get a unique solution of the remaining q variables. The q variables (that are not bound to zero) are called the basic variables of the basic solution. The $p - q$ variables at zero are called non-basic variables. Any basic solution to this linear system is referred to as a basic feasible solution (BFS) if all its variables are non-negative. The important property of the basic feasible solutions is that there is a unique corner point (extreme point) for each basic feasible solution, and there is at least one basic feasible solution for each corner or extreme point. These corner or extreme points are points on the intersection of two adjacent boundary lines such as A and B in Figure 4.2. Two basic feasible solutions are said to be adjacent if they have $q - 1$ basic variables in common in the standard form.

Suppose $q = 500$, even the simplest integer equalities $x_i + x_j = 1$ where $i, j = 1, 2, ..., 500$, would give a huge number of combinations 2^{500}. Thus the number of basic feasible solutions will be the order of $2^{500} \approx 3 \times 10^{150}$, which is larger than the number of particles in the whole universe. This huge number of basic feasible solutions and extreme points necessitates a systematic and efficient search method. Simplex method is a powerful method to carry out such a mathematical programming task.

4.5.2 Augmented Form

The linear optimization problem (4.42) can usually be converted into the following standard augmented form or the canonical form

$$\begin{pmatrix} 1 & -\boldsymbol{\alpha}^T \\ 0 & \boldsymbol{A} \end{pmatrix} \begin{pmatrix} Z \\ \boldsymbol{x} \end{pmatrix} = \begin{pmatrix} 0 \\ \boldsymbol{b} \end{pmatrix}, \tag{4.44}$$

with the objective to maximize Z. In this canonical form, all the constraints are expressed as equalities for all non-negative variables. All the right-hand sides for all constraints are also non-negative, and each constraint equation has a single basic variable. The intention of writing in this canonical form is to identify basic feasible solutions, and move from one basic feasible solution to another via a so-called pivot operation. Geometrically speaking, this means to find all the corner or extreme points first, then evaluate the objective function by going through the extreme points so as to determine if the current basic feasible solution can be improved or not.

In the framework of the canonical form, the basic steps of the simplex method are: 1) to find a basic feasible solution to start the algorithm. Sometimes, it might be difficult to start, which may either imply there is no feasible solution or that it is necessary to reformulate the problem in a slightly different way by changing the canonical form so that a basic feasible solution can be found; 2) to see if the current basic feasible solution can be improved (even marginally) by increasing the non-basic variables from zero to non-negative values; 3) stop the process if the current feasible solution cannot be improved, which means that it is optimal. If the current feasible solution is not optimal, then move to an adjacent basic feasible solution. This adjacent basic feasible solution can be obtained by changing the canonical form via elementary row operations.

The pivot manipulations are based on the fact that a linear system will remain an equivalent system by multiplying a non-zero constant with a row and adding it to the other row. This procedure continues by going to the second step and repeating the evaluation of the objective function. The optimality of the problem will be reached, or we can stop the iteration if the solution becomes unbounded in the event that you can improve the objective indefinitely.

Now we come back to our ISP example, if we use $\alpha = 2$, $\beta = 3$, $n_1 = 16$, $n_2 = 10$ and $n = 20$, we then have

$$\begin{pmatrix} 1 & -2 & -3 & 0 & 0 & 0 \\ 0 & 1 & 1 & 1 & 0 & 0 \\ 0 & 1 & 0 & 0 & 1 & 0 \\ 0 & 0 & 1 & 0 & 0 & 1 \end{pmatrix} \begin{pmatrix} Z \\ x_1 \\ x_2 \\ s_1 \\ s_2 \\ s_3 \end{pmatrix} = \begin{pmatrix} 0 \\ 20 \\ 16 \\ 10 \end{pmatrix}, \qquad (4.45)$$

where $x_1, x_2, s_1, ..., s_3 \geq 0$. Now the first step is to identify a corner point or basic feasible solution by setting non-isolated variables $x_1 = 0$ and $x_2 = 0$ (thus the basic variables are s_1, s_2, s_3). We now have

$$s_1 = 20, \; s_2 = 16, \; s_3 = 10. \qquad (4.46)$$

The objective function $Z = 0$, which corresponds to the corner point O in Figure 4.2. In the present canonical form, the corresponding column associated with each basic variable has only one non-zero entry (marked by a box) for each constraint equality, and all other entries in the same column are zero. The non-zero value is usually converted into 1 if it is not unity. This is shown as follows:

$$\begin{matrix} Z & x_1 & x_2 & s_1 & s_2 & s_3 \end{matrix} \qquad (4.47)$$
$$\begin{pmatrix} 1 & -2 & -3 & 0 & 0 & 0 \\ 0 & 1 & 1 & \boxed{1} & 0 & 0 \\ 0 & 1 & 0 & 0 & \boxed{1} & 0 \\ 0 & 0 & 1 & 0 & 0 & \boxed{1} \end{pmatrix}$$

When we change the set or the bases of basic variables from one set to another, we will aim to convert to a similar form using pivot row operations. There are two ways of numbering this matrix. One way is to call the first row [1 −2 −3 0 0 0] as the zeroth row, so that all other rows correspond to their corresponding constraint equation. The other way is simply to use its order in the matrix, so [1 −2 −3 0 0 0] is simply the first row. We will use this standard notation.

Now the question is whether we can improve the objective by increasing one of the non-basic variables x_1 and x_2? Obviously, if we increase x_1 by a unit, then Z will also increase by 2 units. However, if we increase x_2 by a unit, then Z will increase by 3 units. Since our objective is to increase Z as much as possible, we choose to increase x_2. As the requirement of the non-negativeness of all variables, we cannot increase x_2 without limit. So we increase x_2 while holding $x_1 = 0$, and we have

$$s_1 = 20 - x_2, \; s_2 = 16, \; s_3 = 10 - x_2. \qquad (4.48)$$

Thus, the highest possible value of x_2 is $x = 10$ when $s_1 = s_3 = 0$. If x_2 increases further, both s_1 and s_3 will become negative; thus, it is no longer a basic feasible solution.

The next step is either to set $x_1 = 0$ and $s_1 = 0$ as non-basic variables or to set $x_1 = 0$ and $s_3 = 0$. Both cases correspond to the point A in our example, so we simply choose $x_1 = 0$ and $s_3 = 0$ as non-basic variables, and the basic variables are thus x_2, s_1 and s_2. Now we have to do some pivot operations so that s_3 will be replaced by x_2 as a new basic variable. Each constraint equation has only a single basic variable in the new canonical form. This means that each column corresponding to each basic variable should have only a single non-zero entry (usually 1). In addition, the right-hand sides of all the constraints are non-negative and increase the value of the objective function at the same time. In order to convert the third column for x_2 to the form with only a single non-zero entry 1 (all other coefficients in the column should be zero), we first multiply the fourth row by 3 and add it to the first row, and the first row becomes

$$Z - 2x_1 + 0x_2 + 0s_1 + 0s_2 + 3s_3 = 30. \tag{4.49}$$

Then, we multiply the fourth row by -1 and add it to the second row, and we have

$$0Z + x_1 + 0x_2 + s_1 + 0s_2 - s_3 = 10. \tag{4.50}$$

So the new canonical form becomes

$$\begin{pmatrix} 1 & -2 & 0 & 0 & 0 & 3 \\ 0 & 1 & 0 & 1 & 0 & -1 \\ 0 & 1 & 0 & 0 & 1 & 0 \\ 0 & 0 & 1 & 0 & 0 & 1 \end{pmatrix} \begin{pmatrix} Z \\ x_1 \\ x_2 \\ s_1 \\ s_2 \\ s_3 \end{pmatrix} = \begin{pmatrix} 30 \\ 10 \\ 16 \\ 10 \end{pmatrix}, \tag{4.51}$$

where the third, fourth, and fifth columns (for x_2, s_1 and s_2, respectively) have only one non-zero coefficient. All the values on the right-hand side are non-negative. From this canonical form, we can find the basic feasible solution by setting non-basic variables equal to zero. This is to set $x_1 = 0$ and $s_3 = 0$. We now have the basic feasible solution

$$x_2 = 10, \; s_1 = 10, \; s_2 = 16, \tag{4.52}$$

which corresponds to the corner point A. The objective $Z = 30$.

Now again the question is whether we can improve the objective by increasing the non-basic variables. As the objective function is

$$Z = 30 + 2x_1 - 3s_3, \tag{4.53}$$

Z will increase 2 units if we increase x_1 by 1, but Z will decrease -3 if we increase s_3. Thus, the best way to improve the objective is to increase x_1. The question is what the limit of x_1 is. To answer this question, we hold s_3 at 0, we have

$$s_1 = 10 - x_1, \; s_2 = 16 - x_1, x_2 = 10. \tag{4.54}$$

We can see if x_1 can increase up to $x_1 = 10$, after that s_1 becomes negative, and this occurs when $x_1 = 10$ and $s_1 = 0$. This also suggests that the new adjacent basic feasible solution can be obtained by choosing s_1 and s_3 as the non-basic variables. Therefore, we have to replace s_1 with x_1 so that the new basic variables are x_1, x_2 and s_2.

Using these basic variables, we have to make sure that the second column (for x_1) has only a single non-zero entry. Thus, we multiply the second row by 2 and add it to the first row, and the first row becomes

$$Z + 0x_1 + 0x_2 + 2s_1 + 0s_2 + s_3 = 50. \tag{4.55}$$

We then multiply the second row by -1 and add it to the third row, and we have

$$0Z + 0x_1 + 0x_2 - s_1 + s_2 + s_3 = 6. \tag{4.56}$$

Thus we have the following canonical form

$$\begin{pmatrix} 1 & 0 & 0 & 2 & 0 & 1 \\ 0 & 1 & 0 & 1 & 0 & -1 \\ 0 & 0 & 0 & -1 & 1 & 1 \\ 0 & 0 & 1 & 0 & 0 & 1 \end{pmatrix} \begin{pmatrix} Z \\ x_1 \\ x_2 \\ s_1 \\ s_2 \\ s_3 \end{pmatrix} = \begin{pmatrix} 50 \\ 10 \\ 6 \\ 10 \end{pmatrix}, \tag{4.57}$$

whose basic feasible solution can be obtained by setting non-basic variables $s_1 = s_3 = 0$. We have

$$x_1 = 10, \; x_2 = 10, \; s_2 = 6, \tag{4.58}$$

which corresponds to the extreme point B in Figure 4.2. The objective value is $Z = 50$ for this basic feasible solution. Let us see if we can improve the objective further. Since the objective becomes

$$Z = 50 - 2s_1 - s_3, \tag{4.59}$$

any increase of s_1 or s_3 from zero will decrease the objective value. Therefore, this basic feasible solution is optimal. Indeed, this is the same solution as that obtained earlier from the graph method. We can see that a major advantage is that we have reached the optimal solution after searching a certain number of extreme points, and there is no need to evaluate other extreme points. This is exactly why the simplex method is so efficient.

The case study we used here is relatively simple, but it is useful to show how the basic procedure works in linear programming. For more practical applications, there are well-established software packages that will do the work for you once you have properly set up the objective and constraints.

4.6 NONLINEAR OPTIMIZATION

As most real-world problems are nonlinear, nonlinear mathematical programming forms an important part of mathematical optimization methods. A broad class of nonlinear programming problems is about the minimization or maximization of $f(\boldsymbol{x})$ subject to no constraints, and another important class is the minimization of a quadratic objective function subject to nonlinear constraints. There are many other nonlinear programming problems as well.

Nonlinear programming problems are often classified according to the convexity of the defining functions. An interesting property of a convex function f is that the vanishing of the gradient $\nabla f(\boldsymbol{x}_*) = 0$ guarantees that the point x_* is a global minimum or maximum of f. If a function is not convex or concave, then it is much more difficult to find its global minimum or maximum.

4.7 PENALTY METHOD

For a simple function optimization problem with equality and inequality constraints, a common method is the penalty method. For the optimization problem

$$\begin{aligned} &\underset{\boldsymbol{x} \in \Re^n}{\text{minimize}} f(\boldsymbol{x}), \quad \boldsymbol{x} = (x_1, ..., x_n)^T \in \Re^n, \\ &\text{subject to } \phi_i(\boldsymbol{x}) = 0, \ (i = 1, ..., M), \\ &\psi_j(\boldsymbol{x}) \le 0, \ \ (j = 1, ..., N), \end{aligned} \tag{4.60}$$

the idea is to define a penalty function so that the constrained problem is transformed into an unconstrained problem. Now we define $\Pi(\boldsymbol{x}, \mu_i, \nu_j)$

$$\Pi(\boldsymbol{x}, \mu_i, \nu_j) = f(\boldsymbol{x}) + \sum_{i=1}^{M} \mu_i \phi_i^2(\boldsymbol{x}) + \sum_{j=1}^{N} \nu_j \psi_j^2(\boldsymbol{x}), \tag{4.61}$$

where $\mu_i \gg 1$ and $\nu_j \ge 0$ which should be large enough, depending on the solution quality needed. However, a more general method is to use the Lagrange multipliers to convert the nonlinear constrained problem into an unconstrained one, which is usually easier to solve.

4.8 LAGRANGE MULTIPLIERS

Another powerful method without the above limitation of using large μ_i and ν_j is the method of Lagrange multipliers. If we want to minimize a function

$$\underset{\boldsymbol{x} \in \Re^n}{\text{minimize}} f(\boldsymbol{x}), \qquad \boldsymbol{x} = (x_1, ..., x_n)^T \in \Re^n, \tag{4.62}$$

subject to the following nonlinear equality constraint

$$g(\boldsymbol{x}) = 0, \tag{4.63}$$

then we can combine the objective function $f(\boldsymbol{x})$ with the equality to form a new function, called the Lagrangian

$$L(x, \lambda) = f(\boldsymbol{x}) + \lambda g(\boldsymbol{x}), \tag{4.64}$$

where λ is the Lagrange multiplier, which is an unknown scalar to be determined. This again converts the constrained optimization into an unstrained problem for $L(\boldsymbol{x}, \lambda)$, which is exactly the beauty of this method. The question is now what value of λ should be used? To demonstrate the importance of λ, we now look at a simple example.

■ EXAMPLE 4.2

To solve the nonlinear constrained problem

$$\text{maximize } f(x, y) = 10 - x^2 - (y-2)^2,$$

$$\text{subject to } x + 2y = 5,$$

we can reformulate it using a Lagrange multiplier. Obviously, if there is no constraint at all, the optimal solution is simply $x = 0$ and $y = 2$. To incorporate the equality constraint, we can write

$$L(x, y, \lambda) = 10 - x^2 - (y-2)^2 + \lambda(x + 2y - 5),$$

which becomes an unconstrained optimization problem for $L(x, y, \lambda)$. Obviously, if $\lambda = 0$, then the constraint has no effect at all. If $|\lambda| \gg 1$, then we may have put more weight on the constraint. For various values of λ, we can try to maximize L using the standard stationary conditions

$$\frac{\partial L}{\partial x} = -2x + \lambda = 0, \qquad \frac{\partial L}{\partial y} = -2(y-2) + 2\lambda = 0,$$

or

$$x = \frac{\lambda}{2}, \qquad y = \lambda + 2.$$

Now we have

- $\lambda = -1$, we have $x = -1/2$ and $y = 1$;
- $\lambda = 0$, we have $x = 0$ and $y = 2$ (no constraint at all);
- $\lambda = 1$, we have $x = 1/2$ and $y = 3$
- $\lambda = 100$, we get $x = 50$ and $y = 102$.

We can see that none of these solutions can satisfy the equality $x + 2y = 5$. However, if we try $\lambda = 2/5$, we got $x = 1/5$ and $y = 12/5$ and equality constraint is indeed satisfied. Therefore, $\lambda = 2/5$ is the right value for

λ. Now the question is how to calculate λ. If we take the derivative of L with respect to λ, we get

$$\frac{\partial L}{\partial \lambda} = x + 2y - 5,$$

which becomes the original constraint if we set it to zero. This means that a constrained problem can be converted into an unconstrained one if we consider λ as an additional parameter or independent variable. Then, the standard stationary conditions apply.

This method can be extended to nonlinear optimization with multiple equality constraints. If we have M equalities,

$$g_j(\boldsymbol{x}) = 0, \qquad (j = 1, ..., M), \tag{4.65}$$

then we need M Lagrange multipliers $\lambda_j (j = 1, ..., M)$. We thus have

$$L(x, \lambda_j) = f(\boldsymbol{x}) + \sum_{j=1}^{M} \lambda_j g_j(\boldsymbol{x}). \tag{4.66}$$

The requirement of stationary conditions leads to

$$\frac{\partial L}{\partial x_i} = \frac{\partial f}{\partial x_i} + \sum_{j=1}^{M} \lambda_j \frac{\partial g_j}{\partial x_i}, \quad (i = 1, ..., n), \tag{4.67}$$

and

$$\frac{\partial L}{\partial \lambda_j} = g_j = 0, \quad (j = 1, ..., M). \tag{4.68}$$

These $M + n$ equations will determine the n-component of $\boldsymbol{x}$ and M Lagrange multipliers. As $\frac{\partial L}{\partial g_j} = \lambda_j$, we can consider λ_j as the rate of the change of the quantity $L(\boldsymbol{x}, \lambda_j)$ as a functional of g_j.

■ EXAMPLE 4.3

To solve the optimization problem

$$\min_{(x,y) \in \Re^2} f(x, y) = x + y^2,$$

subject to the conditions

$$g_1(x, y) = x^2 + 2y^2 - 1 = 0, \qquad g_2(x, y) = x - y + 1 = 0,$$

we can now define

$$L(x, y, \lambda_1, \lambda_2) = x + y^2 + \lambda_1(x^2 + 2y^2 - 1) + \lambda_2(x - y + 1).$$

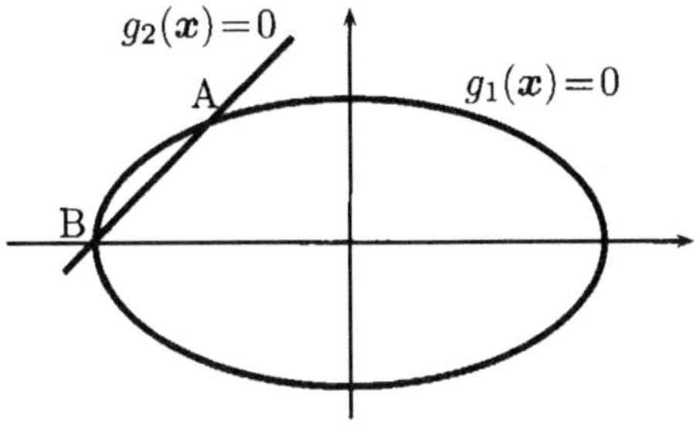

Figure 4.3: Minimization of a function with the two equality constraints.

The stationary conditions are

$$\frac{\partial L}{\partial x} = 1 + 2\lambda_1 x + \lambda_2 = 0,$$

$$\frac{\partial L}{\partial y} = 2y + 4y\lambda_1 - \lambda_2 = 0,$$

$$\frac{\partial L}{\partial \lambda_1} = x^2 + 2y^2 - 1 = 0,$$

$$\frac{\partial L}{\partial \lambda_2} = x - y + 1 = 0.$$

The first two conditions give

$$x = \frac{-(\lambda_2 + 1)}{2\lambda_1}, \qquad y = \frac{\lambda_2}{2 + 4\lambda_1}.$$

Substituting them into the last two conditions, we have

$$\frac{(\lambda_2+1)^2}{4\lambda_1^2} + \frac{2\lambda_2^2}{(2+4\lambda_1)^2} = 1, \quad \frac{(\lambda_2+1)}{2\lambda_1} + \frac{\lambda_2}{2+4\lambda_1} = 1.$$

After some straightforward algebraic manipulations, we have the solutions

$$\lambda_1 = -\frac{7}{6}, \ \lambda_2 = -\frac{16}{9}, \quad \text{or} \quad \lambda_1 = \frac{1}{2}, \ \lambda_2 = 0.$$

These correspond to the two points A$(-1/3, 2/3)$ and B$(-1, 0)$. The objective function takes the value $1/9$ and -1 at A and B, respectively. So the the best solution is $f_{\min} = -1$ at $x_* = -1$ and $y_* = 0$ (see point B in Figure 4.3).

4.9 KARUSH-KUHN-TUCKER CONDITIONS

There is a counterpart of the Lagrange multipliers for the nonlinear optimization with constraint inequalities. The Karush-Kuhn-Tucker (KKT) conditions concern the requirement for a solution to be optimal in nonlinear programming. For the nonlinear optimization problem

$$\underset{\boldsymbol{x}\in\Re^n}{\text{minimize}}\, f(\boldsymbol{x}),$$

$$\text{subject to } \phi_i(\boldsymbol{x}) = 0, \; (i = 1, ..., M),$$

$$\psi_j(\boldsymbol{x}) \leq 0, \;\; (j = 1, ..., N). \tag{4.69}$$

If all the functions are continuously differentiable, at a local minimum $\boldsymbol{x}_*$, there exist constants $\lambda_1, ..., \lambda_M$ and $\mu_0, \mu_1, ..., \mu_N$ such that

$$\mu_0 \nabla f(\boldsymbol{x}_*) + \sum_{i=1}^{M} \lambda_i \nabla \phi_i(\boldsymbol{x}_*) + \sum_{j=1}^{N} \mu_j \nabla \psi_j(\boldsymbol{x}_*) = 0, \tag{4.70}$$

and

$$\psi_j(\boldsymbol{x}_*) \leq 0, \qquad \mu_j \psi_j(\boldsymbol{x}_*) = 0, \;\; (j = 1, 2, ..., N), \tag{4.71}$$

where

$$\mu_j \geq 0, \;\; (j = 0, 1, ..., N). \tag{4.72}$$

This is the non-negativity condition for μ_j, though there is no constraint on the sign of λ_i.

The constants satisfy the following condition

$$\sum_{j=0}^{N} \mu_j + \sum_{i=1}^{M} |\lambda_i| \geq 0. \tag{4.73}$$

This is essentially a generalized method of the Lagrange multipliers. However, there is a possibility of degeneracy when $\mu_0 = 0$ under certain conditions.

There are two possibilities: A) there exist vectors $\boldsymbol{\lambda}^* = (\lambda_1^*, ..., \lambda_M^*)^T$ and $\boldsymbol{\mu}^* = (\mu_1^*, .., \mu_N^*)^T$ such that above equations hold, or B) All the vectors $\nabla \phi_1(\boldsymbol{x}_*), \nabla \phi_2(\boldsymbol{x}_*), ..., \nabla \psi_1(\boldsymbol{x}_*), ..., \nabla \psi_N(\boldsymbol{x}_*)$ are linearly independent, and in this case, the stationary conditions $\frac{\partial L}{\partial x_i}$ do not necessarily hold. As the case B is a special case, we will not discuss this further.

The condition $\mu_j^* \psi_j(\boldsymbol{x}_*) = 0$ in (4.71) is often called the complementarity condition or complementary slackness condition. It either means $\mu_j^* = 0$ or $\psi_j(\boldsymbol{x}_*) = 0$. The later case $\psi_j(\boldsymbol{x}_*) = 0$ for any particular j means the inequality becomes tight, and thus becoming an equality. For the former case $\mu_j^* = 0$, the inequality for a particular j holds and is not tight; however, $\mu_j^* = 0$ means that this corresponding inequality can be ignored. Therefore, those inequalities that are not tight are ignored, while inequalities which are tight become equalities. Consequently, the constrained problem with equality

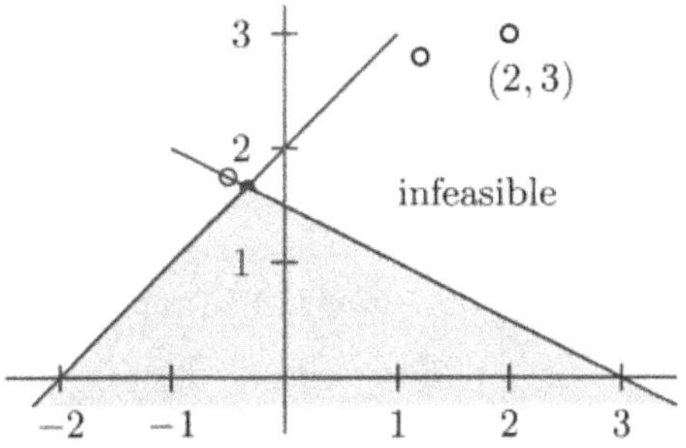

Figure 4.4: The feasible region, infeasible solutions (marked with ∘) and the optimal point (marked with •).

and inequality constraints now essentially becomes a modified constrained problem with selected equality constraints. This is the beauty of the KKT conditions. The main issue remains to identify which inequality becomes tight, and this depends on the individual optimization problem.

EXAMPLE 4.4

Let us solve this constrained problem

$$\text{minimize } f(x,y) = (x-2)^2 + 4(y-3)^2,$$

subject to the following constraints

$$-x + y \leq 2, \qquad x + 2y \leq 3.$$

First, we write the above problem in terms of the Lagrangian

$$L = (x-2)^2 + 4(y-3)^2 + \mu_1(-x+y-2) + \mu_2(x+2y-3).$$

The stationary conditions and KKT conditions become

$$\frac{\partial L}{\partial x} = 2(x-2) - \mu_1 + \mu_2 = 0,$$

$$\frac{\partial L}{\partial y} = 8(y-3) + \mu_1 + 2\mu_2 = 0,$$

$$\mu_1(-x+y-2) = 0, \tag{4.74}$$

$$\mu_2(x+2y-3) = 0, \tag{4.75}$$

$$-x+y-2 \leq 0,$$

$$x+2y-3 \leq 0,$$

and

$$\mu_1 \geq 0, \qquad \mu_2 \geq 0.$$

The first two conditions give

$$x = \frac{\mu_1 - \mu_2}{2} + 2, \qquad y = -\frac{(\mu_1 + 2\mu_2)}{8} + 3. \tag{4.76}$$

The two complementarity equations (4.74) and (4.75) imply four possibilities:

- If $\mu_1 = \mu_2 = 0$, we have from (4.76),

  $$x = 2, \qquad y = 3,$$

  which is essentially the solution without any constraint. However, $x = 3$ and $y = 3$ do not satisfy the constraint $x + 2y - 3 \leq 0$. So this solution is not feasible.

- If $\mu_1 = 0$ but $\mu_2 \neq 0$, from (4.75), we have

  $$x + 2y - 3 = 0.$$

  Using (4.76), we have

  $$\mu_2 = 5, \qquad x = -1/2, \qquad y = 7/4.$$

  This set of solution does not satisfy $-x + y \leq 2$. So this is not a feasible solution.

- If $\mu_2 = 0$ but $\mu_1 \neq 0$, we have $x = \mu_1/2 + 2$ and $y = -\mu_1/8 + 3$. Then, equation (4.74) implies that

  $$-x + y = 2.$$

  This gives $\mu_1 = -8/5$ and subsequently leads to $x = 6/5$ and $y = 14/5$. However, this is not a feasible solution because it does not satisfy the other constraint $x + 2y \leq 3$.

- If $\mu_1 \neq 0$ and $\mu_2 \neq 0$, we have two equalities as the inequalities become tight

  $$-x + y = 2, \qquad x + 2y = 3,$$

  whose solution is $x = -1/3$ and $y = 5/3$. From (4.76), we have

  $$\mu_1 - \mu_2 = -\frac{8}{3}, \qquad \mu_1 + 2\mu_2 = \frac{32}{3}.$$

  Its solution can be obtained easily as

  $$\mu_1 = 16/9, \qquad \mu_2 = 40/9,$$

which indeed satisfy $\mu_1, \mu_2 \geq 0$. This is also a feasible solution, and its objective function has a value of $f(-1/3, 5/3) = 113/9 \approx 12.556$.

Therefore, the solution $x = -1/3$ and $y = 5/3$ with $f_{\min} = 12.5$ is the optimal solution. The feasible region, infeasible solutions and the optimal solution are shown in Figure 4.4.

EXERCISES

4.1 Use a penalty function to solve the following problem of Gill-Murray-Wright type

$$\underset{x \in \Re}{\text{minimize}} \; f(x) = 100(x - b)^2 + \pi,$$
$$\text{subject to } g(x) = x - a \geq 0, \tag{4.77}$$

where $a > b$ is a given value.

4.2 Write a simple program using two different methods to find the optimal solution of De Jong's function

$$f(\boldsymbol{x}) = \sum_{i=1}^{n} x_i^2, \qquad -5.12 \leq x_i \leq 5.12,$$

for the cases of $n = 5$ and $n = 50$.

4.3 Write a program to search the shortest distance to travel 20 cities of your choice exactly once, and return to your starting city. Provide a simple case where the naive strategy of going to the nearest neighboring city does not necessarily lead to the global optimal route.

4.4 Find the maximum value of

$$f(x, y) = (|x) + |y|) \exp[-x^2 - y^2],$$

using a suitable method introduced in this chapter.

4.5 Using linear programming to solve a linear system $\boldsymbol{Au} = \boldsymbol{b}$.

4.6 Write a simple program in Matlab/Octave to implement the simplest version of the conjugate gradient method so as to solve $\boldsymbol{Au} = \boldsymbol{b}$.

REFERENCES

1. A. Antoniou and W. S. Lu, *Practical Optimization: Algorithms and Engineering Applications*, Springer, 2007.
2. M. Bartholomew-Biggs, *Nonlinear Optimization with Engineering Applications*, Springer, 2008.

3. B. A. Cipra, "The best of the 20th century: editors name top 10 algorithms", *SIAM News*, **33**, Number 4, 2000.
4. S. Boyd and L. Vandenberghe, *Convex Optimization*, Cambridge University Press, 2004.
5. M. Celis, J. E. Dennis, and R. A. Tapia."A trust region strategy for nonlinear equality constrained optimization", in *Numerical Optimization 1994*" (Eds. P. Boggs, R. Byrd and R. Schnabel), Philadelphia: SIAM, 1985, pp. 71-82.
6. G. B. Dantzig, *Linear Programming and Extensions*, Princeton University Press, 1959.
7. R. Fletcher, *Practical Methods of Optimization*, 2nd Edition, John Wiley & Sons, 2000.
8. P. E. Gill, W. Murray, and M. H. Wright, *Practical Optimization*, Academic Press, 1982.
9. M. R. Hestenes and E. Stiefel, "Methods of conjugate gradients for solving linear systems", *Journal of Research of the National Bureaus of Standards*, **49**(6),409-436 (1952).
10. Y. Saad and M. H. Schultz, "GMRES: A generalized minimal residual algorithm for solving nonsymmetric linear systems", *SIAM J. Sci. Stat. Comput.*, **7**, 856-869 (1986).
11. A. Schrijver, *Theory of Linear and Integer Programming*, John Wiley & Sons, 1998.
12. X. S. Yang, *Introduction to Computational Mathematics*, World Scientific, 2008.

CHAPTER 5

CLASSIC OPTIMIZATION METHODS II

The optimization methods we introduced in the last chapter are all well-established. In this chapter, we will continue to introduce more widely used, but relatively modern methods.

5.1 BFGS METHOD

The widely used BFGS method is an abbreviation of the Broydon-Fletcher-Goldfarb-Shanno method, and it is a quasi-Newton method for solving unconstrained nonlinear optimization. It is based on the basic idea of replacing the full Hessian matrix $\boldsymbol{H}$ by an approximate matrix $\boldsymbol{B}$ in terms of an iterative updating formula with rank-one matrices as its increment. Briefly speaking, a rank-one matrix is a matrix which can be written as $\boldsymbol{r} = \boldsymbol{a}\boldsymbol{b}^T$ where $\boldsymbol{a}$ and $\boldsymbol{b}$ are vectors, which has at most one non-zero eigenvalue and this eigenvalue can be calculated by $\boldsymbol{b}^T \boldsymbol{a}$.

To minimize a function $f(\boldsymbol{x})$ with no constraint, the search direction $\boldsymbol{s}_k$ at each iteration is determined by

$$\boldsymbol{B}_k \boldsymbol{s}_k = -\nabla f(\boldsymbol{x}_k), \tag{5.1}$$

By Xin-She Yang
Copyright © 2010 John Wiley & Sons, Inc.

The BFGS Method

Choose an initial guess $\boldsymbol{x}_0$ and approximate $\boldsymbol{B}_0$ (e.g., $\boldsymbol{B}_0 = \boldsymbol{I}$)
while (criterion)
  Calculate $\boldsymbol{s}_k$ by solving $\boldsymbol{B}_k \boldsymbol{s}_k = -\nabla f(\boldsymbol{x}_k)$
  Find an optimal step size β_k by a line search method
  Update $\boldsymbol{x}_{k+1} = \boldsymbol{x}_k + \beta_k \boldsymbol{s}_k$
  Calculate $\boldsymbol{u}_k, \boldsymbol{v}_k$ and update $\boldsymbol{B}_{k+1}$ using (5.3) and (5.4)
end for while
Set $k = k + 1$

Figure 5.1: The pseudocode of the BFGS method.

where $\boldsymbol{B}_k$ is the approximation to the Hessian matrix at kth iteration. Then, a line search is performed to find the optimal stepsize β_k so that the new trial solution is determined by

$$\boldsymbol{x}_{n+1} = \boldsymbol{x}_k + \beta_k \boldsymbol{s}_k. \tag{5.2}$$

Introducing two new variables

$$\boldsymbol{u}_k = \boldsymbol{x}_{k+1} - \boldsymbol{x}_k = \beta_k \boldsymbol{s}_k, \qquad \boldsymbol{v}_k = \nabla f(\boldsymbol{x}_{k+1}) - \nabla f(\boldsymbol{x}_k), \tag{5.3}$$

we can update the new estimate as

$$\boldsymbol{B}_{k+1} = \boldsymbol{B}_k + \frac{\boldsymbol{v}_k \boldsymbol{v}_k^T}{\boldsymbol{v}_k^T \boldsymbol{u}_k} - \frac{(\boldsymbol{B}_k \boldsymbol{u}_k)(\boldsymbol{B}_k \boldsymbol{u}_k)^T}{\boldsymbol{u}_k^T \boldsymbol{B}_k \boldsymbol{u}_k}. \tag{5.4}$$

The procedure of the BFGS method is outlined in Figure 5.1.

5.2 NELDER-MEAD METHOD

5.2.1 A Simplex

In the n-dimensional space, a simplex, which is a generalization of a triangle on a plane, is a convex hull with $n+1$ distinct points. For simplicity, a simplex in the n-dimension space is referred to as n-simplex. Therefore, 1-simplex is a line segment, 2-simplex is a triangle, a 3-simplex is a tetrahedron (see Figure 5.2), and so on.

5.2.2 Nelder-Mead Downhill Simplex

The Nelder-Mead method is a downhill simplex algorithm for unconstrained optimization without using derivatives, and it was first developed by J. A. Nelder and R. Mead in 1965. This is one of the most widely used methods

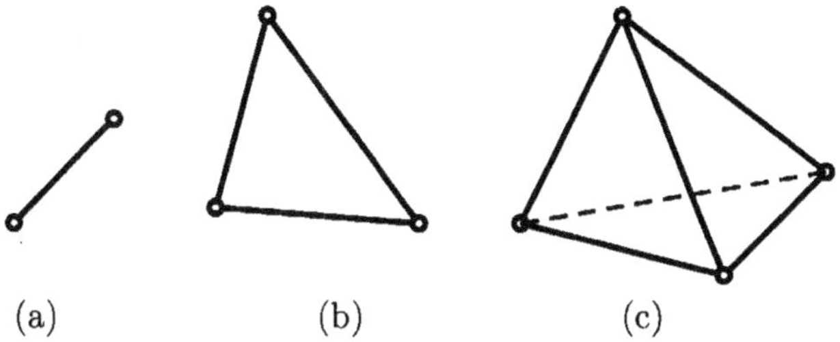

Figure 5.2: The concept of a simplex: (a) 1-simplex, (b) 2-simplex, and (c) 3-simplex.

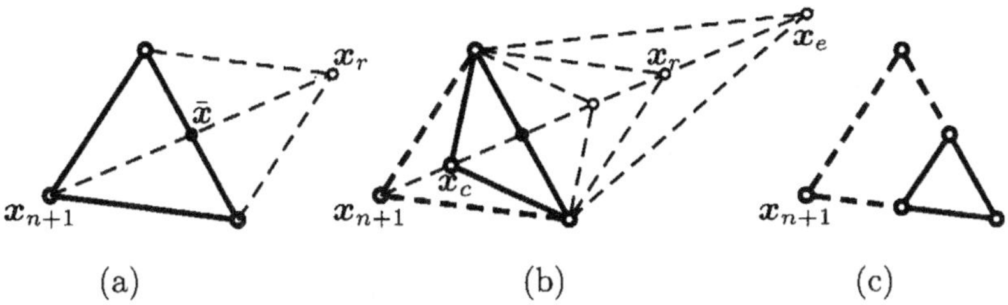

Figure 5.3: Simplex manipulations: (a) reflection with fixed volume (area), (b) expansion or contraction along the line of reflection, (c) reduction.

since its computational effort is relatively small and is something to get a quick grasp of the optimization problem. The basic idea of this method is to use the flexibility of the constructed simplex via amoeba-style manipulations by reflection, expansion, contraction and reduction (see Figure 5.3). In some books such as the best known *Numerical Recipes*, it is also called the 'Amoeba Algorithm'. It is worth pointing out that this downhill simplex method has nothing to do with the simplex method for linear programming.

There are a few variants of the algorithm that use slightly different ways of constructing initial simplex and convergence criteria. However, the fundamental procedure is the same (see Figure 5.4).

The first step is to construct an initial n-simplex with $n+1$ vertices and to evaluate the objective function at the vertices. Then, by ranking the objective values and re-ordering the vertices, we have an ordered set so that

$$f(\boldsymbol{x}_1) \leq f(\boldsymbol{x}_2) \leq ... \leq f(\boldsymbol{x}_{n+1}), \tag{5.5}$$

at $\boldsymbol{x}_1, \boldsymbol{x}_2, ..., \boldsymbol{x}_{n+1}$, respectively. As the downhill simplex method is for minimization, we use the convention that $\boldsymbol{x}_{n+1}$ is the worse point (solution), and $\boldsymbol{x}_1$ is the best solution. Then, at each iteration, similar ranking manipulations are carried out.

Then, we have to calculate the centroid $\boldsymbol{x}$ of simplex excluding the worst vertex $\boldsymbol{x}_{n+1}$

$$\bar{\boldsymbol{x}} = \frac{1}{n}\sum_{i=1}^{n} \boldsymbol{x}_i. \tag{5.6}$$

Using the centroid as the basis point, we try to find the reflection of the worse point $\boldsymbol{x}_{n+1}$ by

$$\boldsymbol{x}_r = \bar{\boldsymbol{x}} + \alpha(\bar{x} - \boldsymbol{x}_{n+1}), \qquad (\alpha > 0), \tag{5.7}$$

though the typical value of $\alpha = 1$ is often used.

Whether the new trial solution is accepted or not and how to update the new vertex, depends on the objective function at $\boldsymbol{x}_r$. There are three possibilities:

- If $f(\boldsymbol{x}_1) \leq f(\boldsymbol{x}_r) < f(\boldsymbol{x}_n)$, then replace the worst vertex $\boldsymbol{x}_{n+1}$ by $\boldsymbol{x}_r$, that is $\boldsymbol{x}_{n+1} \leftarrow \boldsymbol{x}_r$.

- If $f(\boldsymbol{x}_r) < f(\boldsymbol{x}_1)$ which means the objective improves, we then seek a more bold move to see if we can improve the objective even further by moving/expanding the vertex further along the line of reflection to a new trial solution

$$\boldsymbol{x}_e = \boldsymbol{x}_r + \beta(\boldsymbol{x}_r - \bar{\boldsymbol{x}}), \tag{5.8}$$

  where $\beta = 2$. Now we have to test if $f(\boldsymbol{x}_e)$ improves even better. If $f(\boldsymbol{x}_e) < f(\boldsymbol{x}_r)$, we accept it and update $\boldsymbol{x}_{n+1} \leftarrow \boldsymbol{x}_e$; otherwise, we can use the result of the reflection. That is $\boldsymbol{x}_{n+1} \leftarrow \boldsymbol{x}_r$.

- If there is no improvement or $f(\boldsymbol{x}_r) > f(\boldsymbol{x}_n)$, we have to reduce the size of the simplex whiling maintaining the best sides. This is contraction

$$\boldsymbol{x}_c = \boldsymbol{x}_{n+1} + \gamma(\bar{\boldsymbol{x}} - \boldsymbol{x}_{n+1}), \tag{5.9}$$

  where $0 < \gamma < 1$, though $\gamma = 1/2$ is usually used. If $f(\boldsymbol{x}_c) < f(\boldsymbol{x}_{n+1})$ is true, we then update $\boldsymbol{x}_{n+1} \leftarrow \boldsymbol{x}_c$.

If all the above steps fail, we should reduce the size of the simplex towards the best vertex $\boldsymbol{x}_1$. This is the reduction step

$$\boldsymbol{x}_i = \boldsymbol{x}_1 + \delta(\boldsymbol{x}_i - \boldsymbol{x}_1), \qquad (i = 2, 3, ..., n+1). \tag{5.10}$$

Then, we go to the first step of the iteration process and start over again.

5.3 TRUST-REGION METHOD

The fundamental ideas of the trust-region have developed over many years with many seminal papers by a dozen of pioneers. A good history review of the trust-region methods can be found in the book by Conn, Gould and Toint (2000). Briefly speaking, the first important work was due to Levenberg

Nelder-Mead (Downhill Simplex) Method

Initialize a simplex with $n+1$ vertices in n dimension.
while (stop criterion is not true)
(1) Re-order the points such that $f(\boldsymbol{x}_1) \leq f(\boldsymbol{x}_2) \leq ... \leq f(\boldsymbol{x}_{n+1})$
with $\boldsymbol{x}_1$ being the best and $\boldsymbol{x}_{n+1}$ being the worse (highest value)
(2) Find the centroid $\bar{\boldsymbol{x}}$ using $\bar{\boldsymbol{x}} = \sum_{i=1}^{n} \boldsymbol{x}_i/n$ excluding $\boldsymbol{x}_{n+1}$.
(3) Generate a trial point via the reflection of the worse vertex
$\boldsymbol{x}_r = \bar{\boldsymbol{x}} + \alpha(\bar{\boldsymbol{x}} - \boldsymbol{x}_{n+1})$ where $\alpha > 0$ (typically $\alpha = 1$)
(a) **if** $f(\boldsymbol{x}_1) \leq f(\boldsymbol{x}_r) < f(\boldsymbol{x}_n)$, $\boldsymbol{x}_{n+1} \leftarrow \boldsymbol{x}_r$; **end**
(b) **if** $f(\boldsymbol{x}_r) < f(\boldsymbol{x}_1)$,
Expand in the direction of reflection $\boldsymbol{x}_e = \boldsymbol{x}_r + \beta(\boldsymbol{x}_r - \bar{\boldsymbol{x}})$
if $f(\boldsymbol{x}_e) < f(\boldsymbol{x}_r)$, $\boldsymbol{x}_{n+1} \leftarrow \boldsymbol{x}_e$; **else** $\boldsymbol{x}_{n+1} \leftarrow \boldsymbol{x}_r$; **end**
end
(c) **if** $f(\boldsymbol{x}_r) > f(\boldsymbol{x}_n)$, Contract by $\boldsymbol{x}_c = \boldsymbol{x}_{n+1} + \gamma(\bar{\boldsymbol{x}} - \boldsymbol{x}_{n+1})$;
if $f(\boldsymbol{x}_c) < f(\boldsymbol{x}_{n+1})$, $\boldsymbol{x}_{n+1} \leftarrow \boldsymbol{x}_c$;
else Reduction by $\boldsymbol{x}_i = \boldsymbol{x}_1 + \delta(\boldsymbol{x}_i - \boldsymbol{x}_1)$, $(i = 2, ..., n+1)$; **end**
end
end

Figure 5.4: Pseudocode of Nelder-Mead's downhill simplex method.

in 1944, which proposed the usage of addition of a multiple of the identity matrix to the Hessian matrix as a damping measure to stabilize the solution procedure for nonlinear least-squares problems. Later, Marquardt in 1963 independently pointed out the link between such damping in the Hessian and the reduction of the step size in a restricted region. Slightly later in 1966, Goldfelt, Quandt and Trotter essentially set the stage for trust-region methods by introducing an explicit updating formula for the maximum step size. Then, in 1970, Powell proved the global convergence for the trust-region method, though it is believed that the term 'trust region' was coined by Dennis in 1978, as earlier literature used various terminologies such as region of validity, confidence region, and restricted step method.

In the trust-region algorithm, a fundamental step is to approximate the nonlinear objective function by using truncated Taylor expansions, often the quadratic form

$$\phi_k(\boldsymbol{x}) \approx f(\boldsymbol{x}_k) + \nabla f(\boldsymbol{x}_k)^T \boldsymbol{u} + \frac{1}{2}\boldsymbol{u}^T \boldsymbol{H}_k \boldsymbol{u}, \tag{5.11}$$

in a so-called trust region Ω_k defined by

$$\Omega_k = \{\boldsymbol{x} \in \Re^d \Big| \|\boldsymbol{\Gamma}(\boldsymbol{x} - \boldsymbol{x}_k)\| \leq \Delta_k\}, \tag{5.12}$$

where Δ_k is the trust-region radius. Here $\boldsymbol{H}_k$ is the local Hessian matrix. $\boldsymbol{\Gamma}$ is a diagonal scaling matrix that is related to the scalings of the optimization

Trust-Region Method

Start at an initial guess $\boldsymbol{x}_0$ and radius Δ_0 of the trust region Ω_0.
Initialize algorithm constants: $0 < \alpha_1 \leq \alpha_2 < 1$, $0 < \beta_1 \leq \beta_2 < 1$.
while (stop criterion)
Construct an approximate model $\phi_k(\boldsymbol{u})$ for the objective $f(\boldsymbol{x}_k)$ in Ω_k.
Find a trial point $\boldsymbol{x}_{k+1}$ with a sufficient model decrease inside Ω_k.
Calculate the ratio γ_k of the achieved versus predicted decrease:
    $\gamma_k = \frac{f(\boldsymbol{x}_k)-f(\boldsymbol{x}_{k+1})}{\phi_k(\boldsymbol{x}_k)-\phi_k(\boldsymbol{x}_{k+1})}$.
    if $\gamma_k \geq \alpha_1$,
        Accept the move and update the trust-region: $\boldsymbol{x}_k \leftarrow \boldsymbol{x}_{k+1}$;
        if $\gamma_k \geq \alpha_2$, $\Delta_{k+1} \in [\Delta_k, \infty)$; **end if**
        if $\gamma_k \in [\alpha_1, \alpha_2)$, $\Delta_{k+1} \in [\beta_2 \Delta_k, \Delta_k]$; **end if**
    else
        Discard the move and reduce the trust-region radius Δ_{k+1};
        $\Delta_{k+1} \in [\beta_1 \Delta_k, \beta_2 \Delta_k]$.
        end
Update $k = k + 1$
end

Figure 5.5: Pseudocode of a trust region method.

problem. Thus, the shape of the trust region is a hyperellipsoid, and an elliptical region in 2D centered at $\boldsymbol{x}_k$. If the parameters are equally scaled, then $\boldsymbol{\Gamma} = \boldsymbol{I}$ can be used.

The approximation to the objective function in the trust region will make it simpler to find the next trial solution $\boldsymbol{x}_{k+1}$ from the current solution $\boldsymbol{x}_k$. Then, we intend to find $\boldsymbol{x}_{k+1}$ with a sufficient decrease in the objective function. How good the approximation ϕ_k is to the actual objective $f(\boldsymbol{x})$ can be measured by the ratio of the achieved decrease to the predicted decrease

$$\gamma_k = \frac{f(\boldsymbol{x}_k) - f(\boldsymbol{x}_{k+1})}{\phi_k(\boldsymbol{x}_k) - \phi_k(\boldsymbol{x}_{k+1})}. \tag{5.13}$$

If this ratio is close to unity, we have a good approximation and then should move the trust region to $\boldsymbol{x}_{k+1}$.

Now the question is what radius should we use for the newly updated trust region centered at $\boldsymbol{x}_{k+1}$? Since the move is successful, and the decrease is significant, we should be bold enough to expand the trust region a little. A standard measure of such significance in decrease is to use a parameter $\alpha_1 \approx 0.01$. If $\gamma_k > \alpha_1$, the achieved decrease is noticeable, so we should accept the move (i.e., $\boldsymbol{x}_{k+1} \leftarrow \boldsymbol{x}_k$). What radius should we now use? Conventionally, we use another parameter $\alpha_2 > \alpha_1$ as an additional criterion. If γ_k is about $O(1)$ or $\gamma_k \geq \alpha_2 \approx 0.9$, we say that decrease is significant, and we can boldly increase the trust-region radius. Typically, we choose a value $\Delta_{k+1} \in [\Delta_k, \infty)$.

The actual choice may depend on the problem, though typically $\Delta_{k+1} \approx 2\Delta_k$. If the decrease is noticeable but not so significant, that is $\alpha_1 < \gamma_k \leq \alpha_2$, we should shrink the trust region so that $\Delta_{k+1} \in [\beta_2 \Delta_k, \Delta_k]$ or

$$\beta_2 \Delta_k < \Delta_{k+1} < \Delta_k, \qquad (0 < \beta_2 < 1). \tag{5.14}$$

Obviously, if the decrease is too small or $\gamma_k < \alpha_1$, we should abandon the move as the approximation is not good enough over this larger region. We should seek a better approximation on a smaller region by reducing the trust-region radius

$$\Delta_{k+1} \in [\beta_1 \Delta_k, \beta_2 \Delta_k], \tag{5.15}$$

where $0 < \beta_1 \leq \beta_2 < 1$, and typically $\beta_1 = \beta_2 = 1/2$, which means half the original size is used first.

To summarize, the typical values of the parameters are:

$$\alpha_1 = 0.01, \;\; \alpha_2 = 0.9, \qquad \beta_1 = \beta_2 = \frac{1}{2}. \tag{5.16}$$

5.4 SEQUENTIAL QUADRATIC PROGRAMMING

5.4.1 Quadratic Programming

A special type of nonlinear programming is quadratic programming whose objective function is a quadratic form

$$f(\boldsymbol{x}) = \frac{1}{2}\boldsymbol{x}^T \boldsymbol{Q} \boldsymbol{x} + \boldsymbol{b}^T \boldsymbol{x} + \boldsymbol{c}, \tag{5.17}$$

where $\boldsymbol{b}$ and $\boldsymbol{c}$ are constant vectors. $\boldsymbol{Q}$ is a symmetric square matrix. The constraints can be incorporated using Lagrange multipliers and KKT formulation.

5.4.2 Sequential Quadratic Programming

Sequential (or successive) quadratic programming (SQP) represents one of the state-of-art and most popular methods for nonlinear constrained optimization. It is also one of the robust methods. For a general nonlinear optimization problem

$$\text{minimize } f(\boldsymbol{x}), \tag{5.18}$$

$$\text{subject to } \; h_i(\boldsymbol{x}) = 0, \; (i = 1, ..., p), \tag{5.19}$$

$$g_j(\boldsymbol{x}) \leq 0, \; (j = 1, ..., q). \tag{5.20}$$

The fundamental idea of sequential quadratic programming is to approximate the computationally extensive full Hessian matrix using a quasi-Newton updating method. Subsequently, this generates a subproblem of quadratic programming (called QP subproblem) at each iteration, and the solution to this subproblem can be used to determine the search direction and next trial solution.

Using the Taylor expansions, the above problem can be approximated, at each iteration, as the following problem

$$\text{minimize } \frac{1}{2}\boldsymbol{s}^T \nabla^2 L(\boldsymbol{x}_k)\boldsymbol{s} + \nabla f(\boldsymbol{x}_k)^T \boldsymbol{s} + f(\boldsymbol{x}_k), \tag{5.21}$$

$$\text{subject to } \nabla h_i(\boldsymbol{x}_k)^T \boldsymbol{s} + h_i(\boldsymbol{x}_k) = 0, \ (i = 1, ..., p), \tag{5.22}$$

$$\nabla g_j(\boldsymbol{x}_k)^T \boldsymbol{s} + g_j(\boldsymbol{x}_k) \leq 0, \ (j = 1, ..., q), \tag{5.23}$$

where the Lagrange function, also called merit function, is defined by

$$L(\boldsymbol{x}) = f(\boldsymbol{x}) + \sum_{i=1}^{p} \lambda_i h_i(\boldsymbol{x}) + \sum_{j=1}^{q} \mu_j g_j(\boldsymbol{x})$$

$$= f(\boldsymbol{x}) + \boldsymbol{\lambda}^T \boldsymbol{h}(\boldsymbol{x}) + \boldsymbol{\mu}^T \boldsymbol{g}(\boldsymbol{x}), \tag{5.24}$$

where $\boldsymbol{\lambda} = (\lambda_1, ..., \lambda_p)^T$ is the vector of Lagrange multipliers, and $\boldsymbol{\mu} = (\mu_1, ..., \mu_q)^T$ is the vector of KKT multipliers. Here we have used the notation $\boldsymbol{h} = (h_1(\boldsymbol{x}), ..., h_p(\boldsymbol{x}))^T$ and $\boldsymbol{g} = (g_1(\boldsymbol{x}), ..., g_q(\boldsymbol{x}))^T$.

To approximate the Hessian $\nabla^2 L(\boldsymbol{x}_k)$ by a positive definite symmetric matrix $\boldsymbol{H}_k$, the standard Broydon-Fletcher-Goldfarbo-Shanno (BFGS) approximation of the Hessian can be used, and we have

$$\boldsymbol{H}_{k+1} = \boldsymbol{H}_k + \frac{\boldsymbol{v}_k \boldsymbol{v}_k^T}{\boldsymbol{v}_k^T \boldsymbol{u}_k} - \frac{\boldsymbol{H}_k \boldsymbol{u}_k \boldsymbol{u}_k^T \boldsymbol{H}_k^T}{\boldsymbol{u}_k^T \boldsymbol{H}_k \boldsymbol{u}_k}, \tag{5.25}$$

where

$$\boldsymbol{u}_k = \boldsymbol{x}_{k+1} - \boldsymbol{x}_k, \tag{5.26}$$

and

$$\boldsymbol{v}_k = \nabla L(\boldsymbol{x}_{k+1}) - \nabla L(\boldsymbol{x}_k). \tag{5.27}$$

The QP subproblem is solved to obtain the search direction

$$\boldsymbol{x}_{k+1} = \boldsymbol{x}_k + \alpha \boldsymbol{s}_k, \tag{5.28}$$

using a line search method by minimizing a penalty function, also commonly called merit function,

$$\Phi(\boldsymbol{x}) = f(\boldsymbol{x}) + \rho \Big[\sum_{i=1}^{p} |h_i(\boldsymbol{x})| + \sum_{j=1}^{q} \max\{0, g_j(\boldsymbol{x})\} \Big], \tag{5.29}$$

where ρ is the penalty parameter.

It is worth pointing out that any SQP method requires a good choice of $\boldsymbol{H}_k$ as the approximate Hessian of the Lagrangian L. Obviously, if $\boldsymbol{H}_k$ is exactly

Sequential Quadratic Programming

Choose a starting point $\boldsymbol{x}_0$ and approximation H_0 to the Hessian.
repeat $k = 1, 2, ...$
  Solve a QP subproblem: QP_k to get the search direction $\boldsymbol{s}_k$
  Given $\boldsymbol{s}_k$, find α so as to determine $\boldsymbol{x}_{k+1}$
  Update the approximate Hessian $\boldsymbol{H}_{k+1}$ using the BFGS scheme
  $k = k + 1$
until (stop criterion)

Figure 5.6: Procedure of sequential quadratic programming.

calculated as $\nabla^2 L$, SQP essentially becomes Newton's method for solving the optimality condition. A popular way to approximate the Lagrangian Hessian is to use a quasi-Newton scheme as we used the BFGS formula described earlier.

In this chapter, we have outlined several widely used algorithms without providing any examples. The main reason is that the description of such an example may be lengthy, which also depends on the actual implementation. However, there are both commercial and open source software packages for all these algorithms. For example, the Matlab optimization toolbox implemented all these algorithms.

EXERCISES

5.1 Implement the BFGS method in a programming language of your choice so as to find the global minimum of Easom's function

$$f(x) = \cos(x) e^{-(x-\pi)^2}, \qquad x \in [-5, 5].$$

Investigate the effect of starting from different initial solutions, say $x_0 = -5$ and $x_0 = 5$.

5.2 Write a simple program to implement the standard Nelder-Mead method and then use it to find the minimum of

$$f(\boldsymbol{x}) = \Big\{ \sum_{i=1}^{n} [x_i - (i - 1/i)]^{2p} \Big\}^{1/2p}, \qquad (p = 1, 2, ...),$$

where $-n \leq x_i \leq n$ and $(n = 1, 2, ...)$.

5.3 Stochastic tunneling is a good idea for changing the response surface of an objective function f, which suppresses certain (visited) modes while retaining the modes of unvisited regions. Show that

$$\psi = 1 - \exp[-\gamma(f - f_{\min})], \qquad \gamma > 0,$$

preserves that the location of the global minimum. Here $f_{\min}$ is the current minimum found so far during iterations, and $\gamma > 0$ is a scaling factor which can be set $\gamma = O(1/f)$.

REFERENCES

1. C. G. Broyden, "The convergence of a class of double-rank minimization algorithms", *IMA J. Applied Math.*, **6**, 76-90 (1970).
2. A. R. Conn, N. I. M. Gould, P. L. Toint, *Trust-region methods*, SIAM & MPS, 2000.
3. J. E. Dennis, "A brief introduction to quasi-Newton methods", in: *Numerical Analysis: Proceedings of Symposia in Applied Mathematics* (Eds G. H. Golub and J. Oliger, pp. 19-52 (1978).
4. A. V. Fiacco and G. P. McCormick, *Nonlinear Porgramming: Sequential Unconstrained Minimization Techniques*, John Wiley & Sons, 1969.
5. R. Fletcher, "A new approach to variable metric algorithms", *Computer Journal*, **13** 317-322 (1970).
6. D. Goldfarb, "A family of variable metric update derived by variational means", *Mathematics of Computation*, **24**, 23-26 (1970).
7. S. M. Goldeldt, R. E. Quandt, and H. F. Trotter, "Maximization by quadratic hill-climbing", *Econometrica*, **34**, 541-551, (1996).
8. N. Karmarkar, "A new polynomial-time algorithm for linear programming", *Combinatorica*, **4** (4), 373-395 (1984).
9. K. Levenberg, "A method for the solution of certain problems in least squares", *Quart. J. Applied Math.*, **2**, 164-168, (1944).
10. D. Marquardt, "An algorithm for least-squares estimation of nonlinear parameters", *SIAM J. Applied Math.*, **11**, 431-441 (1963).
11. J. A. Nelder and R. Mead, "A simplex method for function optimization", *Computer Journal*, **7**, 308-313 (1965).
12. M. J. D. Powell, "A new algorithm for unconstrained optimization", in: *Nonlinear Programming* (Eds J. B. Rosen, O. L. Mangasarian, and K. Ritter), pp. 31-65 (1970).
13. D. F. Shanno, "Conditioning of quasi-Newton methods for function minimization", *Mathematics of Computation*, **25**, 647-656 (1970).

CHAPTER 6

CONVEX OPTIMIZATION

Convex optimization is a special class of nonlinear optimization that has become a central part of engineering optimization. The reasons are twofold: any local optimum is also the global optimum for convex optimization, and many engineering optimization problems can be reformulated as a convex optimization problem. In addition, there are rigorous theoretical basis and important theorems for convex optimization. Most algorithms such as interior-point methods are not only theoretically sound, but also computationally efficient.

6.1 KKT CONDITIONS

From earlier chapters, we know that KKT conditions play a central role in nonlinear optimization as they are optimality conditions. For a generic nonlinear optimization

$$\underset{\boldsymbol{x} \in \Re^n}{\text{minimize}} f(\boldsymbol{x}),$$

$$\text{subject to } h_i(\boldsymbol{x}) = 0, \ (i = 1, ..., M),$$

$$g_j(\boldsymbol{x}) \leq 0, \ \ (j = 1, ..., N), \tag{6.1}$$

Engineering Optimization: An Introduction with Metaheuristic Applications.
By Xin-She Yang
Copyright © 2010 John Wiley & Sons, Inc.

its corresponding Karush-Kuhn-Tucker (KKT) optimality conditions can be written as

$$\nabla f(\boldsymbol{x}_*) + \sum_{i=1}^{M} \mu_i \nabla h_i(\boldsymbol{x}_*) + \sum_{j=1}^{N} \lambda_j \nabla g_j(\boldsymbol{x}_*) = 0, \tag{6.2}$$

and

$$g_j(\boldsymbol{x}_*) \leq 0, \qquad \lambda_j g_j(\boldsymbol{x}_*) = 0, \;\; (j = 1, 2, ..., N), \tag{6.3}$$

with the non-negativity condition

$$\lambda_j \geq 0, (i = 0, 1, ..., N). \tag{6.4}$$

However, there is no restriction on the sign of μ_i. Using componentwise notation for simplicity and letting $\boldsymbol{\lambda} = (\lambda_1, ..., \lambda_N)^T$, the above non-negativity condition can be written as

$$\boldsymbol{\lambda} \succeq 0. \tag{6.5}$$

It is worth pointing out that we have used slightly different notation conventions here that are different from what we used in Section 4.9 by swapping λ and μ. Though various notation conventions are used, the main reason here is to try to use similar notations to those used in Boyd and Vandenberghe's excellent book on convex optimization, and makes things easier if you wish to pursue further studies in these areas.

Under some special requirements, the KKT condition can guarantee a global optimality. That is, if $f(\boldsymbol{x})$ is convex and twice-differentiable, all $g_j(\boldsymbol{x})$ are convex function and twice-differentiable, and all $h_i(\boldsymbol{x})$ are affine, then the Hessian matrix

$$\boldsymbol{H} = \nabla^2 \mathcal{L}(\boldsymbol{x}_*) = \nabla^2 f(\boldsymbol{x}_*) + \sum_i \mu_i \nabla^2 h_(\boldsymbol{x}_*) + \sum_j \lambda_j \nabla^2 g_j(\boldsymbol{x}_*), \tag{6.6}$$

is positive definite. We then have a convex optimization problem

$$\begin{aligned} &\underset{\boldsymbol{x} \in \Re^n}{\text{minimize}} \; f(\boldsymbol{x}), \\ &\text{subject to} \;\; g_i(\boldsymbol{x}) \leq 0, \;\; (i = 1, 2, ..., N), \\ &\boldsymbol{A}\boldsymbol{x} = \boldsymbol{b}, \end{aligned} \tag{6.7}$$

where $\boldsymbol{A}$ and $\boldsymbol{b}$ are an $M \times M$ matrix and an M-vector, respectively. Then the KKT conditions are both sufficient and necessary conditions. Consequently there exists a solution $\boldsymbol{x}_*$ that satisfies the above KKT optimality conditions, and this solution $\boldsymbol{x}_*$ is in fact the global optimum. In this special convex case, the KKT conditions become

$$\boldsymbol{A}\boldsymbol{x}_* = \boldsymbol{b}, \qquad g_i(\boldsymbol{x}_*) \leq 0, \;\; (i = 1, 2, ..., N), \tag{6.8}$$

$$\lambda_i g_i(\boldsymbol{x}_*) = 0, \;\; (i = 1, 2, ..., N), \tag{6.9}$$

$$\nabla f(\boldsymbol{x}_*) + \sum_{i=1}^{N} \lambda_i \nabla g_i(\boldsymbol{x}_*) + \boldsymbol{A}^T \boldsymbol{\mu} = 0, \tag{6.10}$$

$$\boldsymbol{\lambda} \succeq 0. \tag{6.11}$$

This optimization problem (6.7) or its corresponding KKT conditions can be solved efficiently by using the interior-point methods that employ Newton's method to solve a sequence of either equality constrained problems or modified versions of the above KKT conditions. The interior-point methods are a class of methods, and the convergence of a popular version, called the barrier method, has been proved mathematically. Newton's method can also solve the equality constrained problem directly, rather than transforming it into an unconstrained one.

So let us first see a few examples before we proceed to discuss Newton's method for equality constrained problem and the idea of barrier functions.

6.2 CONVEX OPTIMIZATION EXAMPLES

Convex optimization has become increasing important in engineering and many other disciplines. This is largely due to the significant development in polynomial time algorithms such as the interior-point methods to be discussed later. In addition, many problem in science and engineering can be reformulated as some equivalent convex optimization problems. A vast list of detailed examples can be found in more advanced literature such as Boyd and Vandenberghe's book. Here we only briefly introduce a few examples.

As a first example, we have to solve a linear system

$$\boldsymbol{A}\boldsymbol{u} = \boldsymbol{b}, \tag{6.12}$$

where $\boldsymbol{A}$ is an $m \times n$ matrix, $\boldsymbol{b}$ is an n-vector, and $\boldsymbol{u}$ is the unknown vector. The uniqueness of the solution requires $m \geq n$. To solve this equation, we can try to minimize the residual

$$\boldsymbol{r} = \boldsymbol{A}\boldsymbol{u} - \boldsymbol{b}, \tag{6.13}$$

which will lead to the exact solution if $\boldsymbol{r} = 0$. Therefore, this is equivalent to the following minimization problem

$$\text{minimize} \quad ||\boldsymbol{A}\boldsymbol{u} - \boldsymbol{b}||. \tag{6.14}$$

In the special case when $\boldsymbol{A}$ is symmetric and positive semidefinite, we can rewrite it as a convex quadratic function

$$\text{minimize} \quad \frac{1}{2}\boldsymbol{u}^T \boldsymbol{A}\boldsymbol{u} - \boldsymbol{b}^T \boldsymbol{u}. \tag{6.15}$$

In many applications, we often have to carry out some curve-fitting using some experimental data set. A very widely used method is the method of least squares. For a set of m observations (x_i, y_i) where $i = 1, 2, ..., m$, we often try to best-fit the data to a polynomial function

$$p(x) = \alpha_1 + \alpha_2 x + ... + \alpha_n x^{n-1}, \tag{6.16}$$

by minimizing the errors or residuals

$$\boldsymbol{r} = [r_1, ..., r_m]^T = [p(x_1) - y_1, ..., p(x_n) - y_m]^T, \tag{6.17}$$

so as to determine the coefficients

$$\boldsymbol{\alpha} = \big(\alpha_1, ..., \alpha_n\big)^T. \tag{6.18}$$

That is to minimize the square of the ℓ_2-norm

$$\text{minimize } ||\boldsymbol{r}||_2^2 = ||\boldsymbol{A\alpha} - \boldsymbol{b}||_2^2 = r_1^2 + ... + r_m^2. \tag{6.19}$$

This is equivalent to the minimization of the following convex quadratic function

$$f(\boldsymbol{x}) = \boldsymbol{\alpha}^T \boldsymbol{A}^T \boldsymbol{A\alpha} - 2\boldsymbol{b}^T \boldsymbol{A\alpha} + \boldsymbol{b}^T \boldsymbol{b}. \tag{6.20}$$

The stationary condition leads to

$$\nabla f(\boldsymbol{x}) = 2\boldsymbol{A}^T \boldsymbol{A\alpha} - 2\boldsymbol{A}^T \boldsymbol{b} = 0, \tag{6.21}$$

which gives the standard normal equation commonly used in the method of least-squares

$$\boldsymbol{A}^T \boldsymbol{A\alpha} = \boldsymbol{A}^T \boldsymbol{b}, \tag{6.22}$$

whose solution can be written as

$$\boldsymbol{\alpha} = (\boldsymbol{A}^T \boldsymbol{A})^{-1} \boldsymbol{A}^T \boldsymbol{b}. \tag{6.23}$$

For a given data set, the goodness of best-fit typically increases as increase the degree of the polynomial (or the model complexity), but this often introduces strong oscillations. Therefore, there is a fine balance between the goodness of the fit and the complexity of the mathematical model, and this balance can be achieved by using penalty or regularization. The idea is to choose the simpler model if the goodness of the best-fit is essentially the same. Here, the goodness of the fit is represented by the sum of the residual squares or the ℓ_2-norm. For example, Tikhonov regularization method intends to minimize

$$\text{minimize } \quad ||\boldsymbol{A\alpha} - \boldsymbol{b}||_2^2 + \gamma ||\boldsymbol{\alpha}||_2^2, \qquad \gamma \in (0, \infty), \tag{6.24}$$

where the first term of the objective is the standard least-squares, while the second intends to penalize the complex models. Here γ is called the penalty or regularization parameter. This problem is equivalent to the following convex quadratic optimization

$$\text{minimize } \quad f(\boldsymbol{x}) = \boldsymbol{\alpha}^T (\boldsymbol{A}^T \boldsymbol{A} + \gamma \boldsymbol{I}) \boldsymbol{\alpha} - 2\boldsymbol{b}^T \boldsymbol{A\alpha} + \boldsymbol{b}^T \boldsymbol{b}, \tag{6.25}$$

where $\boldsymbol{I}$ is the identity matrix. The optimality condition requires that

$$\nabla f(\boldsymbol{x}) = 2(\boldsymbol{A}^T\boldsymbol{A} + \gamma\boldsymbol{I})\boldsymbol{\alpha} - 2\boldsymbol{A}^T\boldsymbol{b} = 0, \tag{6.26}$$

which leads to

$$(\boldsymbol{A}^T\boldsymbol{A} + \gamma\boldsymbol{I})\boldsymbol{\alpha} = \boldsymbol{A}^T\boldsymbol{b}, \tag{6.27}$$

whose solution is

$$\boldsymbol{\alpha} = (\boldsymbol{A}^T\boldsymbol{A} + \gamma\boldsymbol{I})^{-1}\boldsymbol{A}^T\boldsymbol{b}. \tag{6.28}$$

It is worth pointing out that the inverse does exist due to the fact that $\boldsymbol{A}^T\boldsymbol{A}+\gamma\boldsymbol{I} \succ 0$ for all $\gamma > 0$.

As a final example, the minimax or Chebyshev optimization uses the ℓ_∞-norm and intends to minimize

$$\text{minimize} \quad \|\boldsymbol{A\alpha} - \boldsymbol{b}\|_\infty = \max\{|r_1|, ..., |r_m|\}. \tag{6.29}$$

Here the name is clearly from the minimization of the maximum absolute value of the residuals. We can essentially rewrite this problem in terms of a linear programming problem with a parameter $\theta \in \Re$

$$\begin{aligned} &\text{minimize} \quad \theta \\ &\text{subject to} \quad \boldsymbol{A\alpha} - \boldsymbol{b} \preceq \theta\boldsymbol{1} \\ &\qquad\qquad -\theta\boldsymbol{1} \preceq \boldsymbol{A\alpha} - \boldsymbol{b}, \end{aligned} \tag{6.30}$$

where $\boldsymbol{1}$ is a matrix with the same size as $\boldsymbol{b}$ with each entry being one. Obviously, this problem can be solved using linear programming techniques.

6.3 EQUALITY CONSTRAINED OPTIMIZATION

For an equality constrained convex optimization problem, Newton's method is very efficient. For simplicity, we now discuss the case when there is no inequality constraint at all (or $N = 0$) in (6.7), and we have the following equality constrained problem

$$\begin{aligned} &\text{minimize} \quad f(\boldsymbol{x}) \\ &\text{subject to} \quad \boldsymbol{Ax} = \boldsymbol{b}, \end{aligned} \tag{6.31}$$

where the objective function $f(\boldsymbol{x})$ is convex, continuous and twice differentiable. We often assume that there are fewer equality constraints than the number of design variables, that is, it is required that the $M \times n$ matrix $\boldsymbol{A}$ has a rank rank($\boldsymbol{A}$)=$M < n$. The KKT conditions (6.8) to (6.11) for this problem simply become

$$\boldsymbol{Ax}_* = \boldsymbol{b}, \qquad \nabla f(\boldsymbol{x}_*) + \boldsymbol{A}^T\boldsymbol{\mu} = 0, \tag{6.32}$$

which forms a set of $M + n$ equations for $M + n$ variables $\boldsymbol{x}_*$ (n variables) and $\boldsymbol{\mu}$ (M variables). We can solve this set of KKT conditions directly.

Newton's method with equality

Initial guess $\boldsymbol{x}_0$ and accuracy ϵ
while $(\theta^2/2 > \epsilon)$
  Calculate $\Delta\boldsymbol{x}_k$ using (6.34)
  Using a line search method to determine α
  Update $\boldsymbol{x}_{k+1} = \boldsymbol{x}_k + \alpha\Delta\boldsymbol{x}_k$
  Estimate θ using (6.35)
end

Figure 6.1: Newton's method for the equality constrained optimization.

However, instead of solving the KKT conditions, it is often more desirable to solve the original problem (6.31) directly, especially for the case when the matrix $\boldsymbol{A}$ is sparse. This is because the sparsity is often destroyed by converting the equality constrained problem into an unconstrained one in terms of, for example, the Lagrange multipliers.

Newton's method often provides the superior convergence over many other algorithms; however, it uses second derivative information directly. The move or descent direction is determined by the first derivative, and descent step or Newton step Δx_k has to be determined by solving an approximate quadratic equation. In order to solve this equality constrained convex optimization problem directly using Newton's method, we now reformulate it as the following

$$\begin{pmatrix} \nabla^2 f(\boldsymbol{x}) & \boldsymbol{A}^T \\ \boldsymbol{A} & 0 \end{pmatrix} \begin{pmatrix} \Delta\boldsymbol{x}_k \\ \boldsymbol{\xi} \end{pmatrix} = \begin{pmatrix} -\nabla f(\boldsymbol{x}) \\ 0 \end{pmatrix}, \tag{6.33}$$

where $\boldsymbol{\xi}$ is an optimal dual variable associated with the quadratic problem.

Similar to an unconstrained minimization problem, the Newton step can be determined by

$$\Delta\boldsymbol{x}_k = -\frac{\nabla f(\boldsymbol{x})}{\nabla^2 f(\boldsymbol{x})} = -[\nabla^2 f(\boldsymbol{x})]^{-1}\nabla f(\boldsymbol{x}). \tag{6.34}$$

However, for the equality constraint, the feasible descent direction has to be modified slightly. A direction $\boldsymbol{s}$ is feasible if $\boldsymbol{A}\boldsymbol{s} = 0$. Consequently, $\boldsymbol{x} + \alpha\boldsymbol{s}$ where $\alpha > 0$ is also feasible under the constraint $\boldsymbol{A}\boldsymbol{x} = \boldsymbol{b}$, and α can be obtained using a line search method. Boyd and Vandenberghe has proved that a good estimate of $f(\boldsymbol{x}) - f_*$ where f_* is the optimal value at $\boldsymbol{x} = \boldsymbol{x}_*$, and thus a stopping criterion, is $\theta^2(\boldsymbol{x})/2$ with $\theta(\boldsymbol{x})$ being given by

$$\theta(\boldsymbol{x}) = [\Delta\boldsymbol{x}_k^T \nabla^2 f(\boldsymbol{x}) \Delta\boldsymbol{x}_k]^{1/2}. \tag{6.35}$$

Finally, we can summarize Newton's method for the equality constrained optimization as shown in Figure 6.1.

Now for the optimization with inequality and equality constraints; if we can somehow remove the inequality constraints by incorporating them in the objective, then the original nonlinear convex optimization problem (6.7) can be converted into a convex one with equality constraints only. Subsequently, we can use the above Newton algorithm to solve it. This conversion typically requires some sort of barrier functions.

6.4 BARRIER FUNCTIONS

In optimization, a barrier function is essentially a continuous function that becomes singular or unbounded when approaching the boundary of the feasible region. Such singular behavior, often difficult to deal with in many mathematical problems, can be used as an advantage to impose constraints by significantly penalizing any potential violation of such constraints.

In order to apply standard optimization techniques such as Newton's method for equality constrained optimization, we often intend to reformulate the constrained optimization with inequality constraints by directly writing the inequality constraints in the objective function. One way to achieve this is to use the indicator function I_- which is defined by

$$I_-[u] = \begin{cases} 0 & \text{if } u \leq 0 \\ \infty & \text{if } u > 0 \end{cases}. \tag{6.36}$$

Now the convex optimization problem (6.7) can be rewritten as

$$\begin{cases} \text{minimize} \;\; f(\boldsymbol{x}) + \sum_{i=1}^{N} I_-[g_i(\boldsymbol{x})] \\ \\ \text{subject to} \;\; \boldsymbol{A}\boldsymbol{x} = \boldsymbol{b}. \end{cases} \tag{6.37}$$

It seems that the problem is now solvable, but the indicator has singularity and is not differentiable. Consequently, the objective function is not differentiable in general, and many derivative-based methods including Newton's method cannot be used for such optimization problems.

The main difficulty we have to overcome is to use a continuous function with similar property as the indicator function when approaching the boundary. That is why barrier functions serve this purpose well. A good approximation to an indicator function should be smooth enough, and we often use the following function

$$\bar{I}_-(u) = -\frac{1}{t}\log(-u), \qquad u < 0, \tag{6.38}$$

where $t > 0$ is an accuracy parameter for the approximation. Indeed, the function $\bar{I}_-$ is convex, non-decreasing and differentiable. Now we can rewrite the above convex optimization as

$$\text{minimize } f(\boldsymbol{x}) + \sum_{i=1}^{N} -\frac{1}{t}\log[-g_i(\boldsymbol{x})], \tag{6.39}$$

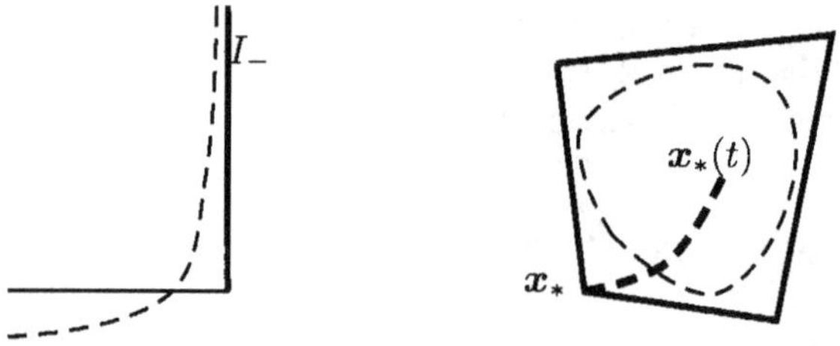

(a) (b)

Figure 6.2: The barrier method: (a) log barrier near a boundary, and (b) central path for $n = 2$ and $N = 4$.

$$\text{subject to} \quad \boldsymbol{A}\boldsymbol{x} = \boldsymbol{b}, \tag{6.40}$$

which can be solved using Newton's method. Here, the function

$$\psi(\boldsymbol{x}) = -\sum_{i=1}^{N} \log[-g_i(\boldsymbol{x})], \tag{6.41}$$

is called the logarithmic barrier function for this optimization problem, or simply the log barrier (see Figure 6.2). Its gradient can be expressed as

$$\nabla\psi(\boldsymbol{x}) = \sum_{i=1}^{N} \frac{1}{-g_i(\boldsymbol{x})} \nabla g_i(\boldsymbol{x}), \tag{6.42}$$

and its Hessian is

$$\nabla^2\psi(\boldsymbol{x}) = \sum_{i=1}^{N} \frac{1}{[g_i(\boldsymbol{x})]^2} \nabla g_i(\boldsymbol{x}) \nabla g_i(\boldsymbol{x})^T + \sum_{i=1}^{N} \frac{1}{-g_i(\boldsymbol{x})} \nabla^2 g_i(\boldsymbol{x}). \tag{6.43}$$

These are frequently used in optimization literature. It is worth pointing out that $-g_i(\boldsymbol{x})$ is always non-negative due to the fact that $g_i(\boldsymbol{x}) \leq 0$. In addition, since $g_i(\boldsymbol{x})$ are convex, and $-\log()$ is convex, from the composition rule of convex functions, it is straightforward to check that $\Phi(\boldsymbol{x})$ is indeed convex, and also twice continuously differentiable. Using this log barrier, we can write the convex optimization problem (6.37) as

$$\begin{aligned} &\text{minimize} \quad f(\boldsymbol{x}) + \tfrac{1}{t}\psi(\boldsymbol{x}) \\ &\text{subject to} \quad \boldsymbol{A}\boldsymbol{x} = \boldsymbol{b}. \end{aligned} \tag{6.44}$$

From the above equations, we can see that $t \to \infty$, the modified objective approaches the original objective as the second term becomes smaller and smaller, as the log barrier is an approximation whose accuracy is controlled by a parameter t. As t increases, the approximation improves; however, the Hessian varies more dramatically near the boundary, and thus makes it difficult to use Newton's method. Now the question is how large t should be enough?

To get around this problem, we can use a sequence of problems (6.44) by increasing the values of t gradually, and starting the optimization from the previous solution at the previous value of t. As $t > 0$, the multiplication of the objective by t will not affect the optimality of the problem. So we can rewrite the above problem as

$$\begin{aligned} &\text{minimize} \;\; tf(\boldsymbol{x}) + \psi(\boldsymbol{x}) \\ &\text{subject to} \;\; \boldsymbol{A}\boldsymbol{x} = \boldsymbol{b}. \end{aligned} \tag{6.45}$$

For any given t, there should be an optimal solution $\boldsymbol{x}_*$ to this problem. Due to the convexity of the problem, we can assume that this solution is unique for a specific value of t. As a result, the solution $\boldsymbol{x}_*$ is a function of t, that is $\boldsymbol{x}_*(t)$. As we have seen earlier in linear programming, the search is carried out along the boundaries, not the interior of the feasible region. Here for convex optimization (6.7), the solution $\boldsymbol{x}_*(t)$ corresponds a point, usually an interior point, and the set of all the points $\boldsymbol{x}_*(t)$ for $t > 0$ will trace out a path (see Figure 6.2), called central path associated with the original problem (6.7). In Figure 6.2, the central path is sketched for the case of $n = 2$ (two variables) and $N = 4$ (four inequality constraints).

The point $\boldsymbol{x}_*(t)$ is strictly feasible and satisfies the following condition

$$\boldsymbol{A}\boldsymbol{x}_*(t) = \boldsymbol{b}, \qquad g_i(\boldsymbol{x}_*(t)) < 0, \tag{6.46}$$

for $i = 1, ..., N$. It is worth pointing out that we cannot include the equal sign in the inequality $g_i \leq 0$ because $\psi \to \infty$ as $g_i = 0$. This means that the inequalities hold strictly, and consequently the points are interior points. Thus, the name for the interior-point methods.

Applying the optimality condition for the equality constrained problem (6.45) in the same manner as (6.32), we have

$$t\nabla f(\boldsymbol{x}_*(t)) + \nabla\psi(\boldsymbol{x}_*(t)) + \boldsymbol{A}^T\boldsymbol{\mu} = 0, \tag{6.47}$$

or

$$t\nabla f(\boldsymbol{x}_*(t)) + \sum_{i=1}^{N} \frac{1}{-g_i(\boldsymbol{x}_*(t))} \nabla g_i(\boldsymbol{x}_*(t)) + \boldsymbol{A}^T\boldsymbol{\mu} = 0. \tag{6.48}$$

The central path has an interesting property, that is, every central point will yield a dual feasible point with a lower bound on the optimal value f_*. For any given tolerance or accuracy, the optimal value of t can be determined by

$$t = \frac{N}{\epsilon}. \tag{6.49}$$

For an excellent discussion concerning these concepts, readers can refer to Boyd and Vanderberghe's book. Now let us introduce a very popular interior-point method, called the barrier method.

Barrier Method

Start a guess in the strictly feasible region $\boldsymbol{x}_0$
Initialize $t_0 > 0$, $\beta > 1$, and $\epsilon > 0$
while $(N/t > \epsilon)$
  Inner loop:
    Start at $\boldsymbol{x}_k$
    Solve $\min t f(\boldsymbol{x}) + \psi(\boldsymbol{x})$ subject to $\boldsymbol{A}\boldsymbol{x} = \boldsymbol{b}$ to get $\boldsymbol{x}_*(t)$
  end loop
  Update $\boldsymbol{x}_k \leftarrow \boldsymbol{x}_*(t)$
  $t = \beta t$
end

Figure 6.3: The procedure of the barrier method for convex optimization.

6.5 INTERIOR-POINT METHODS

The interior-point methods are a class of algorithms using the idea of central paths, and the huge success of the methods are due to the fact that the algorithm complexity is polynomial. Various names have been used. For example, when the first version of the methods was pioneered by Fiacco and McCormick in the 1960s, the term sequential unconstrained minimization techniques were used. In modern literature, we tend to use the barrier method, path-following method or interior-point method. A significant step in the development of interior-point methods was the seminal work by Karmarkar in 1984 outlining the new polynomial-time algorithm for linear programming. The convergence of the interior-point polynomial methods was proved by Nesterov and Nemirovskii for a wider class of convex optimization problems in 1990s. Here we will introduce the simplest well-known barrier method.

The basic procedure of the barrier methods consists of two loops: an inner loop and an outer loop as shown in Figure 6.3. The method starts with a guess which is strictly inside the feasible region at $t = t_0$. For a given accuracy or tolerance ϵ and a parameter $\beta > 1$ for increasing t, a sequence of optimization problems are solved by increasing t gradually. At each step, the main task of the inner loop tries to find the best solution $\boldsymbol{x}_*(t)$ to the problem (6.45) for a given t, while the outer loop updates the solutions and increases t sequentially. The whole iteration process stops once the prescribed accuracy is reached.

Now the question is what value of β should be used. If β is too large, then the change of t at each step in the outer loop is large, then the initial $\boldsymbol{x}_k$ is far from the actual optimal solution for a given t. It takes more steps to find the solution $\boldsymbol{x}_*(t)$. On the other hand, if β is near 1, then the change in t is small, and the initial solution $\boldsymbol{x}_k$ for each value of t is good, so a smaller number of steps is needed to find $\boldsymbol{x}_*(t)$. However, as the increment in t is small, to reach the same accuracy, more jumps or increments in t in the outer loop are needed. Therefore, it seems that a fine balance is needed in choosing β such that the overall number of steps are minimal. This itself is an optimization problem. In practice, any value $\beta = 3$ to 100 are acceptable, though $\beta = 10$ to 20 works well for most applications as suggested in Boyd and Vandenberghe's book.

One of the important results concerning the barrier method is the convergence estimate, and it has been proved that for given β, ϵ and N, the number of steps for the outer loop required is exactly

$$K = \left\lceil \frac{\log(\frac{N}{t_0 \epsilon})}{\log \beta} \right\rceil, \tag{6.50}$$

where $\lceil k \rceil$ means to take the next integer greater than k. For example, for $N = 500$, $\epsilon = 10^{-9}$, $t_0 = 1$ and $\beta = 10$, we have

$$K = \left\lceil \frac{\log(500/10^{-9})}{\log 10} \right\rceil = 12. \tag{6.51}$$

This is indeed extremely efficient.

There are many other methods concerning the interior-point methods, including geometric programming, phase I method, phase II method, and primal-dual interior-point methods. Interested readers can refer to more advanced literature at the end of this chapter.

6.6 STOCHASTIC AND ROBUST OPTIMIZATION

The optimization problems we have discussed so far implicitly assume that the parameters of objective or constraints are exact and deterministic; however, in reality there is always some uncertainty. In engineering optimization, this is often the case. For example, material properties always have certain degree of inhomogeneity and stochastic components. The objective and the optimal solution may depend on some uncertainty, and they are often very sensitive to such variations and uncertainty. For example, the optimal solution at B in Figure 6.4 is very sensitive to uncertainty. If there is any small change in the solution, the global minimum is no longer the global minimum as the perturbed solution may be not so good as the solution at A where the solution is more stable. We say the optimal solution at B is not robust. In practical applications, we are trying to find not only the best solution, but also the most robust solution. Because only robust solutions can have realistic engineering applications. In this present case, we usually seek good quality solutions around A, rather than the non-robust optimal solution at B.

In this case, we are dealing with stochastic optimization. In general, we can write a stochastic optimization problem as

$$\text{minimize} \;\; f(\boldsymbol{x}, \xi), \tag{6.52}$$

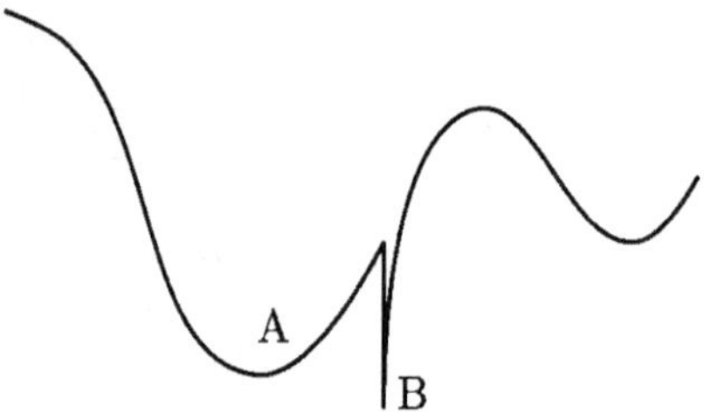

Figure 6.4: Robustness of the optimal solution.

$$\text{subject to} \quad G(\boldsymbol{x}, \xi) \in \Omega, \tag{6.53}$$

where $G(\boldsymbol{x}, \xi)$ consists of all the constraints in the well-structure subset Ω. Here ξ is a random variable with certain, probably unknown, probability distribution.

There are three ways to approach this optimization problem: sensitivity analysis, stochastic programming/optimization, and robust optimization. Sensitivity analysis intends to solve the optimization with a fixed choice of ξ, and then analyze the sensitivity of the solution to the uncertainty by sensitivity techniques such as Monte Carlo methods (to be introduced later in this book). It is relative simple to implement, but the solution may not be optimal or near the true range.

Stochastic programming tries to use the information of the probability density function for ξ while focusing the mean of the objective. If we know the distribution of the random variable ξ such as the Guassian or normal distribution $N(\mu, \sigma^2)$ with a mean of μ and variance of σ^2. We can then try to solve

$$\text{minimize} \quad E_{\xi} f(\boldsymbol{x}, \xi), \tag{6.54}$$

$$\text{subject to} \quad E_{\xi} G(\boldsymbol{x}, \xi) \in \Omega, \tag{6.55}$$

where $E_{\xi} f$ is the expectation of the function f averaging over ξ. Typically, this expectation involves multidimensional integrals which are analytically intractable, and we have to use Monte Carlo integration techniques. It is worth pointing out that we have to modify the objective even more in some cases to include the uncertainty explicitly so that we have $\min E_{\xi} f(\boldsymbol{x}, \xi) + \lambda\sigma$ where $\lambda > 0$ is a parameter. The advantage of stochastic programming is that it has taken accounting the probability information into consideration; however, the implementation is often difficult, especially if there are highly nonlinear constraints.

A good alternative to tackle this problem is to use the robust optimization where it tries to study the optimization problem with uncertainty inputs and intends to seek good solutions with reasonably good performance. Even so, most problems are not tractable and some approximations are needed. However, under some strict conditions such as Ω is a well-structured convex cone

K, and $f(\boldsymbol{x}, .)$ is convex with respect to $\boldsymbol{x}$, this problem is largely tractable and this is the main topics of robust conic optimization. If we further assume $G(\boldsymbol{x})$ does not depend on ξ (no uncertainty in constraints), then the problem is generally solvable. In the rest of section, we will introduce briefly the stochastic robust least-squares as it is very important in engineering applications.

Now we try to solve

$$\text{minimize} \quad ||\boldsymbol{A}\boldsymbol{u} - \boldsymbol{b}||_2^2, \tag{6.56}$$

where $\boldsymbol{A}(\xi)$ depends on the random variable ξ with some uncertainty such that

$$\boldsymbol{A}(\xi) = \bar{\boldsymbol{A}} + \boldsymbol{\epsilon}. \tag{6.57}$$

Here the mean $\bar{\boldsymbol{A}}$ is fixed, and uncertainty or random matrix $\boldsymbol{\epsilon}$ has a zero mean. That is

$$E_\xi(\boldsymbol{\epsilon}) = 0. \tag{6.58}$$

The stochastic robust least-squares problem becomes the minimization of the expectation

$$\text{minimize} \quad E_\xi||\boldsymbol{A}\boldsymbol{u} - \boldsymbol{b}||_2^2. \tag{6.59}$$

From the basic definition of mean, we know that

$$E_\xi||\boldsymbol{A}\boldsymbol{u} - \boldsymbol{b}||_2^2 = E_\xi||(\bar{\boldsymbol{A}} + \boldsymbol{\epsilon})\boldsymbol{u} - \boldsymbol{b}||_2^2 = E_\xi||(\bar{\boldsymbol{A}}\boldsymbol{u} - \boldsymbol{b}) + \boldsymbol{\epsilon}\boldsymbol{u}||_2^2$$

$$= ||\bar{\boldsymbol{A}} - \boldsymbol{b}||_2^2 + E_\xi[\boldsymbol{u}^T\boldsymbol{\epsilon}^T\boldsymbol{\epsilon}\boldsymbol{u}] = ||\bar{\boldsymbol{A}} - \boldsymbol{b}||_2^2 + \boldsymbol{u}^T\boldsymbol{Q}\boldsymbol{u}^T, \tag{6.60}$$

where $\boldsymbol{Q} = E_\xi[\boldsymbol{\epsilon}^T\boldsymbol{\epsilon}]$. Using these results, we can recast the optimization as

$$\text{minimize} \quad ||\bar{\boldsymbol{A}} - \boldsymbol{b}||_2^2 + ||\boldsymbol{Q}^{1/2}\boldsymbol{u}||_2^2, \tag{6.61}$$

which becomes the Tikhonov regularization problem in the case of $\boldsymbol{Q} = \gamma\boldsymbol{I}$. That is

$$\text{minimize} \quad ||\bar{\boldsymbol{A}} - \boldsymbol{b}||_2^2 + \gamma||\boldsymbol{u}||_2^2, \tag{6.62}$$

which is a convex optimization problem and can be solved by the relevant methods such as the interior-point methods.

EXERCISES

6.1 A common technique for some optimization is to convert it to an equivalent problem with a linear objective. Write the following generic convex optimization as a linear programming problem

$$\text{minimize} \quad f(\boldsymbol{x}),$$

$$\text{subject to} \quad g(\boldsymbol{x}) \le 0, \qquad h(\boldsymbol{x}) = 0.$$

6.2 Prove that $f(\boldsymbol{x}) = \sum_{i=1}^n i x_i^2$ is convex.

6.3 Find the optimal solution of the following stochastic function in n dimensions

$$f(x) = \sum_{j=1}^{K} \epsilon_j \exp[-\alpha \sum_{i=1}^{n} (x_i - \epsilon_i)^2] - (K+1) \exp[-\beta \sum_{i=1}^{n} (x_i - i + 1)^2],$$

where $\alpha > 0, \beta > 0$. $K = 1, 2, ...$ is an integer. The domain is $-n \le x_i \le n$ for $i = 1, 2, ..., n$. In addition, ϵ_i, ϵ_j are random numbers drawn from a uniform distribution in $[0, 1]$. Write a simple program to find its global minimum and carry out sensitivity analysis around this global minimum.

REFERENCES

1. C. G. Broyden, "The convergence of a class of double-rank minimization algorithms", *IMA J. Applied Math.*, **6**, 76-90 (1970).
2. S. P. Boyd and L. Vandenberghe, *Convex Optimization*, Cambridge University Press, 2004. Also http://www.stanford.edu/ boyd
3. A. R. Conn, N. I. M. Gould, P. L. Toint, *Trust-region methods*, SIAM & MPS, 2000.
4. J. E. Dennis, "A brief introduction to quasi-Newton methods", in: *Numerical Analysis: Proceedings of Symposia in Applied Mathematics* (Eds G. H. Golub and J. Oliger, pp. 19-52 (1978).
5. A. V. Fiacco and G. P. McCormick, *Nonlinear Porgramming: Sequential Unconstrained Minimization Techniques*, John Wiley & Sons, 1969.
6. R. Fletcher, "A new approach to variable metric algorithms", *Computer Journal*, **13** 317-322 (1970).
7. D. Goldfarb, "A family of variable metric update derived by variational means", *Mathematics of Computation*, **24**, 23-26 (1970).
8. S. M. Goldeldt, R. E. Quandt, and H. F. Trotter, "Maximization by quadratic hill-climbing", *Econometrica*, **34**, 541-551, (1996).
9. N. Karmarkar, "A new polynomial-time algorithm for linear programming", *Combinatorica*, **4** (4), 373-395 (1984).
10. K. Levenberg, "A method for the solution of certain problems in least squares", *Quart. J. Applied Math.*, **2**, 164-168, (1944).
11. D. Marquardt, "An algorithm for least-squares estimation of nonlinear parameters", *SIAM J. Applied Math.*, **11**, 431-441 (1963).
12. J. A. Nelder and R. Mead, "A simplex method for function optimization", *Computer Journal*, **7**, 308-313 (1965).
13. Y. Nesterov and A. Nemirovskii, *Interior-Point Polynomial Methods in Convex Programming*, Society for Industrial and Applied Mathematics, 1994.
14. M. J. D. Powell, "A new algorithm for unconstrained optimization", in: *Nonlinear Programming* (Eds J. B. Rosen, O. L. Mangasarian, and K. Ritter), pp. 31-65 (1970).

15. D. F. Shanno, "Conditioning of quasi-Newton methods for function minimization", *Mathematics of Computation*, **25**, 647-656 (1970).

16. S. J. Wright, *Primal-Dual Interior-Point Methods*, Society for Industrial and Applied Mathematics, 1997.

CHAPTER 7

CALCULUS OF VARIATIONS

The calculus of variations is important in computational sciences and many optimization problems including optimal control. In this chapter, we will briefly touch upon these topics. The main aim of the calculus of variations is to find a function that makes the integral stationary, making the value of the integral a local maximum or minimum. For example, in mechanics we may want to find the shape $y(x)$ of a rope or chain when suspended under its own weight from two fixed points. In this case, the calculus of variations provides a method for finding the function $y(x)$ so that the curve $y(x)$ minimizes the gravitational potential energy of the hanging rope system.

7.1 EULER-LAGRANGE EQUATION

7.1.1 Curvature

Before we proceed to the calculus of variations, let us first discuss an important concept, namely the curvature of a curve. In general, a curve $y(x)$ can be described in a parametric form in terms of a vector $\mathbf{r}(s)$ with a parameter s which is the arc length along the curve measured from a fixed point. The

By Xin-She Yang
Copyright © 2010 John Wiley & Sons, Inc.

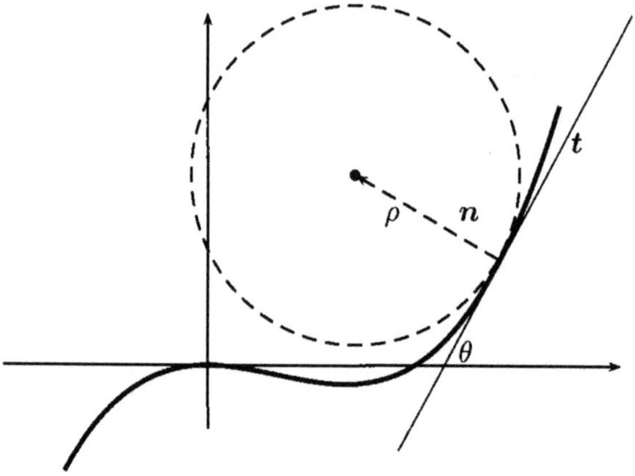

Figure 7.1: Concept of curvature.

curvature κ of a curve is defined as the rate at which the unit tangent $\mathbf{t}$ changes with respect to s. The change of arc length is

$$\frac{ds}{dx} = \sqrt{1 + (\frac{dy}{dx})^2} = \sqrt{1 + y'^2}. \tag{7.1}$$

We have the curvature

$$\frac{d\mathbf{t}}{ds} = \kappa \, \boldsymbol{n} = \frac{1}{\rho} \boldsymbol{n}, \tag{7.2}$$

where ρ is the radius of the curvature, and $\boldsymbol{n}$ is the principal normal shown in Figure 7.1.

As the direction of the tangent is defined by the angle θ made with the x-axis by $\mathbf{t}$, we have $\tan\theta = y'$. Hence, the curvature becomes

$$\kappa = \frac{d\theta}{ds} = \frac{d\theta}{dx} \frac{dx}{ds}. \tag{7.3}$$

From $\theta = \tan^{-1} y'(x)$, we have

$$\frac{d\theta}{dx} = [\tan^{-1}(y')]' = \frac{y''}{(1 + y'^2)}. \tag{7.4}$$

Using the expression for ds/dx, the curvature can be written in terms of $y(x)$, and we get

$$\kappa = \left|\frac{d^2\mathbf{r}}{ds^2}\right| = \left|\frac{y''}{[1 + (y')^2]^{3/2}}\right|. \tag{7.5}$$

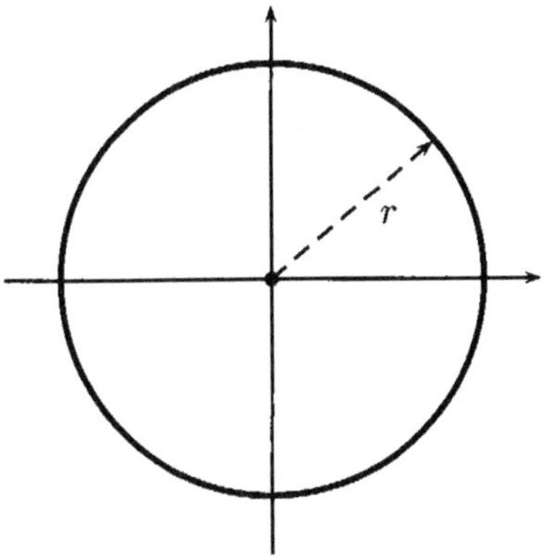

Figure 7.2: The curvature of a circle at any point is $1/r$.

■ EXAMPLE 7.1

We know that the equation for a circle centered at $(0, 0)$ is

$$x^2 + y^2 = r^2,$$

where $r > 0$ is its radius. In order to calculate the curvature at any point (x, y) along the circle, we have to calculate y' and y''. Using implicit differentiation with respect to x, we have

$$2x + 2y'y = 0,$$

which lead to

$$y' = -\frac{x}{y}.$$

The second derivative is

$$y'' = -\frac{x^2 + y^2}{y^3}.$$

Using these expressions and $x^2 + y^2 = r^2$, we have the curvature

$$\kappa = \left|\frac{y''}{[1 + (y')^2]^{3/2}}\right| = \left| -\frac{r^2/y^3}{[1 + (-x/y)^2]^{3/2}}\right|$$

$$= \left| -\frac{1}{y\sqrt{\frac{r^2}{y^2}}}\right| = \frac{1}{r}.$$

Indeed, the curvature of a circle is everywhere $1/r$. Thus, the radius of curvature is the radius of the circle $\rho = 1/\kappa = r$.

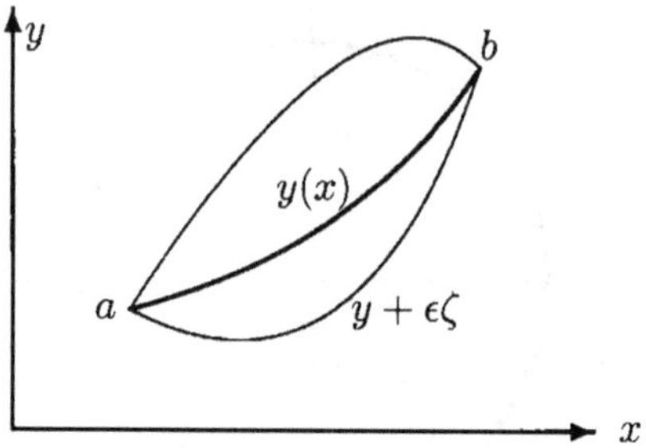

Figure 7.3: Variations in the path $y(x)$.

7.1.2 Euler-Lagrange Equation

Since the calculus of variations is always related to some minimization or maximization, we can in general assume that the integrand ψ of the integral is a function of the shape or curve $y(x)$ (shown in Figure 7.3), its derivative $y'(x)$ and the spatial coordinate x (or time t, depending on the context). For the integral

$$I = \int_a^b \psi(x, y, y')dx, \tag{7.6}$$

where a and b are fixed, the aim is to find the solution of the curve $y(x)$ such that it makes the value of I stationary or optimal. In this sense, $I[y(x)]$ is a function of the function $y(x)$, and thus it is referred to as the functional.

Here, stationary means that the small change of the first order in $y(x)$ will only lead to the second-order changes in values of $I[y(x)]$, and subsequently, the change δI of I should be virtually zero due to the small variation in the function $y(x)$. Translating this into the mathematical language, we suppose that $y(x)$ has a small change of magnitude of ϵ so that

$$y(x) \to y(x) + \epsilon\zeta(x), \tag{7.7}$$

where $\zeta(x)$ is an arbitrary function. The requirement of I to be stationary means that

$$\delta I = 0, \tag{7.8}$$

or more accurately,

$$\frac{dI}{d\epsilon}\bigg|_{\epsilon=0} = 0, \qquad \text{for all} \quad \zeta(x). \tag{7.9}$$

Thus I becomes

$$I(y, \epsilon) = \int_a^b \psi(x, y + \epsilon\zeta, y' + \epsilon\zeta')dx$$

$$= \int_a^b \psi(x, y, y')dx + \int_a^b \epsilon\Big[\zeta\frac{\partial\psi}{\partial y} + \zeta'\frac{\partial\psi}{\partial y'}\Big]dx + O(\epsilon^2). \tag{7.10}$$

The first derivative of I should be zero, and we have

$$\frac{\delta I}{\delta\epsilon} = \int_a^b [\frac{\partial\psi}{\partial y}\zeta + \frac{\partial\psi}{\partial y'}\zeta']dx = 0, \tag{7.11}$$

which is exactly what we mean that the change δI (or the first order variation) in the value of I should be zero. Integrating this equation by parts, we have

$$\int_a^b [\frac{\partial\psi}{\partial y} - \frac{d}{dx}\frac{\partial\psi}{\partial y'}]\zeta dx = -[\zeta\frac{\partial\psi}{\partial y'}]\Big|_a^b. \tag{7.12}$$

If we require that $y(a)$ and $y(b)$ are known at the fixed points $x = a$ and $x = b$, then these requirements naturally lead to $\zeta(a) = \zeta(b) = 0$. This means that the above right hand side of the equation is zero. That is,

$$\Big[\zeta\frac{\partial\psi}{\partial y'}\Big]_a^b = 0, \tag{7.13}$$

which gives

$$\int_a^b \Big[\frac{\partial\psi}{\partial y} - \frac{d}{dx}\frac{\partial\psi}{\partial y'}\Big]\zeta dx = 0. \tag{7.14}$$

As this equation holds for all $\zeta(x)$, the integrand must be zero. Therefore, we have the well-known Euler-Lagrange equation

$$\frac{\partial\psi}{\partial y} = \frac{d}{dx}(\frac{\partial\psi}{\partial y'}). \tag{7.15}$$

It is worth pointing out that this equation is very special in the sense that ψ is known and the unknown is $y(x)$. It has many applications in mathematics, natural sciences and engineering.

■ EXAMPLE 7.2

The simplest and classical example is to find the shortest path on a plane joining two points, say, $(0, 0)$ and $(1, 1)$. We know that the total length along a curve $y(x)$ is

$$L = \int_0^1 \sqrt{1 + y'^2}dx. \tag{7.16}$$

Since $\psi = \sqrt{1 + y'^2}$ does not contain y, thus $\frac{\partial\psi}{\partial y} = 0$. From the Euler-Lagrange equation, we have

$$\frac{d}{dx}(\frac{\partial\psi}{\partial y'}) = 0, \tag{7.17}$$

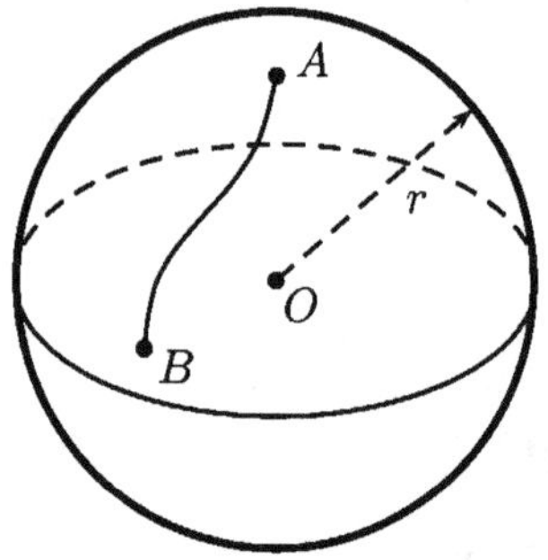

Figure 7.4: Geodesic path on the surface of a sphere.

its integration is

$$\frac{\partial \psi}{\partial y'} = \frac{y'}{\sqrt{1+y'^2}} = A. \tag{7.18}$$

Rearranging it as

$$y'^2 = \frac{A^2}{1-A^2}, \qquad \text{or} \qquad y' = \frac{A}{\sqrt{1-A^2}}, \tag{7.19}$$

and integrating again, we have

$$y = kx + c, \qquad k = \frac{A}{\sqrt{1-A^2}}. \tag{7.20}$$

This is a straight line. That is exactly what we expect from the plane geometry.

Well, you may say, this is trivial and there is nothing new about it. Let us now study a slightly more complicated example to find the shortest path on the surface of a sphere.

■ EXAMPLE 7.3

For any two points A and B on the surface of a sphere with radius r as shown in Figure 7.4, we now use the calculus of variations to find the shortest path connecting A and B on the surface.

Since the sphere has a fixed radius, we need only two coordinates (θ, ϕ) to uniquely determine the position on the sphere. The length element ds can be written in terms of the two spherical coordinate angles

$$ds = r\sqrt{d\theta^2 + \sin^2\theta d\phi^2} = r\sqrt{(\frac{d\theta}{d\phi})^2 + \sin^2\theta} \ \ |d\phi|,$$

where in the second step we assume that $\theta = \theta(\phi)$ is a function of ϕ so that ϕ becomes the only independent variable. This is possible because $\theta(\phi)$ represents a curve on the surface of the sphere just as $y = y(x)$ represents a curve on a plane. Thus, we want to minimize the total length

$$L = \int_A^B ds = \int_{\phi_A}^{\phi_B} \sqrt{\theta'^2 + \sin^2\theta}\, d\phi,$$

where $\theta' = d\theta/d\phi$. Since the integrand

$$\psi = \sqrt{\theta'^2 + \sin^2\theta}$$

does not explicitly depend on ϕ, we can use the simplified form of Euler-Lagrange equation (7.35) to be discussed later

$$\psi - \theta' \frac{\partial \psi}{\partial \theta'} = k,$$

where k is a constant. We have

$$\sqrt{\theta'^2 + \sin^2\theta} - \theta' \frac{\theta'}{\sqrt{\theta' + \sin^2\theta}} = k,$$

or

$$\theta'^2 = (\frac{d\theta}{d\phi})^2 = \frac{\sin^2\theta(\sin^2\theta - k^2)}{k^2}.$$

By taking square root and rearranging the above equation, we get

$$d\phi = \pm \frac{k d\theta}{\sin\theta\sqrt{\sin^2\theta - k^2}} = \pm \frac{\frac{1}{\sin^2\theta} d\theta}{\sqrt{(\frac{1}{k^2} - 1) - \cot^2\theta}}.$$

Its integration gives

$$\phi = \pm \sin^{-1}[\frac{\cot\theta}{\sqrt{\frac{1}{k^2} - 1}}] + \alpha.$$

where α is the integration constant. Taking sin of both sides, we have

$$\sin(\phi - \alpha) = \pm\beta\cot\theta = \sin\phi\cos\alpha - \sin\alpha\cos\phi,$$

where

$$\beta = \frac{1}{\sqrt{\frac{1}{k^2} - 1}}.$$

Multiplying both sides by $r\sin\theta$ and using $x = r\sin\theta\cos\phi$, $y = r\sin\theta\sin\phi$ and $z = r\cos\theta$, we have

$$y\cos\alpha - x\sin\alpha = \pm\beta z,$$

which corresponds to a plane passing through points A, B, and the origin $(x, y, z) = (0, 0, 0)$. Therefore, the intersection of the plane and sphere produces a great circle on the surface connecting A and B. As the two signs corresponds to the two segments of the great circle, one of which is shorter in general. However, in the special case when A and B are at the opposite points, the two segments will have the same length. Great circles are important in geodesy as straight lines in plane geometry.

The Euler-Lagrange equation is very general and includes many physical laws if the appropriate form of ψ is used. For a point mass m following under the Earth's gravity g, the action (see below) is defined as

$$\psi = \frac{1}{2}mv^2 - mgy = \frac{1}{2}m(\dot{y})^2 - mgy,$$

where $y(t)$ is the path, and now x is replaced by t. $v = \dot{y}$ is the velocity. The Euler-Lagrange equation becomes

$$\frac{\partial \psi}{\partial y} = \frac{d}{dt}(\frac{\partial \psi}{\partial v}),$$

or

$$-mg = \frac{d}{dt}(mv),$$

which is essentially the Newton's second law $F = ma$ because the right hand side is the rate of change of the momentum mv, and the left hand side is the force.

These examples are relatively simple. Let us now study a more complicated case so as to demonstrate the wide applications of the Euler-Lagrange equation. In mechanics, there is a Hamilton's principle that states that the configuration of a mechanical system is such that the action integral I of the Lagrangian

$$\mathcal{L} = T - V, \tag{7.21}$$

is stationary with respect to the variations in the path. That is to say that the configuration can be uniquely defined by its coordinates q_i and time t, when moving from one configuration at time t_0 to another time $t = t*$

$$I = \int_0^{t^*} \mathcal{L}(t, q_i, \dot{q}_i)dt, \qquad i = 1, 2, ..., N, \tag{7.22}$$

where T is the total kinetic energy (usually, a function of $\dot{q}_i$), and V is the potential energy (usually, a function of q). Here $\dot{q}_i$ means

$$\dot{q}_i = \frac{\partial q_i}{\partial t}. \tag{7.23}$$

In analytical mechanics and engineering, the Lagrangian $\mathcal{L}$ (=kinetic energy - potential energy) is often called the action, thus this principle is also called the

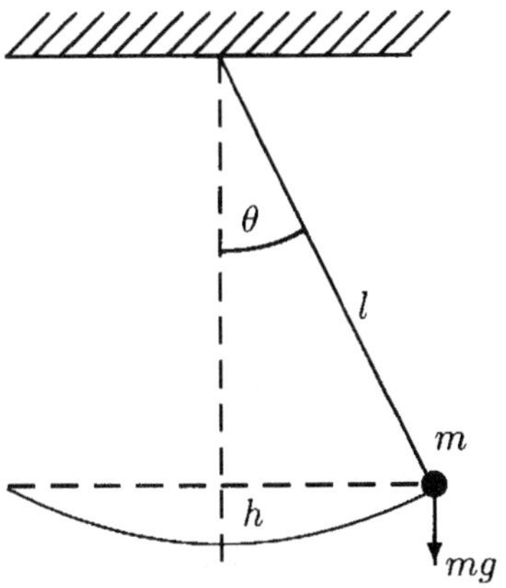

Figure 7.5: A simple pendulum.

principle of least action. The physical configuration or the path of movement follows such a path that makes the action integral stationary.

In the special case, $x \to t$, the Euler-Lagrange equation becomes

$$\frac{\partial \mathcal{L}}{\partial q_i} = \frac{d}{dt}\left(\frac{\partial \mathcal{L}}{\partial \dot{q}_i}\right), \tag{7.24}$$

which is the well-known Lagrange's equation. This seems too abstract. Now let us look at a classic example.

■ EXAMPLE 7.4

For a simple pendulum shown in Figure 7.5, we now try to derive its equation of oscillations. We know the kinetic energy T and the potential energy V are

$$T = \frac{1}{2}ml^2\left(\frac{d\theta}{dt}\right)^2 = \frac{1}{2}ml^2\dot{\theta}^2, \qquad V = mgh = mgl(1-\cos\theta).$$

Using $\mathcal{L} = T - V$, $q = \theta$ and $\dot{q} = \dot{\theta}$, we have

$$\frac{\partial \mathcal{L}}{\partial \theta} - \frac{d}{dt}\left(\frac{\partial \mathcal{L}}{\partial \dot{\theta}}\right) = 0,$$

which becomes

$$-mgl\sin\theta - \frac{d}{dt}(ml^2\dot{\theta}) = 0.$$

Therefore, we have the pendulum equation

$$\frac{d^2\theta}{dt^2} + \frac{g}{l}\sin\theta = 0.$$

This is a nonlinear equation. If the angle is very small ($\theta \ll 1$), $\sin\theta \approx \theta$, we then have the standard equation for the linear harmonic motion

$$\frac{d^2\theta}{dt^2} + \frac{g}{l}\theta = 0.$$

7.2 VARIATIONS WITH CONSTRAINTS

Although the stationary requirement in the calculus of variations leads to the minimization of the integral itself, there is no additional constraint. In this sense, the calculus of variation discussed up to now is unconstrained. However, sometimes these variations have certain additional constraints, for example, the sliding of a bead on a hanging string. Now we want to make the integral I stationary under another constraint integral Q that is constant. We have

$$I = \int_a^b \psi(x, y, y')dx, \tag{7.25}$$

subjected to the constraint

$$Q = \int_a^b \phi(x, y, y')dx. \tag{7.26}$$

As for most optimization problems under additional constraints, the method of Lagrange multipliers can transform the constrained problem into an unconstrained one by using a combined functional $J = I + \lambda Q$ or

$$J = \int_a^b [\psi + \lambda\phi]dx, \tag{7.27}$$

where λ is the undetermined Lagrange multiplier. Replacing ψ by $[\psi + \lambda\phi]$ in the Euler-Lagrange equation or following the same procedure of the derivations, we have

$$[\frac{\partial\psi}{\partial y} - \frac{d}{dx}(\frac{\partial\psi}{\partial y'})] + \lambda[\frac{\partial\phi}{\partial y} - \frac{d}{dx}(\frac{\partial\phi}{\partial y'})] = 0. \tag{7.28}$$

Now we can come back to our example of the hanging rope problem with two fixed points. The total length of the rope is L, and it hangs from two fixed points $(-d, 0)$ and $(d, 0)$. From the geometric consideration, it requires that $2d < L$. In order to find the shape of the hanging rope under gravity, we now define its gravitational potential energy E_p as

$$E_p = \int_{-d}^{d} [\rho g y(x) ds] = \rho g \int_{-d}^{d} y\sqrt{1 + y'^2}dx. \tag{7.29}$$

The additional constraint is that the total length of the rope is a constant (L). Thus,

$$Q = \int_{-d}^{d} \sqrt{1+y'^2}dx = L. \tag{7.30}$$

By using the Lagrange multiplier λ, we have $J = E_p + \lambda Q$, or

$$J = \int_{-d}^{d} [\rho g y + \lambda]\sqrt{1+y'^2}dx. \tag{7.31}$$

Since $\Psi = [\rho g y + \lambda]\sqrt{1+y'^2}$ does not contain x explicitly, or $\frac{\partial \Psi}{\partial x} = 0$, then the Euler-Lagrange equation can be reduced into a simpler form in this special case. Using

$$\frac{d\Psi}{dx} = \frac{\partial \Psi}{\partial x} + \frac{\partial \Psi}{\partial y}\frac{dy}{dx} + \frac{\partial \Psi}{\partial y'}\frac{dy'}{dx}$$

$$= 0 + y'\frac{\partial \Psi}{\partial y} + y''\frac{\partial \Psi}{\partial y'}, \tag{7.32}$$

and the Euler-Lagrange equation $\frac{\partial \Psi}{\partial y} = \frac{d}{dx}(\frac{\partial \Psi}{\partial y'})$, we have

$$\frac{d\Psi}{dx} = y'[\frac{d}{dx}(\frac{\partial \Psi}{\partial y'})] + y''\frac{\partial \Psi}{\partial y'} = \frac{d}{dx}[y'\frac{\partial \Psi}{\partial y'}], \tag{7.33}$$

which can again be written as

$$\frac{d}{dx}[\Psi - y'\frac{\partial \Psi}{\partial y'}] = 0. \tag{7.34}$$

The integration of this equation gives

$$\Psi - y'\frac{\partial \Psi}{\partial y'} = A = \text{const}. \tag{7.35}$$

Substituting the expression of Ψ into the above equation, the stationary values of J requires

$$\sqrt{1+y'^2} - \frac{y'^2}{\sqrt{1+y'^2}} = \frac{A}{\rho g y + \lambda}. \tag{7.36}$$

Multiplying both sides by $\sqrt{1+y'^2}$ and using the substitution $A\cosh\zeta = \rho g y + \lambda$, we have

$$y'^2 = \cosh^2\zeta - 1, \tag{7.37}$$

whose solution is

$$\cosh^{-1}[\frac{\rho g y + \lambda}{A}] = \frac{x\rho g}{A} + K. \tag{7.38}$$

Using the boundary conditions at $x = \pm d$ and the constraint $Q = L$, we have $K = 0$ and implicit equation for A

$$\sinh(\frac{\rho g d}{A}) = \frac{\rho g L}{2A}. \tag{7.39}$$

Finally, the curve for the hanging rope becomes the following catenary

$$y(x) = \frac{A}{\rho g}[\cosh(\frac{\rho g x}{A}) - \cosh(\frac{\rho g d}{A})]. \tag{7.40}$$

■ EXAMPLE 7.5

For the hanging rope problem, what happens if we only fix one end at $(a, 0)$, while allowing the free end of the hanging rope to slide on a vertical pole? Well, this forms a variation problem with variable end-point(s). We assume that free end is at $(0, y)$ where y acts like a free parameter to be determined. Now the boundary condition at the free end is different. Since the variation of $\delta I = 0$, we have

$$\delta J = \int_a^b [\frac{\partial \Psi}{\partial y} - \frac{d}{dx}(\frac{\partial \Psi}{\partial y'})]\zeta dx + \Big[\zeta \frac{\partial \Psi}{\partial y'}\Big]_a^b = 0.$$

As the variation ζ is now non-zero at the free end point, we then have

$$\frac{\partial \Psi}{\partial y'} = 0.$$

From $J = E_p + \lambda Q$, we have $\Psi = (\rho g y + \lambda)\sqrt{1 + y'^2}$. Thus, we get

$$\frac{\partial}{\partial y'}\Big[(\rho g y + \lambda)\sqrt{1 + y'^2}\Big] = 0,$$

or

$$y'(\rho g y + \lambda)/\sqrt{1 + y'^2} = 0, \qquad \text{or} \qquad y' = 0.$$

In other words, the slope is zero at the free end.

Such a boundary condition of $y' = 0$ has the real physical meaning because any non-zero gradient at the free end would have a non-zero vertical component, thus causing the vertical slip along the rope due to the tension in the rope. The zero-gradient leads to the static equilibrium with minimum energy. Thus, the whole curve of the hanging rope with one free end forms half the catenary.

■ EXAMPLE 7.6

Dido's problem concerns the strategy to enclose a maximum area with a fixed length circumference. Legend says that Dido was promised a piece of land on the condition that it was enclosed by an oxhide. She had to cover as much as land as possible using the given oxhide. She cut the oxhide into narrow strips with ends joined, and a whole region of a hill was enclosed.

Suppose the total length of the oxhide strip is L. The enclosed area A to be maximized is

$$A = \int_{x_a}^{x_b} y(x)dx,$$

where x_a and x_b are two end points (of course they can be the same points). We also have the additional constraint

$$\int_{x_a}^{x_b} \sqrt{1+y'^2}dx = L = \text{const.}$$

This forms an isoperimetric variation problem. As L is fixed, thus the maximization of A is equivalent to make $I = A + \lambda L$ stationary. That is

$$I = A + \lambda L = \int_{x_a}^{x_b} [y + \lambda\sqrt{1+y'^2}]dx.$$

Using the Euler-Lagrange equation, we have

$$\frac{\partial I}{\partial y} - \frac{d}{dx}\frac{\partial I}{\partial y'} = 0,$$

or

$$\frac{\partial}{\partial y}[y + \lambda\sqrt{1+y'^2}] + \frac{d}{dx}\frac{\partial}{\partial y'}[y + \lambda\sqrt{1+y'^2}] = 0,$$

which becomes

$$1 - \lambda\frac{d}{dx}\Big(\frac{y'}{\sqrt{1+y'^2}}\Big) = 0.$$

Integrating it once, we get

$$\frac{\lambda y'}{\sqrt{1+y'^2}} = x + K,$$

where K is the integration constant. By rearranging, we have

$$y' = \pm\frac{x+K}{\sqrt{\lambda^2 - (x+K)^2}}.$$

Integrating this equation again, we get

$$y(x) = \mp\sqrt{\lambda^2 - (x+K)^2} + B,$$

where B is another integration constant. This is equivalent to

$$(x+K)^2 + (y-B)^2 = \lambda^2,$$

which is essentially the standard equation for a circle with the centre at $(-K, B)$ and a radius λ. Therefore, the most area that can be enclosed by a fixed length is a circle.

An interesting application is the design of the slides in playgrounds. Suppose we want to design a smooth (frictionless) slide, what is the best curve/shape the slide should take so that a child can slide down in a quickest way? This problem is related to the brachistochrone problem, also called the shortest time problem or steepest descent problem, which initiated the development of the calculus of variations. In 1696, Johann Bernoulli posed a problem to find the curve that minimizes the time for a bead attached to a wire to slide from a point $(0, h)$ to a lower point $(a, 0)$.

From the conservation of energy, we can determine the speed of the bead from the equation $\frac{1}{2}mv^2 + mgy = mgh$, and we have

$$v = \sqrt{2g(h-y)}. \tag{7.41}$$

So the total time taken to travel from $(0, h)$ to $(a, 0)$ is

$$t = \int_0^a \frac{1}{v} ds = \int_0^a \frac{\sqrt{1+y'^2}}{\sqrt{2g(h-y)}} dx. \tag{7.42}$$

Using the simplified Euler-Lagrange equation (7.35) because the integrand $\Psi = \sqrt{1+y'^2}/\sqrt{2g(h-y)}$ does not contain x explicitly, we have

$$\sqrt{\frac{(1+y'^2)}{2g(h-y)}} - y' \frac{\partial}{\partial y'}[\sqrt{\frac{(1+y'^2)}{2g(h-y)}}] = A. \tag{7.43}$$

By differentiation and some rearrangements, we have

$$y'^2 = \frac{B-h+y}{h-y}, \qquad B = \frac{1}{2gA^2}. \tag{7.44}$$

By changing of variables $\eta = h - y = \frac{B}{2}(1 - \cos\theta)$ and integrating, we have

$$x = \frac{B}{2}[\theta - \sin\theta] + k, \tag{7.45}$$

where $\theta < \pi$ and k is an integration constant. As the curve must pass the point $(0, h)$, we get $k = 0$. So the parametric equations for the curve become

$$x = \frac{B}{2}(\theta - \sin\theta), \qquad y = h - \frac{B}{2}(1 - \cos\theta). \tag{7.46}$$

This is a cycloid, not a straight line, which seems a bit surprising, or at least it is rather counter-intuitive. The bead travels a longer distance, thus has a higher average velocity and subsequently falls quicker than traveling in a straight line.

7.3 VARIATIONS FOR MULTIPLE VARIABLES

What we have discussed so far mainly concerns the variations in 2-D, and subsequently the variations are in terms $y(x)$ or curves only. What happens if we want to study a surface in the full 3-D configuration? The principle in the previous sections can be extended to any dimensions with multiple variables, however, we will focus on the minimization of a surface here. Suppose we want to study the shape of a soap bubble, the principle of least action leads to the minimal surface problem. The surface integral of a soap bubble should be stationary. Now we assume that the shape of the bubble is $u(x, y)$, then the total surface area is

$$A(u) = \iint_{\Omega} \Psi dxdy = \iint_{\Omega} \sqrt{1 + (\frac{\partial u}{\partial x})^2 + (\frac{\partial u}{\partial y})^2} dxdy, \tag{7.47}$$

where

$$\Psi = \sqrt{1 + (\frac{\partial u}{\partial x})^2 + (\frac{\partial u}{\partial y})^2} = \sqrt{1 + u_x^2 + u_y^2}. \tag{7.48}$$

In this case, the extended Euler-Lagrangian equation for two variables x and y becomes

$$\frac{\partial \Psi}{\partial u} - \frac{\partial}{\partial x}(\frac{\partial \Psi}{\partial u_x}) - \frac{\partial}{\partial y}(\frac{\partial \Psi}{\partial u_y}) = 0. \tag{7.49}$$

Substituting Ψ into the above equation and using $\frac{\partial \Psi}{\partial u} = \Psi_u = 0$ since Ψ does not contain u explicitly, we get

$$-\frac{\partial}{\partial x}[\frac{1}{\Psi}\frac{\partial u}{\partial x}] - \frac{\partial}{\partial y}[\frac{1}{\Psi}\frac{\partial u}{\partial y}] = 0, \tag{7.50}$$

or

$$(1 + u_y^2)u_{xx} - 2u_x u_y + (1 + u_x^2)u_{yy} = 0. \tag{7.51}$$

This is a nonlinear equation and its solution is out of the scope of this book. This nonlinear equation has been one of the active research topics for more than a century. It has been proved that the fundamental solution to this equation is a sphere, and in fact we know that all bubbles are spherical. For some problems, we can approximately assume that u_x and u_y are small, thus the above equation becomes Laplace's equation

$$u_{xx} + u_{yy} = 0. \tag{7.52}$$

The calculus of variations has many applications. The other classical examples include Fermat's principle in optics, Sturm-Liouville problem, surface shape optimization, the action principle, and the finite element analysis.

7.4 OPTIMAL CONTROL

When the objective is an integral and the constraints are differential equation, the optimization problem becomes an optimal control problem. The methods discussed so far cannot be applied directly to solve optimal control problems. Optimal control is an important branch of optimization and control research, especially in engineering design and economics. For example, in order to reach from point A to B on a road $u(t)$, we can vary the speed $\alpha(t)$ of a car so as to minimize the fuel consumption ψ is an optimal control problem. Similarly, to design a railway path on a hilly landscape with the constraint of slope or gradient so as to minimize the distance between any two stations also requires optimal control. Furthermore, a company intends to vary their investment so as to maximize their financial return is again an optimal control problem.

7.4.1 Control Problem

For a simple ordinary differential equation

$$\dot{u} \equiv \frac{du(t)}{dt} = f(u(t)), \tag{7.53}$$

with an initial condition $u(0) = u_0$, we can simply integrate it to obtain a solution for any time $t > 0$, though such an integral may not be easy to calculate. This is a simple dynamical system.

Now suppose that the function $f(u)$ also depends on an extra parameter α so that we have $f(u, \alpha)$. For any given parameter α, we have a solution $u(t)$, and the solution or state is controlled by α for the given system. In a generic case when α depends on time t continuously, that $f(u)$ becomes $F(u(t), \alpha(t), t)$ where we include t in the argument to show its depends on time t explicitly. Now we have a controlled dynamical system

$$\dot{u}(t) = F(u(t), \alpha(t), t), \qquad u(0) = u_0. \tag{7.54}$$

Here $\alpha(t)$ is called the control variable. If we want to get some desired behaviour of the system, we have to find the best control $\alpha(t)$ to achieve the desired characteristics. In addition, in many applications, we often want to optimize some objective which is an integral of $u(t)$ over some time interval. This becomes an optimal control problem.

A generic optimal control problem for a single control $\alpha(t)$ and univariate state $u(t)$ can often be written as

$$\underset{\alpha(t)}{\text{maximize}} \quad \psi(t) = \int_t^{\tau} w(u(t), \alpha(t), t)dt + K(u(\tau)), \tag{7.55}$$

$$\text{subject to} \quad \dot{u}(t) = \frac{du(t)}{dt} = F(u(t), \alpha(t), t), \tag{7.56}$$

where $\alpha(t)$ is call the control variable (or a path or functional), and $u(t)$ is the state variable. $K(u(\tau))$ is the terminal cost (or payoff or reward) at τ. The aim is to find the best control $\alpha(t)$ so as to maximize the objective functional $\psi(t)$ which is the total cost over a period from t to τ of a function w. The decision/control or path taken at each instant of time t will affect the state $u(t)$ in terms of a known function $F(\cdot, \cdot, t)$.

The standard calculus and the method for solving optimization do not work for this type of problem as the standard optimization methods intend to find a single value of the objective function at some optima. Here in optimal control, we intend to find a control path or a function $\alpha(t)$ which optimizes the objective functional. We have to transform the optimal control problem and write the optimality conditions in terms of differentiation, rather than integral. The Pontryagin maximum principle provides a very useful tool for finding the optimal control $\alpha_*(t)$ for the above problem (7.55).

7.4.2 Pontryagin's Principle

First, let us define an auxiliary function, or the Hamiltonian, for this optimal control problem

$$H = w(u(t), \alpha(t), t) + \lambda(t) F(u(t), \alpha(t), t), \tag{7.57}$$

where $\lambda(t)$ is called the costate variable or adjoint variable, which can be considered as a 'Lagrange multiplier' (though now a function) to incorporate the constraint, as we did for the standard constrained optimization problems. Now the Pontryagin maximum principle provides the following optimality conditions:

$$\frac{\partial H}{\partial \alpha} = 0,$$
$$\frac{\partial H}{\partial u} = -\dot{\lambda}, \tag{7.58}$$
$$\frac{\partial H}{\partial \lambda} = \dot{u},$$

and the costate $\lambda_*(t)$ is linked with the terminal payoff function

$$\lambda_*(\tau) = \nabla K(u_*(\tau)), \tag{7.59}$$

which is often referred to as the transversality condition. Here $u_*(t)$ is the optimal state for the optimal control $\alpha_*(t)$. The equation (7.58) is often called the adjoint equation.

Pontryagin's maximum or minimum principle is a necessary condition to find the best possible control. In the case when the final time τ is fixed, the Hamiltonian is independent of time, that is

$$\frac{\partial H_*}{\partial t} = 0, \tag{7.60}$$

where H_* is the optimal value

$$H_*(u_*(t), \alpha_*(t), \lambda_*) = \max H(u(t), \alpha(t), \lambda). \tag{7.61}$$

Under some condition, the above necessary conditions (7.58) can also become sufficient conditions for optimality. For a minimization problem,

$$\text{minimize} \quad \int_{t_0}^{\mathcal{T}} \phi(u(t), \alpha(t), t) dt, \tag{7.62}$$

$$\text{subject to} \quad \dot{u}(t) = F(u(t), \alpha(t), t), \quad u(t_0) = u_0, \tag{7.63}$$

the necessary conditions (7.58) are also sufficient if w and F are differentiable and jointly convex in u and α as well as $\lambda > 0$ for all t. For a maximization problem, ϕ should be concave. This is because the maximization of ϕ is equivalent to the minimization of $-\phi$. For a minimization problem, we have

$$\frac{\partial^2 H}{\partial u^2} > 0. \tag{7.64}$$

Now let us look at a simple example as an application.

$$\text{maximize} \quad \int_0^{\mathcal{T}} [u(t) + \alpha^q(t)] dt, \tag{7.65}$$

$$\text{subject to} \quad \dot{u}(t) = 1 - \alpha^p(t), \tag{7.66}$$

$$u(0) = \tau. \tag{7.67}$$

Here we assume that $p > q > 0$.

The Hamiltonian becomes

$$H = [u(t) + \alpha^q(t)] + \lambda(t)[1 - \alpha^p(t)].$$

The optimality conditions are

$$\frac{\partial H}{\partial \alpha} = q\alpha^{q-1} - p\lambda\alpha^{p-1} = 0, \tag{7.68}$$

$$\frac{\partial H}{\partial u} = 1 = -\dot{\lambda}, \tag{7.69}$$

$$\frac{\partial H}{\partial \lambda} = 1 - \alpha^p = \dot{u}(t). \tag{7.70}$$

From the second condition (7.69), we have $\dot{\lambda} = -1$ and

$$\lambda(t) = \lambda(0) + \int_0^t \dot{\lambda}(t) dt$$

$$= \lambda(0) + \int_0^t (-1) dt = \lambda(0) - t.$$

If we use the transversality condition $\lambda(\tau) = 0$ at $t = \tau$, we have

$$\lambda(\tau) = \lambda(0) - \tau = 0, \quad \text{or} \quad \lambda(0) = \tau,$$

which leads to

$$\lambda(t) = \tau - t. \tag{7.71}$$

From this solution, (7.68) becomes

$$q\alpha^{q-1} - p(\tau - t)\alpha^{p-1} = 0,$$

which gives the optimal solution

$$\alpha = \Big[\frac{q}{p(\tau - t)}\Big]^{\frac{1}{p-q}}. \tag{7.72}$$

Substituting this into (7.70), we have

$$\dot{u}(t) = 1 - \alpha^p = 1 - \Big[\frac{q}{p(\tau - t)}\Big]^{\frac{p}{p-q}}.$$

By simple integration, we have

$$u(t) = u(0) + \int_0^t \dot{u}(t)dt = \tau + \int_0^t \Big\{1 - \Big[\frac{q}{p(\tau - t)}\Big]^{\frac{p}{p-q}}\Big\}dt$$

$$= \tau + t - \frac{(p-q)}{q}\Big(\frac{q}{p}\Big)^{\frac{p}{p-q}}\Big[\Big(\frac{1}{\tau - t}\Big)^{\frac{p}{p-q}} - \Big(\frac{1}{\tau}\Big)^{\frac{q}{p-q}}\Big]. \tag{7.73}$$

For example, in the case of $p = 3$ and $q = 2$, we have

$$u(t) = \tau + t - \frac{4}{27}\Big[\Big(\frac{1}{\tau - t}\Big)^2 - \frac{1}{\tau^2}\Big]. \tag{7.74}$$

7.4.3 Multiple Controls

The optimal control we discussed so far is for a single control variable. This can readily be extended to the multiple control variables in higher dimensions. Let us define the following vector forms

$$\boldsymbol{\alpha}(t) = \begin{pmatrix} \alpha_1(t) \\ \vdots \\ \alpha_n(t) \end{pmatrix}, \quad \boldsymbol{F} = \begin{pmatrix} F_1 \\ \vdots \\ F_m \end{pmatrix}, \quad \boldsymbol{u}(t) = \begin{pmatrix} u_1(t) \\ \vdots \\ u_m(t) \end{pmatrix}, \tag{7.75}$$

then the optimal control problem can be generalized as

$$\underset{\boldsymbol{\alpha}(t)}{\text{maximize}} \quad \psi(\boldsymbol{\alpha}) = \int_0^\tau w(\boldsymbol{u}(t), \boldsymbol{\alpha}(t), t)dt + K(\boldsymbol{u}(\tau)), \tag{7.76}$$

$$\text{subject to} \quad \dot{\boldsymbol{u}}(t) = \boldsymbol{F}(\boldsymbol{u}(t), \boldsymbol{\alpha}(t), t), \qquad (t \geq 0, \ \ \tau > 0), \tag{7.77}$$

$$\boldsymbol{u}(0) = \boldsymbol{u}_0. \tag{7.78}$$

Defining the Hamiltonian as

$$H = w(\boldsymbol{u}(t), \boldsymbol{\alpha}(t), t) + \boldsymbol{\lambda}^T \boldsymbol{F}, \tag{7.79}$$

where $\boldsymbol{\lambda} = (\lambda_1, ..., \lambda_m)^T$, the Pontryagin maximum principle becomes

$$\frac{\partial H}{\partial \boldsymbol{\alpha}} = 0, \tag{7.80}$$

$$\frac{\partial H}{\partial \boldsymbol{u}} = -\dot{\boldsymbol{\lambda}}, \tag{7.81}$$

$$\frac{\partial H}{\partial \boldsymbol{\lambda}} = \dot{\boldsymbol{u}}, \tag{7.82}$$

and

$$\boldsymbol{\lambda}_*(\tau) = \nabla K(\boldsymbol{u}_*(\tau)). \tag{7.83}$$

Here $\frac{\partial H}{\partial \boldsymbol{u}} = -\dot{\boldsymbol{\lambda}}$ is equivalent to m conditions

$$\frac{\partial H}{\partial u_i} = -\dot{\lambda}_i, \qquad (i = 1, 2, ..., m). \tag{7.84}$$

Other equations in the vector form should be expanded similarly.

The above optimal control belongs to a special class where the final time τ is known and fixed, so the optimal control is called fixed-time optimal control. There are other cases when τ is not known, it becomes a free-end point and/or free-time control problem. Furthermore, when $\tau \to \infty$, such an optimal control becomes an infinite horizon problem. As the infinite integral

$$\int_0^{\infty} w(u(t), \alpha(t), t) dt, \tag{7.85}$$

is not necessarily finite. Therefore, we usually need, especially in economics, to use certain discount future benefit parameterized by $\gamma \geq 0$ to reformulate the objective as

$$\text{maximize} \quad \psi = \int_0^{\infty} e^{-\gamma t} w(u(t), \alpha(t), t) dt. \tag{7.86}$$

By choosing appropriate γ, ψ should remain finite. Interested readers can refer to more advanced literature on optimal control and dynamic programming.

7.4.4 Stochastic Optimal Control

So far, we assume that there is no uncertainty in our discussion on optimal control. In reality, there is always some degree of uncertainty. For simplicity, let us assume the function $F(u, \alpha)$ in (7.54) has some uncertainty so that we have

$$\dot{u}(t) = F(u(t), \alpha(t), t) + \sigma\eta(t), \tag{7.87}$$

with the initial condition $u(0) = u_0$. Here $\eta(t)$ is a random variable, and σ is a parameter associated with the random variable η. The above equation is essentially a stochastic differential equation (SDE) or a stochastic dynamical system.

An optimal control problem with an associated constraint (7.87) is called a stochastic optimal control problem. As the objective ψ has randomness, we have to reformulate it in terms of expectation. Therefore, we have

$$\text{maximize} \quad E_\eta\Big[\int_0^\tau w(u(t), \alpha(t), t)dt + K(u(\tau))\Big], \tag{7.88}$$

$$\text{subject to} \quad \dot{u}(t) = F(u(t), \alpha(t), t) + \sigma\eta(t), \qquad u(0) = u_0. \tag{7.89}$$

In general, the solution of a stochastic differential equation requires the Ito integral which is out of the scope of this book. Informally, the above SDE (7.87) can be solved using

$$u(t) = u_0 + \int_0^t F(u(\xi), \alpha(\xi), \xi)d\xi + \sigma B(t), \tag{7.90}$$

where $B(t)$ is the Brownian motion which is Gaussian with zero mean and an ever-increasing variance $\sigma^2 = t$. That is $B(t) \sim N(0, t)$ where $N(\mu, \sigma^2)$ is the Gaussian distribution. Loosely speaking, $dB(t)/dt$ is equivalent to $\eta(\cdot)$. That is why the above integral is possible.

Optimal control is already difficult enough, and stochastic optimal control is even harder. Interested readers can refer to relevant textbooks for more details.

EXERCISES

7.1 Find the optimal solution $x_*(t)$ so that the functional $J = \int_0^1 \dot{x}^2(1+t)dt$ becomes extremal. The boundary conditions are $x(0) = 0$ and $x(1) = 1$.

7.2 Find the extremal of $J = \int_0^\tau \dot{x}^2/x^2 dt$ with the boundary conditions $x(0) = 1$ and $x(\tau) = e$ assuming τ is constant.

7.3 For the well-known moon-landing control problem, the aim is to design a control procedure by varying the thrust $p(t)$ so that the spacecraft can land on the lunar surface safely using the minimum amount of fuel. Formulate this in terms of an optimal control problem.

7.4 Design a cylindrical water tank with height h_0 and base radius r so that the leakage time through any fixed hole size is maximum. The main constraint is that the surface area of the cylinder side and base is fixed, that is, $S_0 = 2\pi r h_0 + \pi r^2$ is constant where we have assumed that top surface is open. What is the optimum of h_0?

7.5 Find the optimal control $u(t)$ so as to minimize the generic energy functional

$$J = \frac{1}{n}\int_0^1 u^n dt, (n > 1),$$

subject to

$$\dot{x} = u, \quad x(0) = 1, \quad x(1) = 0,$$

and

$$\dot{s} = x, \quad s(0) = 0, \quad s(1) = 0.$$

7.6 Extend the optimality conditions for a univariate optimal control to multivariate control with a quadratic objective

$$\text{minimize } J[\boldsymbol{v}(t)] = \frac{1}{2}\boldsymbol{x}^T\boldsymbol{K}\boldsymbol{x} + \frac{1}{2}\int_0^T (\boldsymbol{x}^T\boldsymbol{P}\boldsymbol{x} + \boldsymbol{v}^T\boldsymbol{Q}\boldsymbol{v})dt,$$

with a linear dynamics

$$\dot{\boldsymbol{x}}(t) = \boldsymbol{A}\boldsymbol{x}(t) + \boldsymbol{B}\boldsymbol{v}(t),$$

and the initial condition $\boldsymbol{x}(0) = \boldsymbol{x}_0$. This is essentially a finite-horizon problem with a control $\boldsymbol{v}(t)$. If possible, try to write the conditions in a matrix form.

REFERENCES

1. R. Courant and D. Hilbert, *Methods of Mathematical Physics*, Vol. I, Interscience Press, 1953.
2. B. D. Craven, *Control and Optimization*, Chapman & Hall, 1995.
3. R. Dorfman, "An economic interpretation of optimal control theory", *Amer. Econ. Review*, **59**, 817-831 (1969).
4. L. C. Evans, *An Introduction to Stochastic Differential Equations*, Lecture notes, http://math.berkeley.edu/ evans/SDE.course.pdf
5. W. Fleming and R. Rishel, *Deterministic and Stochastic Optimal Control*, Springer, 1975.
6. I. M. Gelfand and S. V. Fomin, *Calculus of Variations*, Dover Publication, 2000.
7. L. Hocking, *Optimal Control: An Introduction to the Theory with Applications*, Oxford University Press, 1991.
8. J. Macki and A. Strauss, *Introduction to Optimal Control Theory*, Springer, 1982.
9. L. S. Pontryagin, V. G. Boltyanski, R. S. Gamkrelidze and E. F. Mishchenko, *The Mathematical Theory of Optimal Processes*, Interscience, 1962.

CHAPTER 8

RANDOM NUMBER GENERATORS

The random numbers we used in computer programs are not really random as they are generated via algorithms that are deterministic in a way. The true random numbers should be generated using physical methods. Strictly speaking, all random numbers generated by algorithmic generators are pseudo-random numbers. Therefore, such generators are called pseudo-random number generators (PRNG) in the literature. However, modern pseudo-random generators can generate numbers which are almost indistinguishable from true random numbers.

8.1 LINEAR CONGRUENTIAL ALGORITHMS

There are several classes of random generators, including linear congruential generators, Fibonacci generators, inverse transform generators and twisters.

A linear congruential generator, first formulated by D. H. Lehmer, uses an iterative procedure

$$d_i = (ad_{i-1} + c) \bmod m, \tag{8.1}$$

Engineering Optimization: An Introduction with Metaheuristic Applications.
By Xin-She Yang
Copyright © 2010 John Wiley & Sons, Inc.

where a, c and m are all integers; a and m are relatively prime. Here the 'mod' means to take the reminder after division. The starting or initial value d_0 is often called the seed of the generator. In most algorithms, the modulus m takes the form

$$m = 2^k, \ 2^k \pm 1. \tag{8.2}$$

This generator has a possible maximum period m which is attainable only if $c > 0$ and m are relatively prime, and $(a - 1)$ is a multiple of 4. The classic example is the IBM generator with $a = 65539 = 2^{16} + 3$, $c = 0$ and $m = 2^{31}$ or

$$d_i = (2^{16} + 3)d_{i-1} \bmod m. \tag{8.3}$$

However, this generator has many problems because

$$d_i = (2^{16} + 3)d_{i-1} \bmod m = (2^{16} + 3) \cdot [(2^{16} + 3)^2 d_{i-2}] \bmod m$$

$$= (2^{16} + 3)^2 d_{i-2} \bmod m = (2^{32} + 6 \cdot 2^{16} + 9)d_{i-2} \bmod m$$

$$= 6 \cdot (2^{16} + 3)d_{i-2} - 9d_{i-2} = 6d_{i-1} - 9d_{i-2}, \tag{8.4}$$

where we have used $(2^{32} \bmod m) = 0$ due to $m = 2^{32}$. This is a recursive relationship, and there is a strong correlation among any consecutive three numbers d_{i-2}, d_{i-1} and d_i. Consequently, the numbers generated are not so 'random' from the statistical point of view.

Nowadays, modern computers widely use $a = 1,103,515,245$, $c = 12345$ and $m = 2^{32}$.

If the multiplier a and modulus m are carefully chosen, a seemingly simple generator could produce pseudo-random numbers of high quality. Park and Miller's generator use $c = 0$ so that

$$d_i = a d_{i-1} \bmod m, \tag{8.5}$$

where $a = 16807 = 7^5$ and $m = 2^{31} - 1 = 2,147,483,647$.

In the above generators, we have extensively used multiplications and divisions. If we can use additions or subtractions in the formulation, we may increase the efficiency of a generator. In fact, Lagged Fibonacci pseudo-random generator uses the following formula

$$d_i = (d_{i-k} + d_{i-n}) \bmod m, \tag{8.6}$$

where k and n are integers. In the popular Mitchell-Moore generator, $k = 24$, $n = 55$ and $m = 2^{32}$.

8.2 UNIFORM DISTRIBUTION

All the generators we have discussed so far in this chapter generate uniformly-distributed integers from 0 to $m - 1$. In many applications, we are concerned

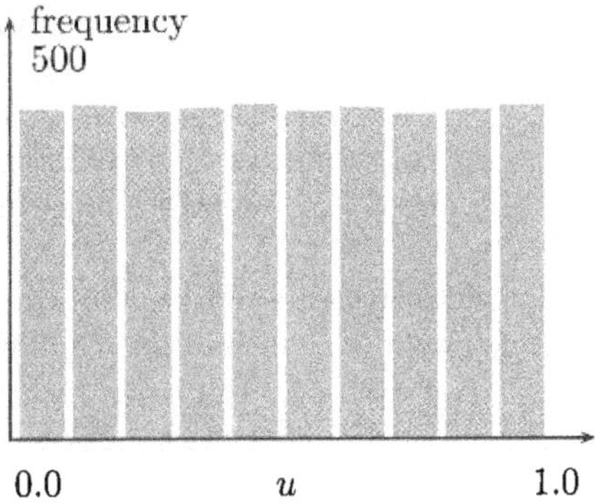

Figure 8.1: Histogram of 5000 random numbers generated from a uniform distribution in the range (0, 1).

with the uniform distribution in a range, say, [0, 1]. In order to obtain scaled floating-point numbers, we have to divide by the period m, so we have d_i/m in the range of [0, 1].

In linear congruential generators and lagged Fibonacci generators, we have to use divisions via mod and multiplications which are not computationally efficient. The better approach is to use additions and subtractions only, and this leads to Marsaglia's generator.

Linear congruential generators produce pseudo-random integers. On the other hand, Marsaglia generators, first introduced by G. Marsaglia, can produce floating-point numbers. For example, in the popular Matlab program, the following 'substract-with-borrow' generator is used. We have

$$d_i = d_{i+20} - d_{i+5} - k, \tag{8.7}$$

where the indices $(i, i+5, i+20)$ only take values mod 32, which means that only their last five bits are used since $2^5 = 32$. Here k is the residual from the previous step so that k is non-negative. That is to say, $k = 0$ if d_i is positive; however, $k = 2^{-53}$ (limited by the machine precision) if d_i is negative and $d_i = d_i + 1$ so that a modified $d_i > 0$ is stored. The period of this generator is about 2^{1492} in theory. However, because of the $k = \epsilon = 2^{-53}$ or $k = 0$, then all the numbers generated are the multiple of $\epsilon = 2^{-53}$ which is a little defect, though it is not noticeable for most applications. Consequently, the generated floating-points are in the range of $[\epsilon, 1-\epsilon]$ or $[2^{-53}, 1-2^{-53}]$. Surprisingly, 0 and 1 cannot be generated by this generator. For example, the histogram of 5000 numbers generated by a typical pseudo-random generator is shown in Figure 8.1.

This so-called machine epsilon (ϵ) is determined by the way that the floating-point numbers are represented. In most modern computers, a double precision real number is represented by a 64-bit string or word in the standard IEEE arithmetic format specified by IEEE standards in 1977 or later 1985.

In general, a 64-bit word consists of 53 bits for a signed fraction in base 2, and 11 bits for a signed exponent. The sign is represented by a single bit. Therefore, the accuracy of the computer is measure by the small quantity $\epsilon = 2^{-53} \approx 1.11 \times 10^{-16}$. Since the exponent of a real number has 11 bits with 1 bit for sign, so the real numbers in this IEEE format are in the range of $2^{\pm 2^{10}} \approx 10^{\pm 308}$. For almost all computations, these representations are sufficient.

Modern algorithms can be exceptionally elaborate and more generators are being invented, with the intention to produce distributions of better quality with higher efficiency. For example, the recent Mersenne twister algorithm could have a period of $2^{19937} - 1$, which is sufficient by any statistical standards.

8.3 OTHER DISTRIBUTIONS

We have seen that most generators just produce pseudo-random numbers for the uniform distribution, often in the range $[0, 1]$; however, we have to deal with other distributions such as normal distributions in many applications.

The simplest method is probably to use the central limit theorem. Let u_i $(i = 1, 2..., n)$ be n independent uniformly-distributed numbers in $[0, 1]$. Then, the central limit theorem implies that the random variable

$$v = \sum_{i=1}^{n} u_i - \frac{n}{2}, \tag{8.8}$$

obeys the standard normal distribution $N(0, 1)$ with a zero mean and a unitary variance. The probability density function of the standard normal distribution is given by

$$p(x) = \frac{1}{\sqrt{2\pi}} e^{-x^2/2}. \tag{8.9}$$

However, this method is very inefficient and with poor tails. So it is rarely used, and more rigorous methods should be used.

The generation of other distributions can be obtained in two major ways: the inverse transform method and the acceptance-rejection method. The former is more computationally extensive but easier to implement, while the latter is not easy to implement though it is usually more efficient. As the speed of modern computers increases, it seems that it does not really matter much which method we choose in most cases.

The basic idea of an acceptance-rejection method is better illustrated in terms of a specific distribution. In order to generate a normal distribution with a bell-shaped probability density function (pdf), we consider the shape of the pdf curve

$$f(x) = \frac{1}{\sqrt{2\pi}} e^{-\frac{x^2}{2}}, \tag{8.10}$$

in a two-dimensional plane, often a truncated box. We first generate random points $P(x, y)$ which distribute uniformly in the plane. If the point P fall under this curve $f(x)$, it is accepted as a valid value, otherwise, it is rejected. The resulting distribution for the accepted points is the normal distribution. This is the basic idea, however, its implementation requires more elaborate algorithms such as the Ziggurat algorithm, which divides the area under the curve into n equal sections, and the generated points are sorted and scaled.

On the other hand, the inverse transform method or sampling intends to generate random numbers by transforming uniformly-distributed numbers. Such transforms use the cumulative distribution function (cdf). The basic procedure of this method of generating a non-uniform distribution $f(x)$ with a given cumulative distribution function $\Phi(v)$ is as follows: First, generate a uniformly-distributed number u, then we try to find the value v so that

$$\Phi(v) = u, \tag{8.11}$$

which is equivalent to finding the inverse

$$v = \Phi^{-1}(u). \tag{8.12}$$

The sample set of v obtained by this inversion obeys the original distribution $f(x)$. Let us look at an example.

■ EXAMPLE 8.1

We know that the probability density of the exponential distribution is

$$f(x) = \lambda e^{-\lambda x}, \qquad x \geq 0,$$

and its cumulative distribution function is

$$\Phi(v) = \int_0^v f(x)dx = 1 - e^{-\lambda v}, \qquad v \geq 0.$$

For a generated uniform distribution $u(0, 1)$ in $(0, 1)$, excluding 0 and 1, we can produce the numbers of the exponential distribution by the inverse transform

$$\Phi(v) = u(0, 1),$$

or

$$v = \Phi^{-1}(u) = -\frac{1}{\lambda} \ln(1 - u).$$

In fact,

$$v = -\frac{1}{\lambda} \ln u,$$

also obeys the same exponential distribution as both u and $1 - u$ are uniformly distributed any way.

For the standard normal distribution $N(0,1)$

$$p(x) = \frac{1}{\sqrt{2\pi}} e^{-x^2/2},$$

its cumulative distribution function is given by

$$\Phi(v) = \frac{1}{\sqrt{2\pi}} \int_{-\infty}^{v} e^{-x^2/2} dx = \frac{1}{2}[1 + \text{erf}(\frac{v}{\sqrt{2}})],$$

where the error function erf is defined by

$$\text{erf}(x) = \frac{2}{\sqrt{\pi}} \int_0^x e^{-\zeta^2} d\zeta. \tag{8.13}$$

From a uniform distribution $u(0,1)$, we have

$$\Phi(v) = u.$$

Thus, the inverse becomes

$$v = \Phi^{-1}(u) = \sqrt{2}\,\text{erf}^{-1}(2u - 1),$$

where erf^{-1} is the inverse error function which is defined by

$$\text{erf}(y) = x, \qquad \text{or} \qquad y = \text{erf}^{-1}(x). \tag{8.14}$$

As the error function and its inverse involve integration, it is not straightforward to estimate their values. Apart from numerical integration, the power series expansions often give good approximations by expanding $e^{-\zeta^2}$ as

$$e^{-\zeta^2} = \sum_{n=0}^{\infty} \frac{(-1)^n \zeta^{2n}}{n!} = 1 - \zeta^2 + \frac{\zeta^4}{2!} - ..., \tag{8.15}$$

Therefore, we have have

$$\text{erf}(x) = \frac{2}{\sqrt{\pi}} \sum_{n=0}^{\infty} \frac{(-1)^n x^{2n+1}}{n!(2n+1)}$$

$$= \frac{2}{\sqrt{\pi}}[x - \frac{x^3}{3} + \frac{x^5}{10} - \frac{x^7}{42} + ...]. \tag{8.16}$$

In a similar way, the inverse error function can be estimated by

$$\text{erf}^{-1}(x) = \frac{\sqrt{\pi}}{2}[x + \frac{\pi x^3}{12} + \frac{7\pi^2 x^5}{480} + \frac{127\pi^3 x^7}{40320} + ...]. \tag{8.17}$$

Let us look at a specific example.

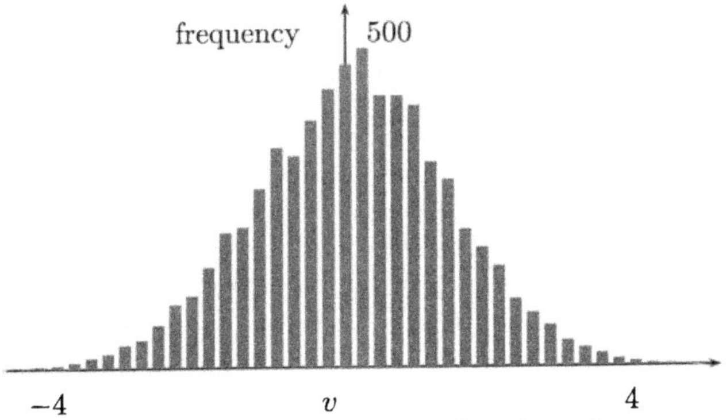

Figure 8.2: Histogram of the normally-distributed numbers generated by the simple inverse transform method.

■ **EXAMPLE 8.2**

Suppose we generate 5000 uniformly-distributed numbers:

$$u = 0.9022,\ 0.4985,\ ...,\ 0.1019,\ ...,\ 0.4990.$$

In order to get a normal distribution, we use

$$v = \sqrt{2}\,\mathrm{erf}^{-1}(2u - 1),$$

to transform the above numbers. For example, the first number 0.9022 will be transformed into

$$v = \sqrt{2}\,\mathrm{erf}^{-1}(2 \times 0.9022 - 1) = 1.2941.$$

We finally have

$$v = 1.2941,\ -0.0037,\ ...,\ -1.2705,\ ...,\ 0.0025.$$

The histograms of the two distributions are shown in Figure 8.1 and Figure 8.2.

We have seen that the standard inverse transform method involves the calculations of erf^{-1} which are difficult to evaluate. Box-Müller method provides a more efficient way to generate a normal distribution.

Let u_1, u_2 be two series of random numbers which are drawn from a uniform distribution in the standard range of $(0, 1)$. The Box-Müller method uses the following transform pair

$$v_1 = \sqrt{-2 \ln u_1}\cos(2\pi u_2), \tag{8.18}$$

$$v_2 = \sqrt{-2 \ln u_1} \sin(2\pi u_2). \tag{8.19}$$

Then, v_1 and v_2 obeys the standard normal distribution $N(0, 1)$

$$dv_1 dv_2 = \frac{1}{2\pi} e^{-(u_1^2+u_2^2)/2} du_1 du_2. \tag{8.20}$$

Here u_1 serves as the radius (with a uniform distribution) of a unit circle, while the angle $2\pi u_2$ distributes uniformly over the range $(0, 2\pi)$.

The disadvantage of this method is that it involves the trigonometric functions $\sin(u)$ and $\cos(u)$, which may slow down the evaluations. A further improvement is to use the polar form. From two uniformly-distributed numbers u_1 and u_2 in $(0, 1)$, we can obtain two uniformly-distributed numbers s_1 and s_2 in the range $(-1, 1)$. We have

$$s_1 = 2u_1 - 1, \qquad s_2 = 2u_2 - 1. \tag{8.21}$$

The condition $r^2 = s_1^2 + s_2^2 \leq 1$ corresponds to the case that the two numbers (s_1, s_2) fall within the unit circle in the region $(-1, 1) \times (-1, 1)$. Inside this unit circle, the angle $\theta = \tan^{-1} s_2/s_1 \in (0, 2\pi)$ and the quantity $p = r^2 = s_1^2 + s_2^2 \in (0, 1)$ are independent, and they both are uniformly distributed. Straightforward trigonometric manipulations lead to

$$v_1 = \sqrt{-2 \ln p} \cos(\theta) = s_1 \sqrt{\frac{-2 \ln p}{p}}, \tag{8.22}$$

$$v_2 = \sqrt{-2 \ln p} \sin(\theta) = s_2 \sqrt{\frac{-2 \ln p}{p}}. \tag{8.23}$$

Again, both v_1 and v_2 are normally distributed with a zero mean and a unitary variance. Other more complicated distributions can be generated in a similar manner.

8.4 METROPOLIS ALGORITHMS

The inverse transform methods can be used to generate random samples with various distributions. However, they are not efficient enough for many applications, especially for multivariate distributions in higher dimensions. In this case, random samples can be generated more efficiently using the Metropolis algorithms first developed in 1953, and the generated samples obey the prescribed probability density function $p(x, y, ..., z)$.

In Metropolis algorithms, the transition probability ϕ from state S_i to state S_j obeys the transition rule

$$p_i \phi(S_i \to S_j) = p_j \phi(S_j \to S_i), \tag{8.24}$$

where $p_i = p(S_i)$ is the probability density function of the random variable concerned. Using the notation $\phi_{i \to j}$ for the transition probability $\phi(S_i \to S_j)$, we have

$$\frac{\phi_{i \to j}}{\phi_{j \to i}} = \frac{p_j}{p_i}. \tag{8.25}$$

This provides an iterative procedure to generate sampling points or states from an initial ϕ_0, and the new candidate sample is accepted or not by an accepting rate. The new states are accepted if the quantity (or energy difference)

$$\Delta E = -\ln \frac{p_j}{p_i}, \tag{8.26}$$

is negative. Otherwise, the new states are only accepted with a probability of p_j/p_i. A major advantage of this method is that we never have to evaluate p_i itself, only the relative probability p_j/p_i matters, which makes many evaluations much simpler. Another advantage is that new sampling points are selected such that more important regions are sampled more intensively. Figuratively speaking, this is equivalent to, say, measuring the depths of a lake. The sampling points are chosen more heavily inside the lake, and the region on the land is rarely measured or not measured at all.

For example, in many applications, we have to evaluate the average

$$<u>= \frac{\sum_i u_i e^{-\gamma E_i}}{\sum_i e^{-\gamma E_i}}, \tag{8.27}$$

where $\gamma = 1/\kappa T$ and κ is the Boltzmann constant and T is the temperature. The states are the energy level E_i and E_j. In this case, we have

$$\frac{p_j}{p_i} = e^{-\gamma(E_j - E_i)}. \tag{8.28}$$

The transition probability $\phi_{i \to j} = 1$ if $p_j > p_i$ or $E_j < E_i$. Otherwise, $\phi_{i \to j} = e^{-\gamma(E_j - E_i)}$ or a random number between 0 and 1. That is

$$\phi_{i \to j} = \begin{cases} 1 & \text{if } p_j > p_i \ (\text{ or } E_j < E_i), \\ e^{-\gamma(E_j - E_i)} & \text{if } E_j \geq E_i. \end{cases} \tag{8.29}$$

In other words, all low-energy states are accepted, while high-energy states are accepted with a small probability. This idea forms the basis of the powerful method of simulated annealing to be introduced later.

EXERCISES

8.1 Write a few lines in Matlab to generate a set of pseudo-random numbers for the Rayleigh distribution

$$p(x; \sigma) = \frac{x}{\sigma^2} \exp[-x^2/2\sigma^2], \qquad x \in [0, \infty),$$

by using the inverse transform method.

8.2 First derive the inverse transform for the Laplace distribution

$$p(x;\mu,\beta) = \frac{1}{2\beta}\exp[-\frac{|x-\mu|}{\beta}], \qquad x \in (-\infty, \infty),$$

where $b > 0$ is the scale parameter and μ is the location parameter. Then, write a program to generate $n = 50,000$ numbers for this distribution.

8.3 Write a program to generate numbers drawn from the Cauchy distribution

$$p(x;\mu,\beta) = \frac{1}{\pi}[\frac{\beta}{(x-\mu)^2+\beta^2}],$$

where $\beta > 0$ is a scale parameter, and μ is the location parameter. If the inverse transform method is used, what is the potential difficulty?

REFERENCES

1. L. Devroye, *Non-Uniform Random Variate Generation*, Springer-Verlag, New York, 1986.
2. W. K. Hastings, "Monte Carlo sampling methods using Markov chains and their applications", *Biometrika*, **57**, 97-109 (1970).
3. D. Knuth, *The Art of Computer Programming*, Vol. 2: *Seminumerical Algorithms*, 3rd Edition, Addison-Wesley, (1997).
4. M. Matsumoto, T. Nishimura, "Mersenne twister: a 623-dimensionally equidistributed uniform pseudo-random number generator", *ACM Trans. Modeling and Computer Simulation*, **8** (1), 3-30 (1998).
5. N. Metropolis, and S. Ulam, "The Monte Carlo method", *J. Amer. Stat. Assoc.*, **44**, 335-341 (1949).
6. N. Metropolis, A. W. Rosenbluth, M. N. Rosenbluth, A. H. Teller, and E. Teller, "Equation of state calculations by fast computing machines", *J. Chem. Phys.*, **21**, 1087-1092 (1953).
7. S. K. Parks and K. W. Miller, "Random number generators: Good ones are hard to find", *Communications of the ACM*, **31** (10), 1192-1201 (1988).
8. I. Peterson, *The Jungle of Randomness: A Mathematical Safari*, John Wiley & Sons, (1998).
9. W. H. Press, S. A. Teukolsky, W. T. Vetterling, B. P. Flannery, *Numerical Recipes – The Art of Scientific Computing*, Cambridge University Press, 3rd Edition, (2007).

CHAPTER 9

MONTE CARLO METHODS

In many applications, the shear number of possible combinations and states is so astronomical that it is impossible to carry out evaluations over all possible combinations systematically. In this case, the Monte Carlo method is the best alternative. Monte Carlo is in fact a class of methods now widely used in computer simulations. Since the pioneer studies in 1940s and 1950s, especially the work by Ulam, von Newmann, and Metropolis, it has been applied in almost all area of simulations, from Ising model to financial market, from molecular dynamics to engineering, and from the routing of the Internet to climate simulations.

9.1 ESTIMATING π

The basic procedure of Monte Carlo is to generate a (large but finite) number of random points so that they distribute uniformly inside the domain Ω of interest. The evaluations of functions (or any quantities of interest) are then carried out over these discrete locations. Then the system characteristics can be expected to be derived or represented by certain quantities averaged over these finite number evaluations.

Engineering Optimization: An Introduction with Metaheuristic Applications.
By Xin-She Yang
Copyright © 2010 John Wiley & Sons, Inc.

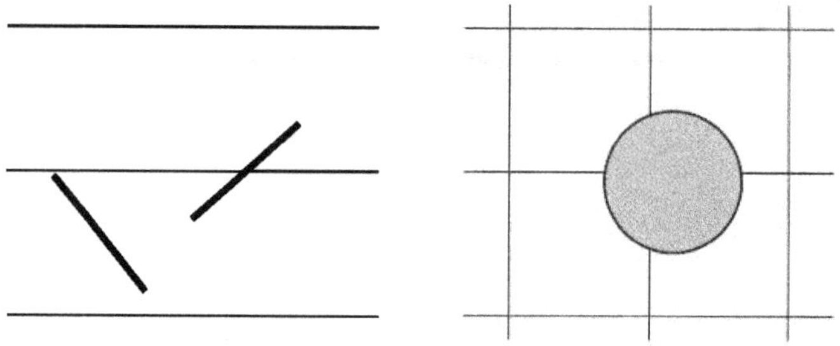

Figure 9.1: Estimating π by repeatedly dropping needles or tossing coins.

A classic example is the geometrical probability problem which provides a method for estimating π using Buffon's needle-throwing technique first described by Compte de Buffon in 1777. The basic idea is to consider a simulation of tossing a needle with a length L (or many needles of the same kind) onto a floor (or a table) with equally-spaced parallel lines with a distance of w apart. You can either drop a needle many times, or drop many needles one by one. What is the probability of the needle(s) crossing these lines (see Figure 9.1)?

Let u be the distance or location of the center of the needle to the nearest line, and θ be the acute angle of the needle formed with the line. For simplicity, we can also assume $L \leq w$ so that a needle can only cross one line.

Since the random nature of this process, both the location u and the angle θ are uniformly distributed. The probability density function of u in the range of 0 and $w/2$ is simply $2du/w$, while the probability density function of θ in the range of 0 and $\pi/2$ is $2d\theta/\pi$. Therefore, the joint probability function is

$$p(u,\theta)dud\theta = \frac{4}{w\pi}dud\theta. \tag{9.1}$$

Since the condition for a needle of crossing a line is $u \leq L/2\sin(\theta)$, we have

$$P = \frac{4}{\pi w}\int_0^{\pi/2}\int_0^{L\sin(\theta)/2} dud\theta = \frac{2L}{\pi w}, \tag{9.2}$$

which is the probability of a needle crossing a line. The interesting feature is that when $L = w$, we have

$$P = \frac{2}{\pi}, \tag{9.3}$$

which is independent of the length of the needle. Suppose we drop N needles with n needles crossing lines, we have

$$\lim_{N\to\infty}\frac{n}{N} = \frac{2}{\pi}, \quad \text{or} \quad \pi = \lim_{N\to\infty}\frac{2N}{n}. \tag{9.4}$$

For example, in a simulation, we dropped $N = 4000$, and found $n = 2547$ needles crossing lines. The estimate of π is

$$\pi = \frac{2 \times 4000}{2547} \approx 3.1410. \tag{9.5}$$

However, the error in such kind of simulation decreases with $1/\sqrt{N}$, and is thus very sensitive to the change of even the last digit. Suppose, in a similar experiment, we get $n = 2546$ instead of 2547, we get

$$\pi = \frac{2 \times 4000}{2546} \approx 3.1422. \tag{9.6}$$

For better accuracy, we have to use large N, otherwise, the results might be suspiciously lucky. For example, if we ran a hypothetical simulation with $N = 22$, we got (luckily) 14 needles crossing the lines, which leads to an estimate $\pi \approx 2 \times 22/14 = 22/7 \approx 3.142857$. However, this estimate is biased because 22/7 is a good approximation of π anyway. Suppose, you get 15 instead of 14, then the new estimate becomes $\pi = 2 \times 22/15 \approx 2.933$. An infamous example is the Lazzarini's estimate in 1901 with $N = 3408$ to get $\pi \approx 355/113$, which seems unlikely to be true.

Furthermore, π can also be estimated by tossing coins of a diameter D onto a chequerboard with the uniform grid spacing of w (see Figure 9.1). For simplicity, we assume that $D \leq w$. For a coin to cross a corner, its center has to be within a quarter of a circle from the corner (or the distance $s \leq D/2$), which means that, for a unit grid, it is the area of a whole circle. So the probability P of a coin crossing a corner is the area of a coin ($\pi D^2/4$) dividing by the area of a unit grid (w^2). That is

$$P = \frac{\pi D^2}{4w^2}. \tag{9.7}$$

In the simple case of $w = D$, we have $P = \pi/4$.

Interestingly, the probability P_n of a coin not crossing any line is $P_n = (1 - D/w)^2$ because a coin has a probability of $(w - D)/w = (1 - D/w)$ of not crossing a line in one direction. The two events of not crossing any line in two directions are independent. Therefore, the probability P_a of a coin crossing any line (not necessarily a corner) is $P_a = 1 - P_n = 1 - (1 - D/w)^2$. In the case of $w = 2D$, we have $P_n = 1/4$ and $P_a = 3/4$.

There are other ways of estimating π using randomized evaluations. The random sampling using the Monte Carlo method is another classic example.

■ **EXAMPLE 9.1**

For a unit square with an inscribed circle, we know that the area of the circle is $A_o = \pi/4$ and the area of the unit square is $A_s = 1$.

If we generate N random points uniformly distributed inside the unit square, there are n points fall within the inscribed circle. Then, the probability of a point falling within the circle is

$$\frac{A_o}{A_s} = \frac{\pi}{4} = \frac{n}{N},$$

when $N \to \infty$. So π can be estimated by $\pi \approx \frac{4n}{N}$. Suppose in a simulation, we have $n = 15707$ points inside the circle among $N = 20000$, we obtain

$$\pi \approx \frac{4n}{N} \approx \frac{4 \times 15707}{20000} \approx 3.1414.$$

We know this method of estimating π is inefficient; however, it does demonstrate simply how the Monte Carlo method works.

9.2 MONTE CARLO INTEGRATION

Monte Carlo integration is an efficient method for numerically estimating multi-dimensional integrals. However, for 1D and 2D integrals, it is less efficient than Gaussian integration. Therefore, Monte Carlo integration is recommended mainly for complicated integrals in higher dimensions.

The fundamental idea of Monte Carlo integration is to randomly sample the domain of integration inside a control volume (often a regular region), and the integral of interest is estimated using the fraction of the random sampling points and the volume of the control region. Mathematical speaking, that is to say,

$$I = \int_\Omega f dV \approx V[\frac{1}{N}\sum_{i=1}^{N} f_i] + O(\epsilon), \tag{9.8}$$

where V is the volume of the domain Ω, and f_i is the evaluation of $f(x, y, z, ...)$ at the sampling point $(x_i, y_i, z_i, ...)$. The error estimate ϵ is given by

$$\epsilon = VS = V\sqrt{\frac{\mu_2 - \mu^2}{N}}, \tag{9.9}$$

where

$$\mu = \frac{1}{N}\sum_{i=1}^{N} f_i, \qquad \mu_2 = \frac{1}{N}\sum_{i=1}^{n} f_i^2. \tag{9.10}$$

Here the sample variance S^2 can be estimated by

$$S^2 = \frac{1}{N-1}\sum_{i=1}^{N}(f_i - \mu)^2 \approx \frac{1}{N}\sum_{i=1}^{N} f_i^2 - \mu^2, \tag{9.11}$$

which is the approximation of the variance σ_f^2

$$\sigma_f^2 = \frac{1}{V}\int (f - \mu)^2 dV. \tag{9.12}$$

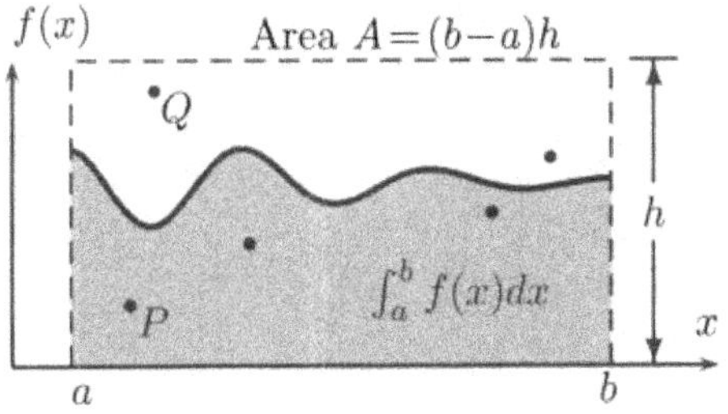

Figure 9.2: Representation of Monte Carlo integration.

The law of large number asymptotics implies that $\epsilon \to 0$ as $N \to \infty$. That is to say

$$V\mu = \lim_{N\to\infty} \frac{V}{N} \sum_{i=1}^{N} \to I. \tag{9.13}$$

In the simplest case in the domain $[0, 1]$, we have

$$\int_0^1 f(x)dx = \lim_{N\to\infty} [\frac{1}{N} \sum_{i=1}^{N} f(x_i)]. \tag{9.14}$$

We can see that the error of Monte Carlo integration decreases with N in a manner of $1/\sqrt{N}$, which is independent of the number (p) of dimensions. This becomes advantageous over other conventional methods for multiple integrals in higher dimensions.

The basic procedure is to generate the random points so that they distribute uniformly inside the domain. In order to calculate the volume in higher dimensions, it is better to use a regular control domain to enclose the domain Ω of the integration.

For simplicity of discussion, we now use the integral of a univariate function $f(x)$ over the interval $[a, b]$ (see Figure 9.2). Let us estimate the integral

$$I = \int_a^b f(x)dx = (b-a)\Big[\frac{1}{N} \sum_{i=1}^{N} f(x_i)\Big], \tag{9.15}$$

where a and b are finite integration limits. Now we first use a regular control volume or a bounding box so that it encloses the interval $[a, b]$ and the curve $f(x)$ itself. As the length of the domain is $(b-a)$ and the height of the box is h, the area A (or more generally the volume in higher dimension) is simply

$$A = (b-a)h. \tag{9.16}$$

Monte Carlo Integration

Define a bounding box/volume $\mathcal{V}$ for $I = \int_\Omega \beta f d\Omega$;
Calculate the area/volume V of $\mathcal{V}$;
Set the counter $S = 0$ and max number N of sampling points;
for $i = 1$ to N;
  Generate a point P randomly inside the volume $\mathcal{V}$;
  if P is inside the domain $\Omega \in \mathcal{V}$;
    $S = S + \beta$;
  end if
end for
Estimate the integral $I = \frac{VS}{N}$;

Figure 9.3: Pseudo code for Monte Carlo integration.

We know the integral I is the shaded area under the curve inside the bounding box, then the fraction or ratio I/A of the integral (area) to the total area of the bounding box is statistically equivalent to the fraction or probability of uniformly-distributed random sampling points falling in the shaded area inside the box. Suppose we generate N sampling points which are uniformly distributed inside the box. If there are K points that are under the curve inside the box (shaded region), then the integral I can be estimated by

$$I \approx A\frac{K}{N} = \frac{(b-a)hK}{N}. \tag{9.17}$$

For the estimation of a multi-dimensional integral in the domain Ω

$$I = \int_\Omega \beta f d\Omega = \int dx \int dy ... \int \beta f dz; \tag{9.18}$$

where $\beta \in \Re$ is a real constant, then the basic steps of Monte Carlo integration can be summarized as the pseudo code shown in Figure 9.3. For simplicity, we can use $\beta = 1$ for most integrals, though it may become a convenient scaling factor for some integrals. Now let us look at a simple example.

■ **EXAMPLE 9.2**

We know that the integral

$$I = \text{erf}(2) = \frac{2}{\sqrt{\pi}} \int_0^2 e^{-x^2} dx \approx 0.995322.$$

Suppose we try to use the Monte Carlo integration to estimate this, and we have $a = 0$, $b = 2$ and $\beta = 1$. The function is simply

$$f(x) = \frac{2}{\sqrt{\pi}} e^{-x^2}.$$

Since the maximum of $f_{\max} = 2/\sqrt{\pi} \approx 1.128$ occurs at $x = 0$, we can use $h = 1.5$. Therefore, the area of the bounding box is $A = (b-a)h = 3$. For $N = 5000$, our simulation suggests that there are $K = 1654$ sampling points falling within the shaded region (under the curve inside the box in Figure 9.2). The estimate of the integral is given by

$$I \approx A\frac{K}{N} \approx 3 \times \frac{1654}{5000} \approx 0.9924,$$

which is within 0.3% of the true value 0.995322.

Well, this is a relatively simple example, and it is not efficient to use Monte Carlo integration. The control region here is regular (a rectangular box). In many applications, the domain of the interest is irregular and it is impossible to calculate the integral analytically. In this case, Monte Carlo integration becomes a powerful alternative. Therefore, Monte Carlo integration is recommended for estimating the integrals with irregular domain and for evaluating multi-dimensional integrals.

9.3 IMPORTANCE OF SAMPLING

In the simple Monte Carlo integration, we have used the uniform sampling. For an integrand that varies rapidly in a narrow region such as a sharp peak (e.g., $f(x) = e^{-(100x)^2}$), the only sampling points that are important are near the peak, the sampling points far outside will contribute less. Thus, it seems that a lot of unnecessary sampling points are wasted. There are two main ways to use the sampling points more effectively, and they are change of variables and importance sampling.

The change of variables uses the integrand itself so that it can be transformed to a more uniform (flat) function. For example, the integral

$$I = \int_a^b f(u)du, \tag{9.19}$$

can be transformed using a known function $u = g(v)$

$$I = \int_{a_v}^{b_v} f[g(v)]\frac{dg}{dv}dv. \tag{9.20}$$

The idea is to make sure that the new integrand is or close to a constant A

$$\phi(v) = f[g(v)]\frac{dg(v)}{dv} = A, \tag{9.21}$$

where $v = g^{-1}(u)$. This means that a uniform sampling can be used for ϕ. The new integration limits are $a_v = g^{-1}(a)$ and $b_v = g^{-1}(b)$.

For example, for an exponential function $f(u) = e^{-\alpha u}$, we can use $u = g(u) = -\ln(\alpha v)/\alpha$ so that $f(u)dg/dv = -1$. We then have

$$I = \int_a^b e^{-\alpha u} du = \int_{e^{-\alpha a}/\alpha}^{e^{-\alpha b}/\alpha} (-1) dv. \tag{9.22}$$

Then, we can use a uniform sampling set to estimate the integral. A slightly different method is to use the stratified sampling; that is, to decompose the domain into subregions. For example, we can use

$$I = \int_a^b f(x)dx = \int_a^d f(x)dx + \int_d^b f(x)dx, \tag{9.23}$$

where d is a known limit which divides the original region using the characteristics of the integrand. In each of the subregions, the function $f(x)$ should be relatively smooth and flat. This can be extended to any number of subregions if required.

Both these methods are limited either to the case when g^{-1} exists uniquely and has explicit expressions in terms of basic functions, or to the case when $f(x)$ are flat in subregions. Otherwise, it is not easy, or even impossible, to make such transformations or domain-decomposition. A far more efficient method is to use the importance sampling.

As the integration is the area (or volume) under a curve (or surface), the region(s) with higher values of integrand will contribute more, thus, we should put more weights on these important points. Importance sampling is just the method for doing such weighted sampling. The integral of interest is often rewritten as the weighted form or a product such that

$$I = \int_a^b f(x)dx = \int_a^b h(x)p(x)dx = \int_a^b \frac{f(x)}{p(x)} p(x)dx, \tag{9.24}$$

where $h(x) = f(x)/p(x)$. Obviously, it is required that $p(x) \neq 0$ in $[a, b]$. Here the function $p(x)$ acts as the probability density function whose integration over the region $[a, b]$ should be always equal to 1. That is

$$\int_a^b p(x)dx = 1. \tag{9.25}$$

The evaluation of the integral becomes the estimation of the expected value of $E = \langle h(x) \rangle = \langle f(x)/p(x) \rangle$. The idea is to choose a function $p(x)$ such that the sampling points become more important when $f(x)$ is in the region with higher values. That is equivalent to a weighted sum with $p(x)$ as the weighting coefficients.

The choice of $p(x)$ should make $h(x) = f(x)/p(x)$ as close to constant as possible. In a special case $p(x) = 1/(b - a)$, equation (9.24) becomes (9.15).

In addition, the error in Monte Carlo integration decreases in the form of $1/\sqrt{N}$ as N increases. As the true randomness of sampling points are not essential as long as the sampling points can be distributed as uniformly as possible. In fact, studies show that it is possible to sample the points in a certain deterministic way so as to minimize the error of the Monte Carlo integration. In this case, the error may decrease in terms of $(\ln N)^p/N$ where p is the dimension if appropriate methods such as Halton sequences are used. In addition, quasi-Monte Carlo methods using low discrepancy pseudo-random numbers are more efficient than standard Monte Carlo methods. Readers interested in such topics can refer to more advanced literature.

EXERCISES

9.1 Write a few lines of Matlab code to simulate the sum distribution of two random variables u and v. If u and v are uniformly distributed in $[0, 1]$, what are the distributions of $u + v$, $u - v$ and uv?

9.2 If two random variables u and v are independent and are drawn from the normal distribution

$$p(x) = \frac{1}{\sigma\sqrt{2\pi}} e^{-x^2/2\sigma^2},$$

what is the sum distribution? Write a program to confirm your conclusion.

9.3 If the distribution of interest is complicated, then analytical methods are not much useful. Monte Carlo simulations are often simple but efficient to obtain estimates. If u is uniformly distributed, what is the distribution $y = f(u) = u^3 - u^2 + u/2$? Write a program, no more than 5 lines to show your results. What happens if u is normally distributed?

9.4 Write a program to simulate throwing $n > 10$ dice (or $n > 40$ coins) and display the distribution of the sum of all face values of n dice or coins.

REFERENCES

1. C. M. Grindstead and J. L. Snell, *Introduction to Probability*, 2nd Edition, American Mathematical Society, 1997.
2. G. S. Fishman, *Monte Carlo: Concepts, Algorithms and Applications*, Springer, New York, 1995.
3. W. K. Hastings, "Monte Carlo sampling methods using Markov chains and their applications", *Biometrika*, **57**(1), 97-109 (1970).
4. N. Metropolis and S. Ulam, "The Monte Carlo method", *J. Am. Stat. Assoc.*, **44**, 335-341 (1949).
5. H. Niederreiter, *Random Number Generation and Quasi-Monte Carlo Methods*, SIAM, 1992.
6. I. M. Sobol, *A Primer for the Monte Carlo Method*, CRC Press, 1994.

CHAPTER 10

RANDOM WALK AND MARKOV CHAIN

Random walk and Markov chain play a central role in modern metaheuristic algorithms and stochastic optimization. In essence, a metaheuristic search algorithm is a procedure with a good combination of some randomization and deterministic components, together with the usage of memory or search history. Randomization is often achieved by some form of random walk in the search space. In this chapter, we will introduce the fundamentals of random walk, Markov chains and their relevance to optimization.

10.1 RANDOM PROCESS

Generally speaking, a random variable can be considered as an expression whose value is the realization or outcome of events associated with a random process such as the noise level on the street. The values of random variables are real, though for some variables such as the number of cars on a road only take discrete values, and such random variables are called discrete random variables. If a random variable such as the noise at a particular location can take any real values in an interval, it is called continuous. If a random variable can have both continuous and discrete values, it is called a mixed

Engineering Optimization: An Introduction with Metaheuristic Applications.
By Xin-She Yang
Copyright © 2010 John Wiley & Sons, Inc.

type. Mathematical speaking, a random variable is a function that maps events to real numbers. The domain of this mapping is called the sample space.

There is a probability distribution function associated with each random variable, and this distribution is often expressed as a so-called probability density function. For example, the number of phone calls per minute, and the number of users of a web server per day all obey the Poisson distribution

$$p(n;\lambda) = \frac{\lambda^n e^{-\lambda}}{n!}, \qquad (n = 0, 1, 2, ...), \tag{10.1}$$

where $\lambda > 0$ is a parameter which is the mean or expectation of the occurrence of the event during a unit interval. This distribution is often called the law for rare events because it is a limit of small probability of the binomial distribution

$$p(k;n,p) = \binom{n}{k} p^k (1-p)^{n-k}, \qquad \binom{n}{k} = \frac{n!}{k!(n-k)!}, \tag{10.2}$$

which is a probability distribution for the number k of success in a sequence of n independent success-or-fail experiments. Here p is the probability of success in each event. Poisson's distribution is the limit of the binomial distribution when $n \to \infty$ while $\lambda = pn$ remains constant.

Sometimes, it is easier to use moment-generating function Φ to calculate mean, variance and other quantities. For example, the moment-generating function for the binomial distribution is

$$\Phi(t) = (pe^t + 1 - p)^n, \tag{10.3}$$

where $t \in \Re$ is the parameter. The mean of the binomial distribution is

$$\mu = \Phi'(t)\Big|_{t=0} = n(pe^t + 1 - p)^n \cdot pe^t\Big|_{t=0} = n(pe^0 + 1 - p)^n \cdot pe^0 = np, \tag{10.4}$$

which is exactly the first moment μ_1'.

The second moment μ_2' is defined as

$$\mu_2' = \Phi''(t)\Big|_{t=0} = n(n-1)[pe^t + (1-p)]^{n-2}(pe^t)^2 + n[pe^t + (1-p)]^{n-1}(pe^t)\Big|_{t=0}$$

$$= n(n-1)p^2 + np. \tag{10.5}$$

The variance is the second moment about the mean, that is

$$\sigma^2 = \mu_2' - \mu^2 = [n(n-1)p^2 + np] - (np)^2 = np(1-p). \tag{10.6}$$

We will use these results later in the discussion of the random walk on a straight line.

Gaussian distribution or normal distribution is by far the most popular distribution because many physical variables including light intensity, and errors/uncertainty in measurements, and many other processes obey the normal distribution

$$p(x;\mu,\sigma^2) = \frac{1}{\sigma\sqrt{2\pi}} \exp[-\frac{(x-\mu)^2}{2\sigma^2}], \qquad -\infty < x < \infty, \tag{10.7}$$

where μ is the mean and $\sigma > 0$ is the standard deviation. This distribution is often denoted by $N(\mu, \sigma^2)$. In the special case when $\mu = 0$ and $\sigma = 1$, it is called a standard normal distribution, denoted by $N(0, 1)$.

For a random variable with discrete values x_i, the entropy of its distribution is given by

$$S = -\sum_{i=1}^{K} p(x_i) \log_b p(x_i), \tag{10.8}$$

where b is the base, often $b = 2$, e or 10. Here K the number of all possible outcomes. If the distribution continuous, then the entropy becomes an integral

$$S = -\int_{\Omega} p(x) \ln p(x)\, dx. \tag{10.9}$$

For example, the entropy of the normal distribution is

$$S = -\int_{-\infty}^{\infty} \frac{1}{\sigma\sqrt{2\pi}} e^{-(x-\mu)^2/2\sigma^2} \ln[\frac{1}{\sigma\sqrt{2\pi}} e^{-(x-\mu)^2/2\sigma^2}]$$

$$= \ln[\sqrt{2\pi e\sigma^2}], \tag{10.10}$$

which is independent of the mean μ.

10.2 RANDOM WALK

A random walk is a random process which consists of taking a series of consecutive random steps. Mathematically speaking, let S_N denotes the sum of each consecutive random step X_i, then S_N forms a random walk

$$S_N = \sum_{i=1}^{N} X_i = X_1 + ... + X_N, \tag{10.11}$$

where X_i is a random step drawn from a random distribution. This relationship can also be written as a recursive formula

$$S_N = \sum_{i=1}^{N-1} + X_N = S_{N-1} + X_N, \tag{10.12}$$

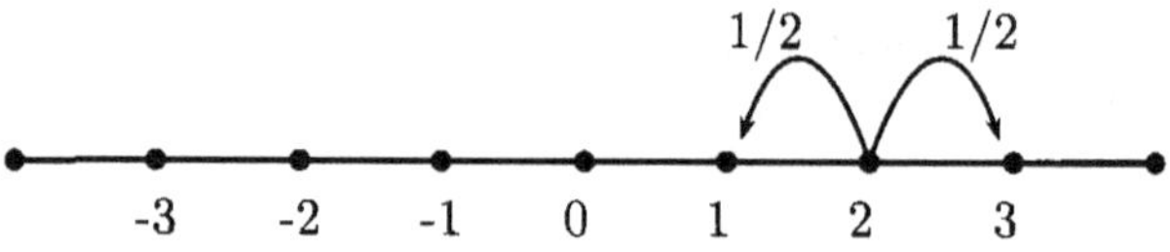

Figure 10.1: Random walk in a one-dimensional line. At any point, the probability moving to the left or right equals to 1/2.

which means the next state S_N will only depend the current existing state S_{N-1} and how the motion or transition X_N from the existing state to the next state. This is typically the main property of a Markov chain to be introduced later.

Here the step size or length in a random walk can be fixed or varying. Random walks have many applications in physics, economics, statistics, computer sciences, environmental science and engineering.

10.2.1 1D Random Walk

Consider a scenario, a drunkard walks on a street, at each step, he can randomly go forward or backward, this forms a one-dimensional random walk. If this drunkard walks on a football pitch, he can walk in any direction randomly, this becomes a 2D random walk. Mathematically speaking, a random walk is given by the following equation

$$S_{t+1} = S_t + w_t, \tag{10.13}$$

where S_t is the current location or state at t, and w_t is a step or random variable with a known distribution.

For the 1D grid line shown in Figure 10.1, a particle can jump to the right or the left with equal probability 1/2, and each jump can only take one step only. This jump probability, often called transition probability, can be written as

$$w_t = \begin{cases} 1/2 & \text{if } S = +1 \\ 1/2 & \text{if } S = -1 \\ 0 & \text{otherwise} \end{cases} . \tag{10.14}$$

A particle starting at $S_0 = 0$ jumps along a straight line, let us follow its first few steps. Suppose if we flip a coin, the particle move to the right (or up) if it is a head; otherwise, the particle moves to the left (or down) when the coin is a tail. First, we flop the coin, we get, say, a head. So the particle moves to the right by a fixed unit step. So $S_1 = S_0 + 1 = 1$. Then, a tail leads a move to the left, that is $S_2 = S_1 - 1 = 0$. By flipping the coin again, we get a tail. So $S_3 = S_2 - 1 = -1$. We continue the process in the similar manner, and the path of 100 steps or jumps is shown in Figure 10.2. It has been proved

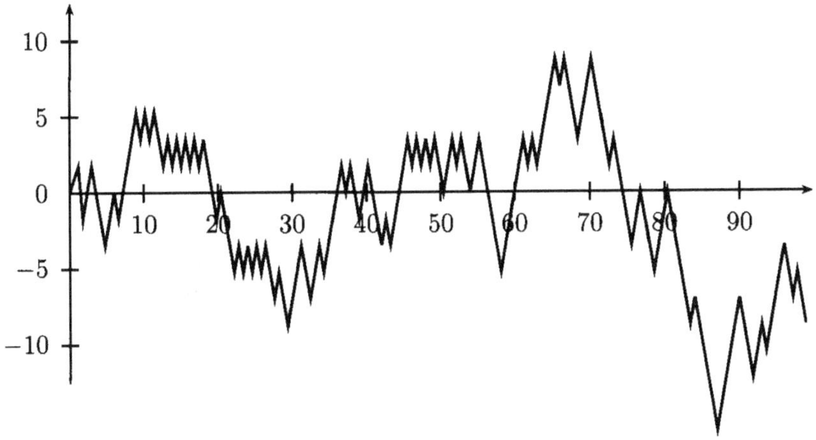

Figure 10.2: Random walk and the path of 100 consecutive steps staring at position 0.

theoretically that the probability of returning to the origin or reaching any point approaches 1 when the number N of steps approaches infinity.

Suppose the probability of moving to the right is p, and thus the probability of moving to the left is $q = 1 - p$. The probability of taking k steps to the right among N steps obeys the the binomial distribution

$$p(k; N, p) = \binom{N}{k} p^k (1-p)^{N-k}. \tag{10.15}$$

From the results (10.4) and (10.6), we know that the mean number of steps to the right is the mean

$$<k> = pN, \tag{10.16}$$

which means the mean number of steps to the left is simply $N - pN = (1-p)N = qN$. The variance associated with k is

$$\sigma_k^2 = p(1-p)N = pqN. \tag{10.17}$$

As time increases, the number of steps also increases. That is $N = t$ if each step or jump takes a unit time. This means that the variance increases linearly with t, or

$$\sigma^2 \propto t. \tag{10.18}$$

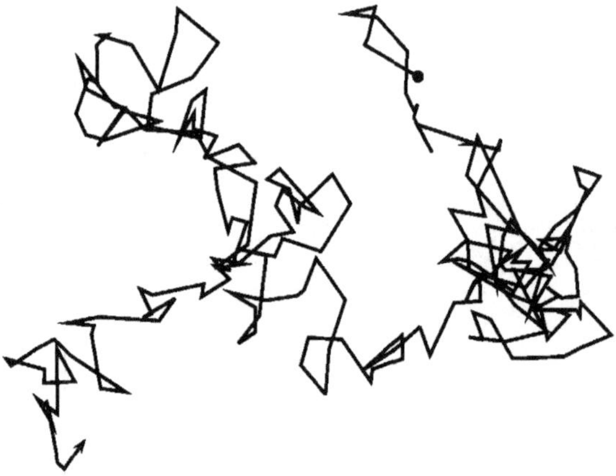

Figure 10.3: Brownian motion in 2D: random walk with a Gaussian step-size distribution and the path of 100 steps starting at the origin $(0,0)$ (marked with •).

10.2.2 Random Walk in Higher Dimensions

If each step or jump is carried out in the n-dimensional space, the random walk discussed earlier

$$S_N = \sum_{i=1}^{N} X_i, \tag{10.19}$$

becomes a random walk in higher dimensions. In addition, there is no reason why each step length should be fixed. In fact, the step size can also vary according to a known distribution. If the step length obeys the Gaussian distribution, the random walk becomes the Brownian motion (see Figure 10.3). Similar to the one-dimensional case, the variance σ^2 also increases linearly with time t or the total number of steps N (see Exercise 10.1 for detail).

In theory, as the number of steps N increases, the central limit theorem implies that the random walk (10.19) should approaches a Gaussian distribution. As the mean of particle locations shown in Figure 10.3 is obviously zero, their variance will increase linearly with t. Therefore, the Brownian motion $B(t)$ essentially obeys a Gaussian distribution with zero mean and time-dependent variance. That is,

$$B(t) \sim N(0, \sigma^2 = t), \tag{10.20}$$

where $\sim$ means the random variance obeys the distribution on the right-hand side, or samples should be drawn from the distribution. The diffusion process can be viewed as a series of Brownian motion, and the motion obeys the Gaussian distribution. For this reason, standard diffusion is often referred to as the Gaussian diffusion. If the motion at each step is not Gaussian, then the diffusion is called non-Gaussian diffusion.

If the step length obeys other distribution, we have to deal with more generalized random walk. A very special case is when the step length obeys the Lévy distribution, such a random walk is called Lévy flights.

10.3 LÉVY FLIGHTS

Broadly speaking, Lévy flights are a random walk whose step length is drawn from the Lévy distribution, often in terms of a simple power-law formula $L(s) \sim |s|^{-1-\beta}$ where $0 < \beta \leq 2$ is an index. Lévy distribution is a distribution of the sum of N identically and independently distribution random variables whose Fourier transform takes the following form

$$F_N(k) = \exp[-N|k|^{\beta}]. \tag{10.21}$$

The inverse to get the actual distribution $L(s)$ is not straightforward, as the integral

$$L(s) = \frac{1}{\pi}\int_0^{\infty} \cos(\tau s)e^{-\alpha\tau^{\beta}}\,d\tau, \qquad (0 < \beta \leq 2), \tag{10.22}$$

does not have analytical form except for a few special cases. Here $L(s)$ is called the Lévy distribution with an index β. For most applications, we can set $\alpha = 1$ for simplicity.

Two special cases are $\beta = 1$ and $\beta = 2$. When $\beta = 1$, the above integral becomes the Cauchy distribution

$$p(s) = \frac{1}{\pi N} \cdot \frac{1}{1 + (\frac{s}{N})^2}. \tag{10.23}$$

When $\beta = 2$, it becomes the normal distribution. In this case, Lévy flights become the standard Brownian motion. However, it is possible to express the integral (10.22) as a series

$$L(s) \sim -\frac{1}{\pi}\sum_{j=1}^{\infty} \frac{(-1)^j}{j!} \sin(j\pi\beta/2)\Gamma(\beta j + 1) \cdot \frac{1}{s^{\beta j+1}}, \tag{10.24}$$

which suggests the leading-order approximation ($j = 1$) for the longer flight length s is a power-law distribution

$$L(s) \sim |s|^{-1-\beta}, \tag{10.25}$$

and it is heavy-tailed. The variance of such a power-law distribution is infinite for $0 < \beta < 2$. Figure 10.4 shows the path of Lévy flights of 100 steps starting from $(0, 0)$ with $\beta = 1$. It is worth pointing out that a power-law distribution is often linked to some scale-free characteristics, and Lévy flights can thus show self-similarity and fractal behavior in the flight patterns.

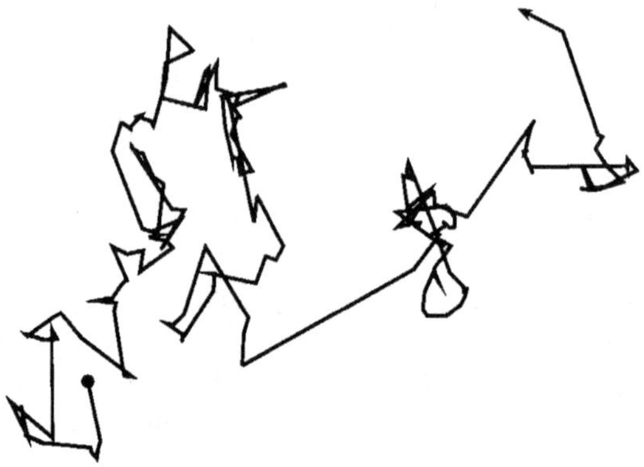

Figure 10.4: Lévy flights in 2D setting starting at the origin $(0, 0)$ (marked with •).

If we calculate the variance of the Lévy flights, though very difficult to do, we have the following relationship

$$\sigma^2(t) \sim \begin{cases} t^2 & \text{if } 0 < \beta < 1 \\ t^{3-\beta} & \text{if } 1 < \beta < 2 \\ t & \text{if } \beta \geq 2 \end{cases}, \tag{10.26}$$

which is in contrast with the linear behavior $\sigma^2 \sim t$ obtained earlier for the Brownian motion and 1D random walk. In the special case of $\beta = 2$, it indeed becomes the standard Brownian motion. It is worth pointing out that in the case of $\beta = 1$, $\sigma^2 \sim t^2/\ln t$ is expected. Therefore, we can say that Lévy flights are the deviation from the behavior governed by the central limit theorem.

Lévy flights have many applications. In fact, Lévy flights have been observed among foraging patterns of albatrosses and fruit flies, spider monkeys, and even humans such as the Ju/'hoansi hunter-gatherers. In addition, many physical phenomena such as the diffusion of fluorenscent molecules, cooling behavior and noise could show Lévy-flight style characteristics under the right conditions.

The non-Gaussian diffusion process for $1 < \beta < 2$ is called superdiffusion, and for example, turbulence can be considered as a case of superdiffusion. The case $\beta > 2$ is called subdiffusion, and obviously, $\beta = 2$ is the standard diffusion. It is worth pointing out that quantum tunneling can be thought as the case of $0 < \beta < 1$. The moments diverge (or are infinite) for $0 < \beta < 2$, which is a stumbling block for mathematical analysis.

If we impose a velocity associated with each flight step or segment, then the potential difficulty may be avoided by looking at the distance wandered by a Lévy walker. Conventionally, the Lévy flight with an associated velocity is often referred to as a Lévy walk. However, as a Lévy flight is a type of random walk, so Lévy flights and Lévy walk are interchangeable in most cases.

10.4 MARKOV CHAIN

Briefly speaking, a random variable U is a Markov process if the transition probability, from state $U_t = S_i$ at time t to another state $U_{t+1} = S_j$, depends only on the current state U_i. That is

$$P(i,j) \equiv P(U_{t+1} = S_j | U_0 = S_p, ..., U_t = S_i)$$
$$= P(U_{t+1} = S_j | U_t = S_i), \tag{10.27}$$

which is independent of the states before t. In addition, the sequence of random variables $(U_0, U_1, ..., U_n)$ generated by a Markov process is subsequently called a Markov chain. The transition probability $P(i,j) \equiv P(i \to j) = P_{ij}$ is also referred to as the transition kernel of the Markov chain.

If we rewrite the random walk relationship (10.12) with a random move governed by w_t which depends on the transition probability P, we have

$$S_{t+1} = S_t + w_t, \tag{10.28}$$

which indeed has the properties of a Markov chain. Therefore, a random walk is a Markov chain.

In order to solve an optimization problem, we can search the solution by performing a random walk starting from a good initial but random guess solution. However, simple or blind random walks are not efficient. To be computationally efficient and effective in searching for new solutions, we have to keep the best solutions found so far, and to increase the mobility of the random walk so as to explore the search space more effectively. Most importantly, we have to find a way to control the walk in such a way that it can move towards the optimal solutions more quickly, rather than wander away from the potential best solutions. These are the challenges for most metaheuristic algorithms, and the same issues are also important for Monte Carlo simulations and Markov chain sampling techniques. Markov chain Monte Carlo methods are a class of sampling techniques by controlling how a random walk behaves.

10.5 MARKOV CHAIN MONTE CARLO

The Markov chain Monte Carlo (MCMC) is a class of sample-generating methods, which attempts to directly draw samples from some highly complex multi-dimensional distribution using a Markov chain with known transition probability. The basic idea of the MCMC methods can trace back to to the classic Metropolis algorithm developed by Metropolis *et al.* in 1953. Since the 1990s, the Markov chain Monte Carlo has become a powerful tool for Bayesian statistical analysis, Monte Carlo simulations, and potentially optimization with high nonlinearity.

We use $\pi_i(t)$ to denote the probability of the chain in the state i (or more accurately S_i) at time t. This means that $\boldsymbol{\pi}(t) = (\pi_1, ..., \pi_m)^T$ is a vector of the state space. At time $t = 0$, $\boldsymbol{\pi}(0)$ is the initial vector.

The k-step transition probability $P_{ij}^{(k)}$ from state i to state j can be calculated by

$$P_{ij}^{(k)} = P(V_{t+k} = S_j | V_t = S_i), \tag{10.29}$$

where $k > 0$ is an integer. The matrix $\boldsymbol{P} = [P_{ij}^{(1)}] = [P_{ij}]$ is the transition matrix which is a right stochastic matrix. A right stochastic matrix is defined as a probability (square) matrix whose entries are non-negative with each row summing to 1. That is

$$P_{ij} > 0, \qquad \sum_{j=1}^{m} P_{ij} = 1, \qquad i = 1, 2, ..., m. \tag{10.30}$$

It is worth pointing the left transition matrix, though less widely used, is a stochastic matrix with each column summing to 1.

A Markov chain is regular if some power of the transition matrix $\boldsymbol{P}$ has only positive elements. That is, there exists a positive integer K such that $P_{ij}^{(K)} > 0$ for $\forall i, j$. This means that there is a non-zero probability to go from any state i to another state j. In other words, every state is accessible in a finite number of steps (not necessarily a few steps). If the number of steps K is not a multiple of some integer, then the chain is called aperiodic. This means there is no fixed-length cycle between certain states of the chain. In addition, a Markov chain is said to be ergodic or irreducible if it is possible to go from every state to every state.

In general, for a Markov chain starting with an initial $\boldsymbol{\pi}_0$ and a transition matrix $\boldsymbol{P}$, we have after n steps

$$\boldsymbol{\pi}_n = \boldsymbol{\pi}_0 \boldsymbol{P}^n, \qquad \text{or} \qquad \boldsymbol{\pi}_n = \boldsymbol{\pi}_{n-1} \boldsymbol{P}, \tag{10.31}$$

where $\boldsymbol{\pi}_n$ is a vector whose jth entry is the probability that the chain is in state S_j after n steps.

There is a fundamental theorem about a regular Markov chain. That is

$$\lim_{n \to \infty} \boldsymbol{P}^n = \boldsymbol{W}, \tag{10.32}$$

where $\boldsymbol{W}$ is a matrix with all rows equal and all entries strictly positive.

As the number of steps n increases, it is possible for a Markov chain to reach a stationary distribution $\boldsymbol{\pi}^*$ defined by

$$\boldsymbol{\pi}^* = \boldsymbol{\pi}^* \boldsymbol{P}. \tag{10.33}$$

From the definition of eigenvalues of a matrix $\boldsymbol{A}$,

$$\boldsymbol{A}\boldsymbol{u} = \lambda \boldsymbol{u}, \tag{10.34}$$

we know that the above equation implies that $\boldsymbol{\pi}^*$ is the eigenvector (of $\boldsymbol{P}$) associated with the eigenvalue $\lambda = 1$. The unique stationary distribution requires the following detailed balance of transition probabilities

$$P_{ij}\pi_i^* = P_{ji}\pi_j^*, \tag{10.35}$$

which is often referred to as the reversibility condition. A Markov chain that satisfies this reversibility condition is said to be reversible.

EXAMPLE 10.1

To see if a Markov chain is regular or not, we have to use the transition matrix $\boldsymbol{P}$. For example, the chain with

$$\boldsymbol{P} = \begin{pmatrix} 0.2 & 0.7 & 0.1 \\ 0.5 & 0.1 & 0.4 \\ 0.5 & 0.2 & 0.3 \end{pmatrix},$$

is regular because all entries of $\boldsymbol{P}$ are positive.

For another transition matrix

$$\boldsymbol{P} = \begin{pmatrix} 1/2 & 0 & 1/2 \\ 0 & 1/4 & 3/4 \\ 1/3 & 2/3 & 0 \end{pmatrix},$$

it has zero entries. However, we have

$$\boldsymbol{P}^2 = \boldsymbol{P}\boldsymbol{P} = \begin{pmatrix} 5/12 & 1/3 & 1/4 \\ 1/4 & 9/16 & 3/16 \\ 1/6 & 1/6 & 2/3 \end{pmatrix},$$

whose entries are all positive. So this is a regular Markov chain. If we have an initial probability vector $\boldsymbol{u}_0 = (1, 0, 0)$, then we have at the next step

$$\boldsymbol{u}_1 = \boldsymbol{u}_0 \boldsymbol{P} = \begin{pmatrix} 1/2 & 0 & 1/2 \end{pmatrix},$$

and

$$\boldsymbol{u}_2 = \boldsymbol{u}_1 \boldsymbol{P} = \boldsymbol{u}_0 \boldsymbol{P}^2 = \begin{pmatrix} 5/12 & 1/3 & 1/4 \end{pmatrix}.$$

In fact, as n increases, we have

$$\lim_{n \to \infty} \boldsymbol{u}_n \approx \begin{pmatrix} 0.26 & 0.35 & 0.39 \end{pmatrix},$$

which is independent of $\boldsymbol{u}_0$. It is easy to check $\boldsymbol{u}_\infty$ will be the same even from $\boldsymbol{u}_0 = (0\ 1\ 0)$ or $(0\ 0\ 1)$.

Furthermore, for

$$\boldsymbol{P} = \begin{pmatrix} 1/2 & 1/2 \\ 1 & 0 \end{pmatrix},$$

there is a zero entry. However, we have

$$\boldsymbol{P}^2 = \begin{pmatrix} 3/4 & 1/4 \\ 1/2 & 1/2 \end{pmatrix},$$

whose entries are strictly positive. So this chain is also regular.

On the other hand,

$$\boldsymbol{P} = \begin{pmatrix} 0 & 1 \\ 1 & 0 \end{pmatrix},$$

is not regular. This is because $\boldsymbol{P}^2 = \boldsymbol{I} = \begin{pmatrix} 1 & 0 \\ 0 & 1 \end{pmatrix}$, $\boldsymbol{P}^3 = \boldsymbol{P}$, and $\boldsymbol{P}^n = I$ if n is even, and $\boldsymbol{P}^n = \boldsymbol{P}$ if n is odd. There are always two entries which are zero.

The above discussion is mainly for the case when the states are discrete. We can generalize the above results to a continuous state Markov chain with a transition probability $P(u, v)$ and the corresponding stationary distribution

$$\pi^*(v) = \int_\Omega \pi^*(u) P(u, v) dv, \tag{10.36}$$

where Ω is the probability state space.

There are many ways to choose the transition probabilities, and different choices will result in different behaviour of the Markov chain. In essence, the characteristics of the transition kernel largely determine how the Markov chain of interest behaves, which also determines the efficiency and convergence of MCMC sampling. There are several widely used sampling algorithms, including Metropolis algorithms, Metropolis-Hasting Algorithms, Independence Sampling, Random-Walk, and of course Gibbs Sampler. We will introduce the most popular Metropolis-Hastings algorithms.

10.5.1 Metropolis-Hastings Algorithms

To draw samples from the target distribution, we may write $\pi(\theta) = \beta p(\theta)$ where β is just a normalising constant which is either difficult to estimate or not known. We will see later that the normalising factor β disappear in the expression of acceptance probability.

The Metropolis-Hastings algorithms essentially expresses an arbitrary transition probability from state θ to state ϕ as the product of an arbitrary transition kernel $q(\theta, \phi)$ and a probability $\alpha(\theta, \phi)$. That is

$$P(\theta, \phi) \equiv P(\theta \to \phi) = q(\theta, \phi)\alpha(\theta, \phi). \tag{10.37}$$

The Metropolis-Hastings algorithm

Begin with any initial θ_0 at time $t \leftarrow 0$ such that $p(\theta_0) > 0$;
for $i = 1$ to n (number of samples),
Generate a candidate $\theta_* \sim q(\theta_t, .)$ from a proposal distribution;
Evaluate the acceptance probability $\alpha(\theta_t, \theta_*)$;
Generate a uniformly-distributed random number $u \sim$ Unif $[0, 1]$,
  if $\alpha \geq u$,
    accept θ_*, that is $\theta_{t+1} \leftarrow \theta_*$;
  else
    $\theta_{t+1} \leftarrow \theta_t$;
  end if
end

Figure 10.5: Metropolis-Hastings algorithm.

Here q is the proposal distribution function, while $\alpha(\theta, \phi)$ can be considered as the acceptance rate from θ to ϕ, and can be determined by

$$\alpha(\theta, \phi) = \min\left[\frac{\pi(\phi)q(\phi, \theta)}{\pi(\theta)q(\theta, \phi)}, 1\right] = \min\left[\frac{p(\phi)q(\phi, \theta)}{p(\theta)q(\theta, \phi)}, 1\right]. \tag{10.38}$$

The essence of Metropolis-Hastings algorithm is to first propose a candidate θ_*, then accept it with a probability α. That is, $\theta_{t+1} \leftarrow \theta_*$ if $\alpha \geq u$ where u is a random value drawn from a uniform distribution in $[0, 1]$, otherwise $\theta_{t+1} \leftarrow \theta_t$. The Metropolis-Hastings algorithms can be summarized as the pseudo code shown in Figure 10.5.

It is straightforward to verify that the reversibility condition is satisfied by the Metropolis-Hastings kernel

$$q(\theta, \phi)\alpha(\theta, \phi)\pi(\theta) = q(\phi, \theta)\alpha(\phi, \theta)\pi(\phi), \tag{10.39}$$

for all θ, ϕ. Consequently, the Markov chain will converge to a stationary distribution which is the target distribution $\pi(\theta)$.

■ EXAMPLE 10.2

Let us try to draw some samples from the Rayleigh distribution

$$f(x) = xe^{-x^2/2}, \qquad x \in [0, \infty),$$

using the Metropolis-Hastings algorithm. Suppose we use a uniform distribution in $(0, 10)$ as the proposal distribution q. Obviously, there is a small probability of x over 10, but such probability is extremely small.

Firstly, starting with $\theta_0 = 1$ at $t = 0$, we now draw a random sample from the uniform distribution $(0, 10)$, and we get a candidate $\theta_* = 0.5$.

Then, we have

$$\alpha = \min\left[\frac{f(\theta_*)}{f(\theta_0)}, 1\right] = \min\left[\frac{0.5 \cdot e^{-0.5^2/2}}{1 \cdot e^{-1^2/2}}, 1\right] \approx 0.7275.$$

Secondly, we now draw a random number u from a uniform distribution $(0, 1)$, suppose we get $u = 0.69$. Since $\alpha > u$, so we accept $\theta_* = 0.5$ as a successfully drawn sample, and we also update that $\theta_1 = 0.5$.

Thirdly, we draw another candidate $\theta_* = 2.40$. Now have

$$\alpha = \min\left[\frac{f(\theta_*)}{f(\theta_1)}, 1\right] = \min\left[\frac{2.40 \cdot e^{-2.40^2/2}}{0.5 \cdot e^{-0.5^2/2}}, 1\right] \approx 0.305.$$

We then draw another uniformly distributed random number, say, we get $u = 0.90$. Since $\alpha < u$, we should reject the sample. We proceed in a similar manner to draw as many samples as we want.

In a special case when the transition kernel is symmetric in its arguments, or

$$q(\theta, \phi) = q(\phi, \theta), \tag{10.40}$$

for all θ, ϕ, then equation (10.38) becomes

$$\alpha = \min[\frac{p(\phi)}{p(\theta)}, 1], \tag{10.41}$$

and the Metropolis-Hastings algorithm reduces to the classic Metropolis algorithm. In this case, the associated Markov chain is called a symmetric chain. In a special case when $\alpha = 1$ is used, that is, the acceptance probability is always 1, then the Metropolis-Hastings degenerates into the classic widely-used Gibbs sampling algorithm. However, Gibbs sampler becomes very inefficient for the distributions that are non-normally distributed or highly nonlinear.

10.5.2 Random Walk

As shown above, the choice of proposal kernels is very important. An efficient variation in generating a candidate sample θ_{t+1} is to use a random walk process. The proposal can be generated from θ_t by

$$\theta_{t+1} = \theta_t + w_t, \tag{10.42}$$

where w_t is a random walk variable with a distribution density Q independent of the Markov chain. The transition kernel becomes a special case

$$q(\theta, \phi) = Q(\phi - \theta), \tag{10.43}$$

which implies that only the difference $\phi - \theta$ matters. If Q is an even function in terms of $\phi - \theta$ (or symmetric around 0), then the kernel is symmetric and thus the Markov chain is also symmetric. In this special case, the classic Metropolis algorithm can be considered as a special case of a random walk.

10.6 MARKOV CHAIN AND OPTIMISATION

An important link between MCMC and optimisation is that most heuristic and metaheuristic search algorithms such as simulated annealing to be introduced later use a trajectory-based approach. They start with some initial (random) state, and propose a new state (solution) randomly. Then, the move is accepted or not depending on some probability. This is similar to a Markov chain. In fact, the standard simulated annealing is a random walk.

In fact, a great leap in understanding metaheuristic algorithms is to view a Markov chain Monte carlo as an optimization procedure. If we want to find the minimum of an objective function $f(\theta)$ at $\theta = \theta_*$ so that $f_* = f(\theta_*) \leq f(\theta)$, we can convert it to a target distribution for a Markov chain

$$\pi(\theta) = e^{-\beta f(\theta)}, \tag{10.44}$$

where $\beta > 0$ is a parameter which acts as a normalized factor. β value should be chosen so that the probability is close to 1 when $\theta \to \theta_*$.

At $\theta = \theta_*$, $\pi(\theta)$ should reach a maximum $\pi_* = \pi(\theta_*) \geq \pi(\theta)$. This often requires that the formulation of $f(\theta)$ should be non-negative, which means that some objective functions can be shifted by a large constant $A > 0$ such as $f \leftarrow f + A$ if necessary.

Using the Markov transition probability $p(\theta, \theta')$ so that

$$\alpha = \min[\frac{\pi(\theta')p(\theta', \theta)}{\pi(\theta)p(\theta, \theta')}, 1], \tag{10.45}$$

we can update the chain if $f(\theta') < f(\theta)$. Ideally, we should choose such transition probabilities so as to make α close to 1 when a new move produces a better solution. This is equivalent to

$$\alpha \to 1, \qquad \text{for } f(\theta') \leq f(\theta), \tag{10.46}$$

and

$$\alpha \to 0, \qquad \text{for } f(\theta') > f(\theta). \tag{10.47}$$

Such optimisation via MCMC can be extended to a generic framework outlined by Ghate and Smith in 2008, as shown in Figure 10.6.

■ EXAMPLE 10.3

Simulated annealing for minimization intends to propose a potential move and then decide to accept by an acceptance probability

$$p_t = \min\left[e^{-\Delta f / k_B T_t}, 1\right],$$

where k_B is the Boltzmann constant which can often be taken as $k_B = 1$. T_t is the current temperature at time t, and $\Delta f = f_{t+1} - f_t$ is the change

Markov Chain Algorithm for Optimization

Objective function $f(\boldsymbol{x})$
Start with $U_0 \in S$, at $t = 0$
while (criterion)
  Generate Y_{t+1} using an appropriate candidate kernel
  Generate a random number $0 \leq P_t \leq 1$

$$U_{t+1} = \begin{cases} Y_{t+1} & \text{with probability } P_t \\ U_t & \text{with probability } 1 - P_t \end{cases} \qquad (10.48)$$

end

Figure 10.6: The Ghate-Smith Markov chain algorithm for optimization.

of the objective function $f(\boldsymbol{x})$ where f_{t+1} and f_t are the values of the objective function at the two consecutive time steps, respectively.

In this framework, simulated annealing and its many variants are simply a special case of a Markov chain with

$$P_t = \begin{cases} \exp[-\frac{\Delta f}{T_t}] & \text{if } f_{t+1} > f_t \\ 1 & \text{if } f_{t+1} \leq f_t \end{cases},$$

In this case, only the difference Δf between the function values is relevant.

In addition, a proper control over the temperature can have a significant effect on how the algorithm converges, and this control is often referred to as the cooling schedule.

Algorithms such as simulated annealing use a single Markov chain, which may not be very efficient. In practice, it is usually advantageous to use multiple Markov chains in parallel to increase the overall efficiency. Algorithms such as simulated tempering uses multiple Markov chains with different temperatures.

Furthermore, there is no reason different chains should not interact. In fact, the algorithms such as particle swarm optimization to be introduced in Part II can be viewed as multiple interacting Markov chains, though such theoretical analysis remains almost intractable. The theory of interacting Markov chains is complicated and yet still under development, however, any progress in such areas will play a central role in the understanding how population- and trajectory-based metaheuristic algorithms perform under various conditions.

It is worth pointing that this generic optimization framework may have good convergence under appropriate conditions. However, the computational efficiency is not always practical for large-scale problems. There is no free lunch in optimization, and consequently, a lot of efforts have been devoted to the development of efficient algorithms to fit for the purpose. In Part II, we will introduce various metaheuristic algorithms in great detail.

EXERCISES

10.1 Consider the case of a 2D random walk where each step is of fixed unit length. Derive the distance traveled after N steps and its variance using the complex numbers on a plane.

10.2 Write a simple program to carry out some Lévy flights and demonstrate how the exponent β affects the distance traveled (10.26).

10.3 Write a simple program to generate of a random walk

10.4 The convergence to the stationary distribution $\boldsymbol{\pi}^*$ of a Markov chain is typically related to the second eigenvalue of the transition matrix $\boldsymbol{P}$. For a simple matrix

$$\boldsymbol{P} = \begin{pmatrix} 1/4 & 3/4 \\ 3/4 & 1/4 \end{pmatrix},$$

what is its second eigenvalue? Show that $\boldsymbol{P}^n$ (where $n = 1, 2, ...$) will converge to

$$\boldsymbol{P}_\infty = \begin{pmatrix} 1/2 & 1/2 \\ 1/2 & 1/2 \end{pmatrix}.$$

10.5 Verify if a Markov chain with the following transition matrix $\boldsymbol{P}$ is regular or not.

$$\text{a) } \boldsymbol{P} = \begin{pmatrix} 0.2 & 0.8 \\ 0.3 & 0.7 \end{pmatrix}; \text{ b) } \boldsymbol{P} = \begin{pmatrix} 1/2 & 1/4 & 1/4 \\ 0 & 2/5 & 3/5 \\ 4/5 & 1/5 & 0 \end{pmatrix}; \text{ c) } \boldsymbol{P} = \begin{pmatrix} 1 & 0 \\ 1/2 & 1/2 \end{pmatrix}.$$

10.6 Markov chains have many applications. For example, the Google's page ranking engine uses a Markov chain with a transition probability $p = \alpha/n_i + (1-\alpha)/N$ for page i with n_i links among N known webpages. Here α is in the range of 0 and 1. How will the choice of α affect the ranking?

REFERENCES

1. W. J. Bell, *Searching Behaviour: The Behavioural Ecology of Finding Resources*, Chapman & Hall, London, 1991.
2. C. Blum and A. Roli, "Metaheuristics in combinatorial optimization: Overview and conceptural comparison", *ACM Comput. Surv.*, **35**, 268-308 (2003).
3. M. G. Cox, A. B. Forbes, P. M. Harris, *Discrete Modelling*, SSfM Best Practice Guide No.4, NPL, UK, 2002.
4. S. R. Finch, *Mathematical Constants*, Cambridge University Press, (2003).
5. D. Gamerman, *Markov Chain Monte Carlo*, Chapman & Hall/CRC, 1997.
6. L. Gerencser, S. D. Hill, Z. Vago, and Z. Vincze, "Discrete optimization, SPSA, and Markov chain Monte Carlo methods", *Proc. 2004 Am. Contr. Conf.*, 3814-3819 (2004).

7. C. J. Geyer, "Practical Markov Chain Monte Carlo", *Statistical Science*, **7**, 473-511 (1992).

8. A. Ghate and R. Smith, "Adaptive search with stochastic acceptance probabilities for global optimization", *Operations Research Lett.*, **36**, 285-290 (2008).

9. W. R. Gilks, S. Richardson, and D. J. Spiegelhalter, *Markov Chain Monte Carlo in Practice*, Chapman & Hall/CRC, 1996.

10. M. Gutowski, "Lévy flights as an underlying mechanism for global optimization algorithms", *ArXiv Mathematical Physics e-Prints*, June, 2001.

11. W. K. Hastings, "Monte Carlo sampling methods using Markov chains and their applications", *Biometrika*, **57**, 97-109 (1970).

12. S. Kirkpatrick, C. D. Gellat and M. P. Vecchi, "Optimization by simulated annealing", *Science*, **220**, 670-680 (1983).

13. E. Marinari and G. Parisi, "Simulated tempering: a new Monte Carlo scheme", *Europhysics Lett.*, **19**, 451-458 (1992).

14. W. H. McCrea and F. J. Whipple, "Random paths in two and three dimensions", *Proc. Roy. Soc. Edinburgh*, **60**, 281-298 (1940).

15. S. P. Meyn, and R. L. Tweedie, *Markov Chains and Stochastic Stability*, Springer-Verlag, London, 1993.

16. N. Metropolis, and S. Ulam, "The Monte Carlo method", *J. Amer. Stat. Assoc.*, **44**, 335-341 (1949).

17. N. Metropolis, A. W. Rosenbluth, M. N. Rosenbluth, A. H. Teller, and E. Teller, "Equation of state calculations by fast computing machines", *J. Chem. Phys.*, **21**, 1087-1092 (1953).

18. D. J. Murdoch and P. J. Green, "Exact sampling from a continuous state space", *Scand. J. Statist.*, **25**, 483-502 (1998).

19. I. Pavlyukevich, "Lévy flights, non-local search and simulated annealing", *J. Computational Physics*, **226**, 1830-1844 (2007).

20. I. Pavlyukevich, "Cooling down Lévy flights", *J. Phys. A:Math. Theor.*, **40**, 12299-12313 (2007).

21. J. Propp and D. Wilson, "Exact sampling with coupled Markov chains and applications to statistical mechanics", *Random Structures and Algorithms*, **9**, 223-252 (1996).

22. G. Ramos-Fernandez, J. L. Mateos, O. Miramontes, G. Cocho, H. Larralde, B. Ayala-Orozco, "Lévy walk patterns in the foraging movements of spider monkeys (*Ateles geoffroyi*)", *Behav. Ecol. Sociobiol.*, **55**, 223-230 (2004).

23. A. M. Reynolds and M. A. Frye, "Free-flight odor tracking in Drosophila is consistent with an optimal intermittent scale-free search", *PLoS One*, **2**, e354 (2007).

24. M. E. Tipping M. E., "Bayesian inference: An introduction to principles and and practice in machine learning", in: *Advanced Lectures on Machine Lerning*, O. Bousquet, U. von Luxburg and G. Ratsch (Eds), pp.41-62 (2004).

25. G. M. Viswanathan, S. V. Buldyrev, S. Havlin, M. G. E. da Luz, E. P. Raposo, and H. E. Stanley, "Lévy flight search patterns of wandering albatrosses", *Nature*, **381**, 413-415 (1996).

26. E. Weisstein, http://mathworld.wolfram.com

PART II

METAHEURISTIC ALGORITHMS

CHAPTER 11

GENETIC ALGORITHMS

11.1 INTRODUCTION

The genetic algorithm (GA), developed by John Holland and his collaborators in the 1960s and 1970s, is a model or abstraction of biological evolution based on Charles Darwin's theory of natural selection. Holland was the first to use the crossover and recombination, mutation, and selection in the study of adaptive and artificial systems. These genetic operators form the essential part of the genetic algorithm as a problem-solving strategy. Since then, many variants of genetic algorithms have been developed and applied to a wide range of optimization problems, from graph colouring to pattern recognition, from discrete systems such as the traveling salesman problem to continuous systems such as the efficient design of airfoil in aerospace engineering, and from financial market to multiobjective engineering optimization.

There are many advantages of genetic algorithms over traditional optimization algorithms, and two most noticeable advantages are: the ability of dealing with complex optimization problems and parallelism. Genetic algorithms can deal with various types of optimization whether the objective (fitness) function is stationary or non-stationary (change with time), linear or nonlinear,

Engineering Optimization: An Introduction with Metaheuristic Applications.
By Xin-She Yang
Copyright © 2010 John Wiley & Sons, Inc.

continuous or discontinuous, or with random noise. As multiple offsprings in a population act like independent agents, the population (or any subgroup) can explore the search space in many directions simultaneously. This feature makes it ideal to parallelize the algorithms for implementation. Different parameters and even different groups of encoded strings can be manipulated at the same time.

However, genetic algorithms also have some disadvantages. The formulation of fitness function, the usage of population size, the choice of the important parameters such as the rate of mutation and crossover, and the selection criteria of new population should carefully be carried out. Any inappropriate choice will make it difficult for the algorithm to converge, or it simply produces meaningless results. Despite these disadvantages, genetic algorithms remain one of the most widely used optimization algorithms in modern nonlinear optimization.

11.2 GENETIC ALGORITHMS

11.2.1 Basic Procedure

The essence of genetic algorithms involves the encoding of an optimization function as arrays of bits or character strings to represent the chromosomes, the manipulation operations of strings by genetic operators, and the selection according to their fitness with the aim to find a solution to the problem concerned. This is often done by the following procedure:

- Encoding the objectives or optimization functions;
- Defining a fitness function or selection criterion;
- Initializing a population of individuals;
- Evaluating the fitness of all the individuals in the population;
- Creating a new population by performing crossover, and mutation, fitness-proportionate reproduction etc;
- Evolving the population until certain stopping criteria are met;
- Decoding the results to obtain the solution to the problem.

These steps can be represented schematically as the pseudo code of genetic algorithms shown in Figure 11.1.

One iteration of creating a new population is called a generation. The fixed-length character strings are used in most of genetic algorithms during each generation although there is substantial research on the variable-length strings and coding structures. The coding of the objective function is usually in the form of binary arrays or real-valued arrays in the adaptive genetic algorithms.

```
Genetic Algorithm
Objective function f(x), x = (x1, ..., xn)^T
Encode the solution into chromosomes (binary strings)
Define fitness F (eg, F ∝ f(x) for maximization)
Generate the initial population
Initial probabilities of crossover (pc) and mutation (pm)
   while ( t <Max number of generations )
       Generate new solution by crossover and mutation
       if pc >rand, Crossover; end if
       if pm >rand, Mutate; end if
       Accept the new solutions if their fitness increase
       Select the current best for new generation (elitism)
   end while
Decode the results and visualization
```

Figure 11.1: Pseudo code of genetic algorithms.

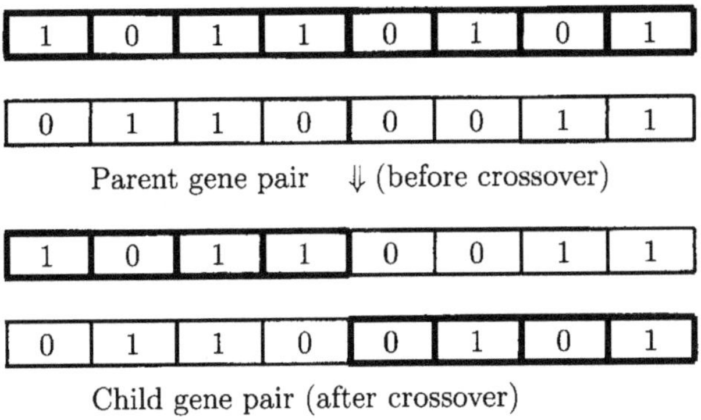

Figure 11.2: Diagram of crossover at a random crossover point (location) in genetic algorithms.

For simplicity, we use binary strings for encoding and decoding. The genetic operators include crossover, mutation, and selection from the population.

The crossover of two parent strings is the main operator with a higher probability p_c and is carried out by swapping one segment of one chromosome with the corresponding segment on another chromosome at a random position (see Figure 11.2). The crossover carried out in this way is a single-point crossover. Crossover at multiple points is also used in many genetic algorithms to increase the efficiency of the algorithms.

The mutation operation is achieved by flopping the randomly selected bits (see Figure 11.3), and the mutation probability p_m is usually small. The selection of an individual in a population is carried out by the evaluation of its fitness, and it can remain in the new generation if a certain threshold of

1	1	0	**1**	1	0	0	1

Original gene ⇓ (before mutation)

1	1	0	**0**	1	0	0	1

New gene (after crossover)

Figure 11.3: Schematic representation of mutation at a single site by flipping a randomly selected bit ($1 \to 0$).

the fitness is reached, we can also use that the reproduction of a population is fitness-proportionate. That is to say, the individuals with higher fitness are more likely to reproduce.

11.2.2 Choice of Parameters

An important issue is the formulation or choice of an appropriate fitness function that determines the selection criterion in a particular problem. For the minimization of a function using genetic algorithms, one simple way of constructing a fitness function is to use the simplest form $F = A - f(\boldsymbol{x})$ where A is a large constant if you wish to obtain a non-negative F. As only the relative fitness matters, in most cases $A = 0$ will do. Then, the objective is to maximize the fitness function and subsequently minimize the objective function $f(\mathbf{x})$. However, there are many different ways of defining a fitness function. For example, we can use the individual fitness assignment relative to the whole population

$$F(x_i) = \frac{f(\xi_i))}{\sum_{i=1}^{N} f(\xi_i)}, \tag{11.1}$$

where N is the population size. Here ξ_i is the phenotypic value of individual i, that is, the solution represented by an individual chromosome. The appropriate form of the fitness function will make sure that the solutions with higher fitness should be selected efficiently. Poor fitness function may result in incorrect or meaningless solutions.

Another important issue is the choice of various parameters. The crossover probability p_c is usually very high, typically in the range of $0.7 \sim 1.0$. On the other hand, the mutation probability p_m is usually small (usually $0.001 \sim 0.05$). If p_c is too small, then the crossover occurs sparsely, which is not efficient for evolution. If the mutation probability is too high, the solutions could still 'jump around' even if the optimal solution is approaching.

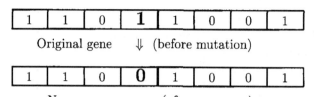

A proper criterion for selecting the best solutions is also important. How to select the current population so that the best individuals with higher fitness should be preserved and passed onto the next generation. That is often carried out in association with certain elitism. The basic elitism is to select the most fit individual (in each generation) which will be carried over to the new generation without being modified by genetic operators. This ensures that the best solution is achieved more quickly.

Other issues include the multiple sites for mutation and the use of various population sizes. The mutation at a single site is not very efficient, mutation at multiple sites will increase the evolution efficiency. However, too many mutants will make it difficult for the system to converge or even make the system go astray to the wrong solutions. In reality, if the mutation rate is too high under high selection pressure, then the whole population might go extinct.

In addition, the choice of the right population size is also very important. If the population size is too small, there is not enough evolution going on, and there is a risk for the whole population to go extinct. In the real world, a species with a small population, ecological theory suggests that there is a real danger of extinction for such species. Even if the system carries on, there is still a danger of premature convergence. In a small population, if a significantly more fit individual appears too early, it may reproduces enough offsprings so that they overwhelm the whole (small) population. This will eventually drive the system to a local optimum, often not the global optimum. On the other hand, if the population is too large, more evaluations of the objective function are needed, which will require extensive computing time.

Furthermore, more complex and adaptive genetic algorithms are under active research and the literature is vast about these topics.

11.3 IMPLEMENTATION

Using the basic procedure described in the above section, we can implement the genetic algorithms in any programming language. For simplicity of demonstrating how it works, we have implemented a function optimization using a simple GA in both Matlab and Octave. There are two main ways to use encoded strings. We can either use a string (or multiple strings) for each individual in a population, or use a long string for all the individuals (or all design variables) shown in Figure 11.4. However, for simplicity, we will use the first approach, though the latter approach is very efficient.

For the generalized De Jong's test function

$$f(\mathbf{x}) = \sum_{i=1}^{n} x_i^{2\alpha}, \qquad |x_i| \le r, \tag{11.2}$$

where $0 < \alpha \in \mathcal{N}$ is a positive integer and $r > 0$ is the half length of the domain. This function has a minimum of $f(\boldsymbol{x}_*) = 0$ at $\boldsymbol{x}_* = 0$. For the values of $\alpha = 2, r = 100$ and $n = 5$ as well as a population size of 40 16-bit strings, an estimate of $f_* \approx 0.0022429$ can be found after about 500 generations.

Figure 11.4: Encode all design variables into a single long string.

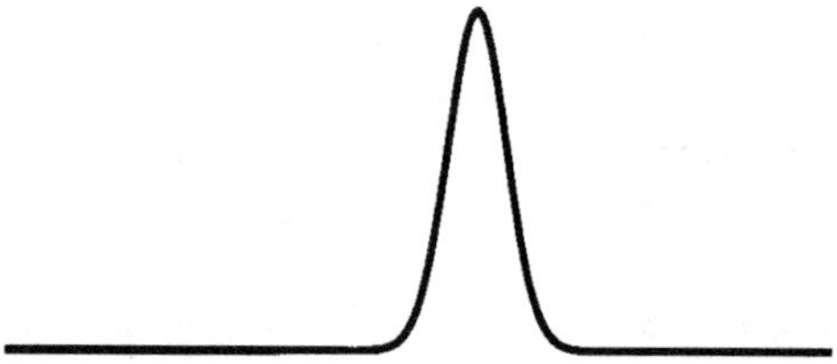

Figure 11.5: Easom's function: $f(x) = -\cos(x)e^{-(x-\pi)^2}$ for $x \in [-10, 10]$ has a unique global maximum $f_{\max} = 1$ at $x_* = \pi$.

Any two runs will give slightly different results due to the stochastic nature of genetic algorithms, but better estimates $f(\mathbf{x}) \to 0$ are obtained as the number of generations increases.

For the well-known Easom function

$$f(x) = -\cos(x)e^{-(x-\pi)^2}, \qquad x \in [-10, 10], \tag{11.3}$$

it has the global maximum $f_{\max} = 1$ at $x_* = \pi$ (see Figure 11.5). Now we can use the Matlab/Octave given in Appendix B to find its global maximum. In our implementation, we have used fixed-length 16-bit strings. The probabilities of crossover and mutation are respectively

$$p_c = 0.95, \qquad p_m = 0.05. \tag{11.4}$$

As it is a maximization problem, we can use the simplest fitness function $F = f(x)$. The outputs from a typical run are shown in Figure 11.6 where the top figure shows the variations of the best estimates as they approach to $x_* \to \pi$ while the lower figure shows the variations of the fitness function.

The simple demo Matlab program provided in Appendix B can easily be extended to higher dimensions. In fact, there is no need to do any programming (if you prefer) because there are many software packages about genetic algorithms, either free or commercial. For example, Matlab has an optimization toolbox with both conventional algorithms and genetic algorithms.

Biology-inspired algorithms have many advantages over traditional optimization methods such as the steepest descent and hill-climbing and calculus-based techniques due to the parallelism and the ability of locating the very good approximate solutions in extremely very large search spaces. Furthermore, more powerful new generation algorithms can be formulated by combining existing and new evolutionary algorithms with classical optimization methods.

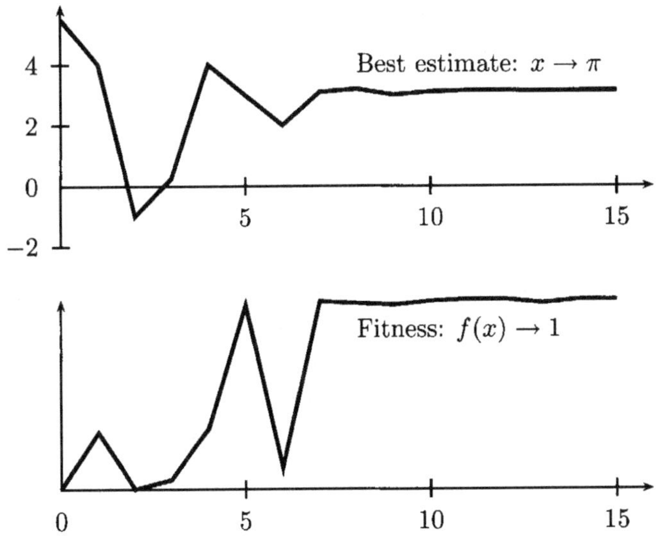

Figure 11.6: Typical outputs from a typical run. The best estimate will approach π while the fitness will approach $f_{\max} = 1$.

EXERCISES

11.1 Write a simple program to find the global minimum of

$$f(x, y) = (4 - 2.1x^2 + x^4/3)x^2 + xy + 4(y^2 - 1)y^2,$$

in the domain of $0 \le x \le 3$ and $-2 \le y \le 2$.

11.2 Write a simple program to implement the genetic algorithm to find the global minimum of the following n-dimensional function

$$f(\boldsymbol{x}) = \sum_{i=1}^{n} |x_i|^{i+1}, \qquad -1 \le x_i \le 1.$$

11.3 Almost all the test functions for validating new optimization algorithms are deterministic. If there is some uncertainty in the objective functions, it makes it much more difficult to find the optimum. For example, the author of this book has designed a stochastic objective function

$$\text{minimize } f(\boldsymbol{x}) = \sum_{i=1}^{n} \epsilon_i \, \big|x_i\big|, \qquad -5 \le x_i \le 5,$$

where the n random variables ϵ_i $(i = 1, 2, ..., n)$ are uniformly distributed. That is

$$\epsilon_i \sim \text{Unif}[0, 1].$$

Modify the genetic algorithm code in Appendix B, and try to see if you can find the minimum of this function.

REFERENCES

1. K. De Jong, *Analysis of the Behaviour of a Class of Genetic Adaptive Systems*, PhD thesis, University of Michigan, Ann Arbor, 1975.
2. L. J. Fogel, A. J. Owens, and M. J. Walsh, *Artificial Intelligence Through Simulated Evolution*, John Wiley & Sons, 1966.
3. D. E. Goldberg, *Genetic Algorithms in Search, Optimisation and Machine Learning*, Reading, Mass., Addison Wesley, 1989.
4. J. Holland, *Adaptation in Natural and Artificial Systems*, University of Michigan Press, Ann Anbor, 1975.
5. Z. Michaelewicz, *Genetic Algorithm + Data Structure = Evolution Progamming*, New York, Springer, 1996.
6. M. Mitchell, *An Introduction to Genetic Algorithms*, Cambridge, Mass: MIT Press, 1996.
7. P. Sirisalee, M. F. Ashby, G. T. Parks, and P. J. Clarkson, "Multi-criteria material selection in engineering design", *Adv. Eng. Mater.*, **6**, 84-92 (2004).
8. E.-G. Talbi, *Metaheuristics: From Design to Implementation*, John Wiley & Sons, 2009.

CHAPTER 12

SIMULATED ANNEALING

12.1 ANNEALING AND PROBABILITY

Simulated annealing (SA) is a random search technique for global optimization problems, and it mimics the annealing process in material processing when a metal cools and freezes into a crystalline state with the minimum energy and larger crystal size so as to reduce the defects in metallic structures. The annealing process involves the careful control of temperature and cooling rate, often called annealing schedule.

The application of simulated annealing into optimization problems was pioneered by Kirkpatrick, Gelatt and Vecchi in 1983. Since then, there have been extensive studies. Unlike the gradient-based methods and other deterministic search methods which have the disadvantage of being trapped into local minima, the main advantage of the simulated annealing is its ability to avoid being trapped in local minima. In fact, it has been proved that the simulated annealing will converge to its global optimality if enough randomness is used in combination with very slow cooling. In fact, simulated annealing algorithm is a search method using a Markov chain which converges under appropriate conditions concerning its transition probability.

By Xin-She Yang
Copyright © 2010 John Wiley & Sons, Inc.

Metaphorically speaking, this is equivalent to dropping some bouncing balls over a landscape, and as the balls bounce and lose energy, they settle down to some local minima. If the balls are allowed to bounce enough times and lose energy slowly enough, some of the balls will eventually fall into the globally lowest locations; hence, the global minimum will be reached.

The basic idea of the simulated annealing algorithm is to use random search in terms of a Markov chain, which not only accepts changes that improve the objective function, but also keeps some changes that are not ideal. In a minimization problem, for example, any better moves or changes that decrease the value of the objective function f will be accepted; however, some changes that increase f will also be accepted with a probability p. This probability p, also called the transition probability, is determined by

$$p = e^{-\frac{\Delta E}{k_B T}}, \tag{12.1}$$

where k_B is the Boltzmann's constant, and for simplicity, we can use k to denote k_B because $k = 1$ is often used. T is the temperature for controlling the annealing process. ΔE is the change in energy levels. This transition probability is based on the Boltzmann distribution in physics.

The simplest way to link ΔE with the change of the objective function Δf is to use

$$\Delta E = \gamma \Delta f, \tag{12.2}$$

where γ is a real constant. For simplicity without losing generality, we can use $k_B = 1$ and $\gamma = 1$. Thus, the probability p simply becomes

$$p(\Delta f, T) = e^{-\frac{\Delta f}{T}}. \tag{12.3}$$

Whether or not we accept a change, we usually use a random number r as a threshold. Thus, if $p > r$ or

$$p = e^{-\frac{\Delta f}{T}} > r, \tag{12.4}$$

the move is accepted.

12.2 CHOICE OF PARAMETERS

Here the choice of the right initial temperature is crucially important. For a given change Δf, if T is too high ($T \to \infty$), then $p \to 1$, which means almost all the changes will be accepted. If T is too low ($T \to 0$), then any $\Delta f > 0$ (worse solution) will rarely be accepted as $p \to 0$ and thus the diversity of the solution is limited, but any improvement Δf will almost always be accepted. In fact, the special case $T \to 0$ corresponds to the gradient-based method because only better solutions are accepted, and the system is essentially climbing up or descending along a hill. Therefore, if T is too high, the system is at a

Simulated Annealing Algorithm

Objective function $f(\boldsymbol{x})$, $\boldsymbol{x} = (x_1, ..., x_p)^T$
Initialize initial temperature T_0 and initial guess $\boldsymbol{x}^{(0)}$
Set final temperature T_f and max number of iterations N
Define cooling schedule $T \mapsto \alpha T$, $(0 < \alpha < 1)$
while ($T > T_f$ and $n < N$)
    Move randomly to new locations: $\boldsymbol{x}_{n+1} = \boldsymbol{x}_n$+randn
    Calculate $\Delta f = f_{n+1}(\boldsymbol{x}_{n+1}) - f_n(\boldsymbol{x}_n)$
    Accept the new solution if better
    if not improved
        Generate a random number r
        Accept if $p = \exp[-\Delta f/T] > r$
    end if
    Update the best $\boldsymbol{x}_*$ and f_*
    $n = n + 1$
end while

Figure 12.1: Simulated annealing algorithm.

high energy state on the topological landscape, and the minima are not easily reached. If T is too low, the system may be trapped in a local minimum, not necessarily the global minimum, and there is not enough energy for the system to jump out the local minimum to explore other minima including the global minimum. So a proper initial temperature should be calculated.

Another important issue is how to control the annealing or cooling process so that the system cools down gradually from a higher temperature to ultimately freeze to a global minimum state. There are many ways of controlling the cooling rate or the decrease of the temperature.

Two commonly used annealing schedules (or cooling schedules) are: linear and geometric. For a linear cooling schedule, we have

$$T = T_0 - \beta t, \tag{12.5}$$

or $T \to T - \delta T$, where T_0 is the initial temperature, and t is the pseudo time for iterations. β is the cooling rate, and it should be chosen in such a way that $T \to 0$ when $t \to t_f$ (or the maximum number N of iterations), this usually gives $\beta = (T_0 - T_f)/t_f$.

On the other hand, a geometric cooling schedule essentially decreases the temperature by a cooling factor $0 < \alpha < 1$ so that T is replaced by αT or

$$T(t) = T_0 \alpha^t, \qquad t = 1, 2, ..., t_f. \tag{12.6}$$

The advantage of the second method is that $T \to 0$ when $t \to \infty$, and thus there is no need to specify the maximum number of iterations. For this reason, we will use this geometric cooling schedule. The cooling process should be slow enough to allow the system to stabilize easily. In practise, $\alpha = 0.7 \sim 0.95$ is commonly used.

In addition, for a given temperature, multiple evaluations of the objective function are needed. If too few evaluations, there is a danger that the system will not stabilize and subsequently will not converge to its global optimality. If too many evaluations, it is time-consuming, and the system will usually converge too slowly as the number of iterations to achieve stability might be exponential to the problem size.

Therefore, there is a fine balance between the number of evaluations and solution quality. We can either do many evaluations at a few temperature levels or do few evaluations at many temperature levels. There are two major ways to set the number of iterations: fixed or varied. The first uses a fixed number of iterations at each temperature, while the second intends to increase the number of iterations at lower temperatures so that the local minima can be fully explored.

12.3 SA ALGORITHM

The simulated annealing algorithm can be summarized as the pseudo code shown in Figure 12.1. In order to find a suitable starting temperature T_0, we can use any available information about the objective function. If we know the maximum change $\max(\Delta f)$ of the objective function, we can use this to estimate an initial temperature T_0 for a given probability p_0. That is

$$T_0 \approx -\frac{\max(\Delta f)}{\ln p_0}.$$

If we do not know the possible maximum change of the objective function, we can use a heuristic approach. We can start evaluations from a very high temperature (so that almost all changes are accepted) and reduce the temperature quickly until about 50% or 60% of the worse moves are accepted, and then use this temperature as the new initial temperature T_0 for proper and relatively slow cooling processing.

For the final temperature, it should be zero in theory so that no worse move can be accepted. However, if $T_f \to 0$, more unnecessary evaluations are needed. In practice, we simply choose a very small value, say, $T_f = 10^{-10} \sim 10^{-5}$, depending on the required quality of the solutions and time constraints.

12.4 IMPLEMENTATION

Based on the guidelines of choosing the important parameters such as the cooling rate, initial and final temperatures, and the balanced number of iterations, we can implement the simulated annealing using both Matlab and

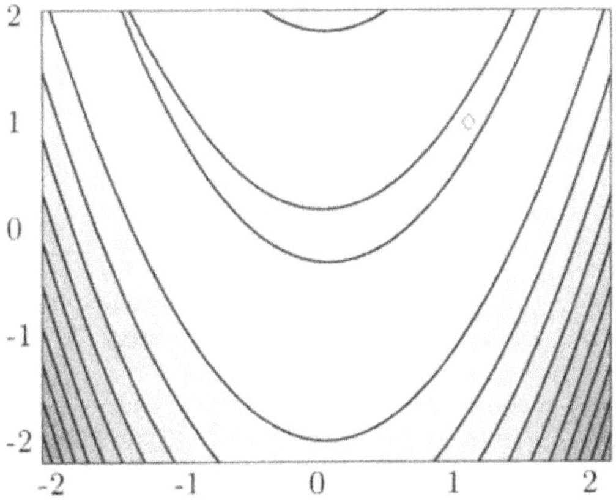

Figure 12.2: Rosenbrock's function with the global minimum $f_* = 0$ at $(1, 1)$.

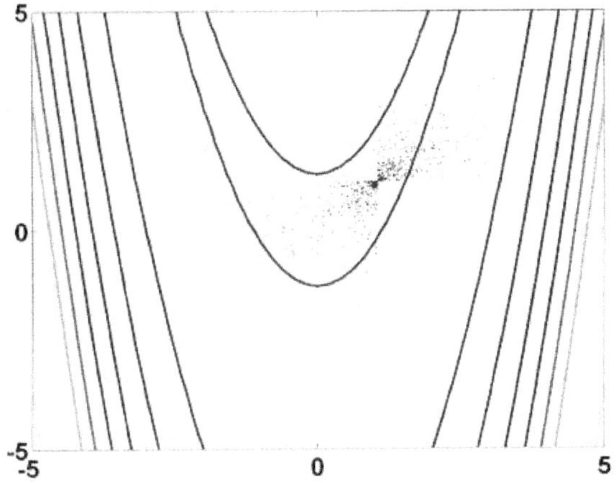

Figure 12.3: 500 evaluations during the simulated annealing. The final global best is marked with •.

Octave. The implemented Matlab and Octave program is given in Appendix B. Some of the simulations of the related test functions are explained in detail in the examples.

■ **EXAMPLE 12.1**

For Rosenbrock's banana function

$$f(x, y) = (1 - x)^2 + 100(y - x^2)^2,$$

we know that its global minimum $f_* = 0$ occurs at $(1, 1)$ (see Figure 12.2). This is a standard test function and quite tough for most conven-

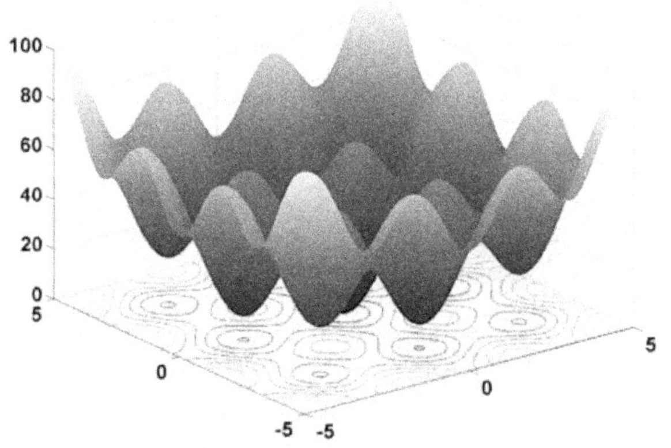

Figure 12.4: The egg crate function with a global minimum $f_* = 0$ at $(0, 0)$.

tional algorithms. However, using the program given in Appendix B, we can find this global minimum easily and the 500 evaluations during the simulated annealing are shown in Figure 12.3.

This banana function is still relatively simple as it has a curved narrow valley. Other functions such as the egg crate function are strongly multimodal and highly nonlinear. It is straightforward to extend the above program to deal with highly nonlinear multimodal functions. Let us look another example.

■ **EXAMPLE 12.2**

The egg crate function

$$f(x, y) = x^2 + y^2 + 25[\sin^2(x) + \sin^2(y)],$$

has the global minimum $f_* = 0$ at $(0, 0)$ in the domain $(x, y) \in [-5, 5] \times [-5, 5]$. The landscape of the egg crate function is shown in Figure 12.4, and the paths of the search during simulated annealing are shown in Figure 12.5. It would takes about 2500 evaluations to get an optimal solution accurate to the third decimal place.

EXERCISES

12.1 Modify the program in Appendix B so as to investigate the rate of convergence of simulated annealing for different cooling schedules such as

$$T(t) = \frac{T_0}{1 + \alpha t}, \qquad \alpha > 0.$$

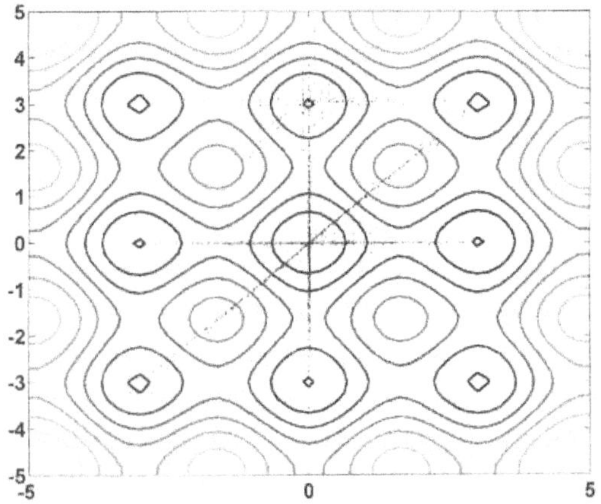

Figure 12.5: The paths of moves of simulated annealing during iterations.

12.2 For standard SA, the cooling schedule is a monotonically decreasing function. There is no reason why we should not use other forms of cooling. For example, we can use

$$T(t) = T_0 \cos^2(t) \exp[-\alpha t], \qquad \alpha > 0.$$

Modify the SA program discussed in this book to study the behavior of various functions as a cooling schedule.

12.3 Write a program to find the minimum of the following function for any $n > 1$

$$f(\boldsymbol{x}) = \Big(\sum_{i=1}^{n} |x_i|\Big) \exp\Big[-\sum_{i=1}^{n} \sin(x_i^2)\Big],$$

where $-2\pi \leq x_i \leq 2\pi$.

12.4 Modify the Matlab program again to solve the following equality-constrained optimization

$$f(\boldsymbol{x}) = -n^{n/2} \prod_{i=1}^{n} x_i, \qquad 0 \leq x_i \leq 1,$$

subject to an equality constraint

$$\sum_{i=1}^{n} x_i^2 = 1.$$

12.5 Modify the program of simulated annealing so as to use multiple parallel chains for simulated annealing.

REFERENCES

1. C. Blum and A. Roli, "Metaheuristics in combinatorial optimization: Overview and conceptual comparison", *ACM Comput. Surv.*, **35**, 268-308 (2003).
2. G. W. Flake, *The Computational Beauty of Nature: Computer Explorations of Fractals, Chaos, Complex Systems, and Adaptation*, Cambridge, Mass.: MIT Press, 1998.
3. L. J. Fogel, A. J. Owens, and M. J. Walsh, *Artificial Intelligence Through Simulated Evolution*, John Wiley & Sons, 1966.
4. S. Kirkpatrick, C. D. Gelatt, and M. P. Vecchi, "Optimization by simulated annealing", *Science*, **220**, No. 4598, 671-680 (1983).
5. M. Molga, C. Smutnicki, "Test functions for optimization needs", http://www.zsd.ict.pwr.wroc.pl/files/docs/functions.pdf
6. I. Pavlyukevich, "Lévy flights, non-local search and simulated annealing", *J. Computational Physics*, **226**, 1830-1844 (2007).
7. W. H. Press, S. A. Teukolsky, W. T. Vetterling, B. P. Flannery, *Numerical Recipes in C++: The Art of Scientific Computing*, Cambridge University Press, 2002.
8. E.-G. Talbi, *Metaheuristics: From Design to Implementation*, John Wiley & Sons, 2009.

CHAPTER 13

ANT ALGORITHMS

From the discussion of genetic algorithms, we know that we can improve the search efficiency by using randomness that will also increase the diversity of the solutions so as to avoid being trapped in local optima. The selection of the best individuals is also equivalent to using memory. In fact, there are other forms of selection such as using chemical messenger (pheromone) which is commonly used by ants, honey bees, and many other insects. In this chapter, we will discuss the nature-inspired ant colony optimization (ACO), which is a metaheuristic method.

13.1 BEHAVIOUR OF ANTS

Ants are social insects in habit and they live together in organized colonies whose population size can range from about 2 to 25 millions. When foraging, a swarm of ants or mobile agents interact or communicate in their local environment. Each ant can lay scent chemicals or pheromone so as to communicate with others, and each ant is also able to follow the route marked with pheromone laid by other ants. When ants find a food source, they will mark it with pheromone and also mark the trails to and from it. From the ini-

Engineering Optimization: An Introduction with Metaheuristic Applications.
By Xin-She Yang

Copyright © 2010 John Wiley & Sons, Inc.

tial random foraging route, the pheromone concentration varies and the ants follow the route with higher pheromone concentration, and the pheromone is enhanced by the increasing number of ants. As more and more ants follow the same route, it becomes the favored path. Thus, some favorite routes emerge, often the shortest or more efficient. This is actually a positive feedback mechanism.

Emerging behavior exists in an ant colony and such emergence arises from simple interactions among individual ants. Individual ants act according to simple and local information (such as pheromone concentration) to carry out their activities. Although there is no master ant overseeing the entire colony and broadcasting instructions to the individual ants, organized behavior still emerges automatically. Therefore, such emergent behavior is similar to other self-organized phenomena which occur in many processes in nature such as the pattern formation in animal skins (tiger and zebra skins).

The foraging pattern of some ant species (such as the army ants) can show extraordinary regularity. Army ants search for food along some regular routes with an angle of about 123° apart. We do not know how they manage to follow such regularity, but studies show that they could move in an area and build a bivouac and start foraging. On the first day, they forage in a random direction, say, the north and travel a few hundred meters, then branch to cover a large area. The next day, they will choose a different direction, which is about 123° from the direction on the previous day and cover a large area. On the following day, they again choose a different direction about 123° from the second day's direction. In this way, they cover the whole area over about 2 weeks and they move out to a different location to build a bivouac and forage in the new region.

The interesting thing is that they do not use the angle of $360°/3 = 120°$, as 120° would mean that, on the fourth day, they will search on the empty area already foraged on the first day. The beauty of this 123° angle is that it leaves an angle of about 10° from the direction on the first day. This means they cover the whole circle in 14 days without repeating a previously-foraged area. This is an amazing phenomenon.

13.2 ANT COLONY OPTIMIZATION

Based on these characteristics of ant behaviour, scientists have developed a number of powerful ant colony algorithms with important progress made in recent years. Marco Dorigo pioneered the research in this area in 1992. Many different variants appear since then.

If we only use some of the foraging behavior of ants and add some new characteristics, we can devise a class of new algorithms. The basic steps of the ant colony optimization (ACO) can be summarized as the pseudo code shown in Figure 13.1.

```
Ant Colony Optimization
---------------------------------------------------------------
Objective function f(x),    x = (x_1, ..., x_n)^T
   [or f(x_ij) for routing problem where (i, j) ∈ {1, 2, ..., n}]
Define pheromone evaporation rate γ
while ( criterion )
   for loop over all n dimensions (or nodes)
   Generate new solutions
   Evaluate the new solutions
   Mark better locations/routes with pheromone δφ_ij
   Update pheromone: φ_ij ← (1 − γ)φ_ij + δφ_ij
   end for
   Daemon actions such as finding the current best
end while
Output the best results and pheromone distribution
```

Figure 13.1: Pseudo code of ant colony optimization.

Apart from the population size, there are two important issues here: the probability of choosing a route, and the evaporation rate of pheromone. There are a few ways of solving these problems although it is still an area of active research. Here we introduce the current best method.

For a network routing problem, the probability of ants at a particular node i to choose the route from node i to node j, among n_d nodes, is given by

$$p_{ij} = \frac{\phi_{ij}^{\alpha} d_{ij}^{\beta}}{\sum_{i,j=1}^{n_d} \phi_{ij}^{\alpha} d_{ij}^{\beta}}, \tag{13.1}$$

where $\alpha > 0$ and $\beta > 0$ are the influence parameters, and their typical values are $\alpha \approx \beta \approx 2$. ϕ_{ij} is the pheromone concentration on the route between i and j, and d_{ij} the desirability of the same route. Some *a priori* knowledge about the route such as the distance s_{ij} is often used so that $d_{ij} \propto 1/s_{ij}$, which implies that shorter routes will be selected due to their shorter traveling time, and thus the pheromone concentrations on these routes are higher. This is because the traveling time is shorter, and thus the less amount of the pheromone has been evaporated during this period.

This probability formula reflects the fact that ants would normally follow the paths with higher pheromone concentrations. In the simpler case when $\alpha = \beta = 1$, the probability of choosing a path by ants is proportional to the pheromone concentration on the path. The denominator normalizes the probability so that it is in the range between 0 and 1.

The pheromone concentration can change with time due to the evaporation of pheromone. Furthermore, the advantage of pheromone evaporation is that the system could avoid being trapped in local optima. If there is no evaporation, then the path randomly chosen by the first ants will become the

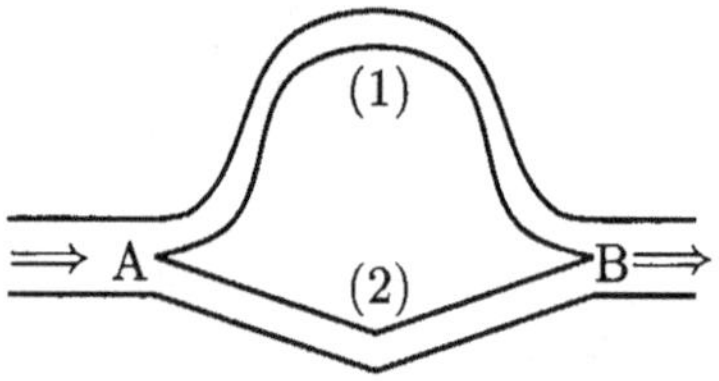

Figure 13.2: The double bridge problem for routing performance: route (2) is shorter than route (1).

preferred path as the attraction of other ants by their pheromone. For a constant rate γ of pheromone decay or evaporation, the pheromone concentration usually varies with time exponentially

$$\phi(t) = \phi_0 e^{-\gamma t}, \tag{13.2}$$

where ϕ_0 is the initial concentration of pheromone and t is time. If $\gamma t \ll 1$, then we have $\phi(t) \approx (1 - \gamma t)\phi_0$. For the unitary time increment $\Delta t = 1$, the evaporation can be approximated by $\phi^{t+1} \leftarrow (1 - \gamma)\phi^t$. Therefore, we have the simplified pheromone update formula:

$$\phi_{ij}^{t+1} = (1 - \gamma)\phi_{ij}^t + \delta\phi_{ij}^t, \tag{13.3}$$

where $\gamma \in [0, 1]$ is the rate of pheromone evaporation. The increment $\delta\phi_{ij}^t$ is the amount of pheromone deposited at time t along route i to j when an ant travels a distance L. Usually $\delta\phi_{ij}^t \propto 1/L$. If there are no ants on a route, then the pheromone deposit is zero.

There are other variations to this basic procedure. A possible acceleration scheme is to use some bounds of the pheromone concentration and only the ants with the current global best solution(s) are allowed to deposit pheromone. In addition, certain ranking of solution fitness can also be used. These are active areas of current research.

13.3 DOUBLE BRIDGE PROBLEM

A standard test problem for ant colony optimization is the simplest double bridge problem with two branches (see Figure 13.2) where route (2) is shorter than route (1). The angles of these two routes are equal at both point A and point B so that the ants have equal chance (or 50-50 probability) of choosing each route randomly at the initial stage at point A.

Initially, fifty percent of the ants would go along the longer route (1) and the pheromone evaporates at a constant rate, but the pheromone concentration

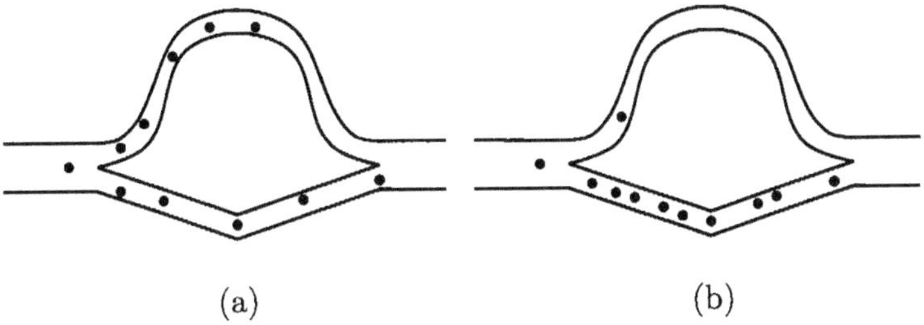

Figure 13.3: Route selection via ACO: (a) initially, ants choose each route with a 50-50 probability, and (b) almost all ants move along the shorter route after 5 iterations.

will become smaller as route (1) is longer and thus takes more time to travel through. Conversely, the pheromone concentration on the shorter route will increase steadily. After some iterations, almost all the ants will move along the shorter route. Figure 13.3 shows the initial snapshot of 10 ants (5 on each route initially) and the snapshot after 5 iterations (or equivalent to the case that 50 ants have moved along this section). In this simple illustration, there are 11 ants, and one has not decided which route to follow as it just comes near to the entrance. Almost all the ants (about 90% in this case) move along the shorter route.

Here we only use two routes at the node, it is straightforward to extend it to the multiple routes at a node. It is expected that only the shortest route will be chosen ultimately. As any complex network system is always made of individual nodes, this algorithms can be extended to solve complex routing problems reasonably efficiently. In fact, the ant colony algorithms have been applied successfully to the Internet routing problem, the traveling salesman problem, combinatorial optimization problems, and other NP-hard problems.

13.4 VIRTUAL ANT ALGORITHM

As we know that ant colony optimization has successfully solved NP-hard problems such as the traveling salesman problem, it can also be extended to solve the standard optimization problems of multimodal functions. The only problem now is to figure out how the ants will move on an n-dimensional hyper-surface. For simplicity, we will discuss the 2-D case which can easily be extended to higher dimensions. On a 2D landscape, ants can move in any direction or $0° \sim 360°$, but this will cause some problems. How to update the pheromone at a particular point as there are infinite number of points. One solution is to track the history of each ant's moves and record the locations consecutively, and the other approach is to use a moving neighborhood or

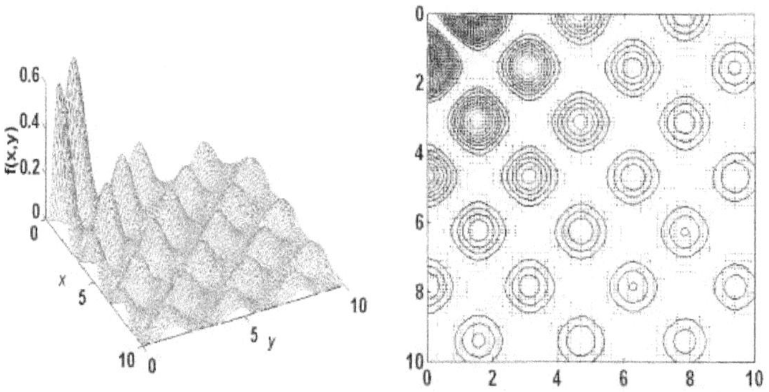

Figure 13.4: Landscape and pheromone distribution of the multi-peak function.

window. The ants 'smell' the pheromone concentration of their neighborhood at any particular location.

In addition, we can limit the number of directions the ants can move by quantizing the directions. For example, ants are only allowed to move left and right, and up and down (only 4 directions). We will use this quantized approach here, which will make the implementation much simpler. Furthermore, the objective function or landscape can be encoded into virtual food so that ants will move to the best locations where the best food sources are. This will make the search process even more simpler. This simplified algorithm is called Virtual Ant Algorithm (VAA) developed by Xin-She Yang and his colleagues in 2006, which has been successfully applied to topological optimization problems in engineering.

The following Keane function with multiple peaks is a standard test function

$$f(x, y) = \frac{\sin^2(x - y)\sin^2(x + y)}{\sqrt{x^2 + y^2}}, \tag{13.4}$$

where

$$(x, y) \in [0, 10] \times [0, 10].$$

This function without any constraint is symmetric and has two highest peaks at $(0, 1.39325)$ and $(1.39325, 0)$. To make the problem harder, it is usually optimized under two constraints:

$$x + y \leq 15, \qquad xy \geq \frac{3}{4}, \tag{13.5}$$

which may make it difficult to find the optimality because the function is now nearly symmetric about $x = y$ in the domain, and the peaks occur in pairs where one is higher than the other. In addition, the true maximum is $f(1.593, 0.471) \approx 0.365$, which is defined by a constraint boundary.

Figure 13.4 shows the surface variations of the multi-peaked function. If we use 50 roaming ants and let them move around for 25 iterations, then the pheromone concentrations (also equivalent to the paths of ants) are displayed in Figure 13.4. We can see that the highest pheromone concentration within the constraint boundary corresponds to the optimal solution.

It is worth pointing out that ant colony algorithms are the right tool for combinatorial and discrete optimization. They may have the advantages over other stochastic algorithms such as genetic algorithms and simulated annealing in dealing with dynamical network routing problems.

For continuous decision variables, its performance is still under active research. For the present example, it took about 1500 evaluations of the objective function so as to find the global optima. This is not as efficient as other metaheuristic methods, especially comparing with particle swarm optimization. This is partly because the handling of the pheromone takes time. Is it possible to eliminate the pheromone and just use the roaming ants? The answer is yes. Particle swarm optimization is just the right kind of algorithm for such further modifications which will be discussed later in detail.

EXERCISES

13.1 Write a simple program to implement the simplest version of the ant algorithms using the exponential decrease of pheromone concentration $\phi(t) = \phi_0 \exp(-\gamma t)$ due to evaporation. Investigate the parameter sensitivity of $\gamma > 0$ and its effect on the rate of convergence.

13.2 Use the program developed in the last question to solve a simple version of the traveling salesman problem by visiting $n = 10$ cities of your favorite cities so as to minimize the overall distance traveled.

13.3 Modify your program again so as to find the minimum overall traveling cost of visiting your 10 favorite cities, either by air or by train.

REFERENCES

1. C. Blum, "Ant colony optimization: introduction and recent trends", *Physics of Life Review*, **2**, 353-373 (2005).

2. C. Blum, A. Roli, "Metaheuristics in combinatorial optimization: Overview and conceptural comparison", *ACM Computing Surveys*, **35** (3), 268-308 (2003).

3. E. Bonabeau, M. Dorigo, G. Theraulaz, *Swarm Intelligence: From Natural to Artificial Systems.* Oxford University Press, 1999.

4. M. Dorigo, *Optimization, Learning and Natural Algorithms*, PhD thesis, Politencnico di Milano, Italy, 1992.

5. M. Dorigo and T. Stützle, *Ant Colony Optimization*, MIT Press, Cambridge, 2004.

6. M. Dorigo, G. Di Caro, L. M. Gambardella, "Ant algorithms for discrete optimization", *Artificial Life*, **5** (2), 137-172 (1999).

7. D. B. Fogel, *Evolutionary Computation*, IEEE Press, 1995.

8. X. S. Yang, J. M. Lees, C. T. Morley, "Application of virtual ant algorithms in the otpimization of CFRP shear strengthened precracked structures", *Lecture Notes in Computer Sciences*, **3991**, 834-837 (2006).

CHAPTER 14

BEE ALGORITHMS

Bee algorithms form another class of algorithms which are closely related to the ant colony optimization. Bee algorithms are inspired by the foraging behaviour of honey bees. Several variants of bee algorithms have been formulated, including the Honey Bee Algorithm (HBA), the Virtual Bee Algorithm (VBA), the Artificial Bee Colony (ABC) optimization, the Honeybee Mating Algorithm (HBMA) and others.

14.1 BEHAVIOR OF HONEY BEES

Honey bees live in a colony and they forage and store honey in their constructed colony. Honey bees can communicate by pheromone and 'waggle dance'. For example, an alarming bee may release a chemical message (pheromone) to stimulate attack response in other bees. Furthermore, when bees find a good food source and bring some nectar back to the hive, they will communicate the location of the food source by performing the so-called waggle dances as a signal system. Such signaling dances vary from species to species, however, they will try to recruit more bees by using directional danc-

Engineering Optimization: An Introduction with Metaheuristic Applications.
By Xin-She Yang
Copyright © 2010 John Wiley & Sons, Inc.

ing with varying strength so as to communicate the direction and distance of the found food resource.

For multiple food sources such as flower patches, studies show that a bee colony seems to be able to allocate forager bees among different flower patches so as to maximize their total nectar intake. In order to survive the winter, a bee colony typically has to collect and store extra nectar, about 15 to 50 kg. The efficiency of nectar collection is consequently very important from the evolution point of view. Experimental studies have also been carried out by researchers, including the important work by S Camazine, T Seeley and J Sney in early 1990s and lately by Quijano and K. Passino and their colleagues. Various algorithms can be designed if we learn from the natural behaviour of bee colonies.

14.2 BEE ALGORITHMS

Over the last decade or so, nature-inspired bee algorithms have started to emerge as a promising and powerful tool. It is difficult to pinpoint the exact dates when the bee algorithms were first formulated. They were developed over a few years independently by several groups of researchers.

From the literature survey, it seems that the Honey Bee Algorithm (HBA) was first formulated in around 2004 by Craig A Tovey at Georgia Tech in collaboration with Sunil Nakrani then at Oxford University to study a method to allocate computers among different clients and web-hosting servers. Later in 2004 and earlier 2005, Xin-She Yang at Cambridge University developed a Virtual Bee Algorithm (VBA) to solve numerical optimization problems, and it can optimize both functions and discrete problems, though only functions with two parameters were given as examples. Slightly later in 2005, Haddad and Afshar and their colleagues presented a Honey-bee mating optimization (HBMO) algorithm which was subsequently applied to reservoir modelling and clustering. Around the same time, D Karabogo in Turkey developed an Artificial Bee Colony (ABC) algorithm for numerical function optimization and a comparison study was carried by in the same group later in 2006 and 2008. These bee algorithms are now becoming more and more popular.

The essence of the bee algorithms are the communication or broadcasting ability of a bee to some neighbourhood bees so they can 'know' and follow a bee to the best source, locations or routes to complete the optimization task. The detailed implementation will depend on the actual algorithms, and they may differ slightly and vary with different variants. However, the essence of all the bee algorithms can be summarized as the pseudo code in Figure 14.1.

14.2.1 Honey Bee Algorithm

In the honey bee algorithm, forager bees are allocated to different food sources (or flower patches) so as to maximize the total nectar intake. The colony has

```
Bee Algorithms
Objective function f(x), x = (x_1, ..., x_n)^T & constraints
Encode f(x) into virtual nectar levels
Define dance routine (strength, direction) or protocol
while ( criterion )
   for loop over all n dimensions
        (or nodes for routing and scheduling problems)
   Generate new solutions
   Evaluate the new solutions
   end for
   Communicate and update the optimal solution set
end while
Decode and output the best results
```

Figure 14.1: Pseudo code of bee algorithms

to 'optimize' the overall efficiency of nectar collection; the allocation of the bees is thus depending on many factors such as the nectar richness and the proximity to the hive. This problem is similar to the allocation of web-hosting servers in the Internet which was in fact one of the first problem solved using bee-inspired algorithms by Nakrani and Tovey in 2004.

Let $w_i(j)$ be the strength of the waggle dance of bee i at time step $t = j$, the probability of an observer bee following the dancing bee to forage can be determined in many ways depending on the actual variant of algorithms. A simple way is given by Guijano and Passino

$$p_i = \frac{w_i^j}{\sum_{i=1}^{n_f} w_i^j}, \tag{14.1}$$

where n_f is the number of bees in foraging process. t is the pseudo time or foraging expedition. The number of observer bees is $N - n_f$ when N is the total number of bees. Alternatively, we can define an exploration probability of a Gaussian type

$$p_e = 1 - p_i = e^{-\frac{w_i^2}{2\sigma^2}}, \tag{14.2}$$

where σ is the volatility of the bee colony, and it controls the exploration and diversity of the foraging sites. If there is no dancing (no food found), then $w_i \to 0$, and $p_e = 1$. So all the bee explore randomly.

In other variant algorithms when applying to discrete problems such as job scheduling, a forager bee will perform waggle dance with a duration $\tau = \alpha f_p$ where f_p is the profitability or the richness of the food site, and α is a scaling factor. The profitability should be related to the objective function.

In addition, the rating of each route is ranked dynamically and the path with the highest number of bees become the preferred path. For a routing problem, the probability of choosing a route between any two nodes can take the form similar to the equation (13.1). That is

$$p_{ij} = \frac{w_{ij}^{\alpha} d_{ij}^{\beta}}{\sum_{i,j=1}^{n_d} w_{ij}^{\alpha} d_{ij}^{\beta}}, \tag{14.3}$$

where $\alpha > 0$ and $\beta > 0$ are the influence parameters. w_{ij} is the dance strength along route i to j, and d_{ij} is the desirability of the same route.

The honey bee algorithm, similar to the ant colony algorithms, is very efficient in dealing with discrete optimization problems such as routing and scheduling. When dealing with continuous optimization problems, it is not straightforward, and some modifications are needed.

14.2.2 Virtual Bee Algorithm

The virtual bee algorithm (VBA), developed by Xin-She Yang in 2005, is an optimization algorithm specially formulated for solving both discrete and continuous problems. It has some similarity to the particle swarm optimization (PSO) to be discussed later, rather than a bee algorithm. In VBA, the continuous objective function is directly encoded as virtual nectar, and the solutions (or decision variables) are the locations of the nectar. The activities such as the waggle dance strength (or similar to the pheromone in the ant algorithms) are combined with the nectar concentration as the 'fitness' of the solutions. For a maximization problem, the objective function can be thought as virtual nectar, while for minimization, the nectar is formulated in such a way that the minimal value of the objective function corresponds to the highest nectar concentration.

For example, we know that the generalized De Jong's test function in the n dimensions has a minimum $f_{\min} = 0$ at the origin $\boldsymbol{x} = (0, 0, ..., 0)$. That is

$$f(\boldsymbol{x}) = \sum_{i=1}^{n} x_i^{2k}, \qquad |x_i| \leq r, \tag{14.4}$$

where k is a positive integer such as $1, 2, ..., m$, and r is the radius of the hypersphere. For $k = 3$, $r = 256$, and $n = 50$, we can use 20 virtual bees after 600 foraging explorations, the best estimate is $f(\boldsymbol{x}) \approx 0.0016616$. Considering the domain size and multiple dimensions of this function, it is much more efficient than genetic algorithms.

For discrete problems, the objective function such as the shorter paths are encoded and linked with the profitability of the nectar explorations, which is in turn linked with the dance strength of forager bees. In this way, the virtual bee algorithm is similar to the honey bee algorithm. However, there is a fundamental difference from other bee algorithms. That is, VBA has a broadcasting ability of the current best. The current best location is 'known' to every bee so that this algorithm is more efficient. In this way, forager bees do not have to come back to the hive to tell other onlooker bees via 'waggle dance' so as to save time. Similar broadcasting ability is also used in the particle swarm optimization, especially in the accelerated PSO algorithms.

For a mixed type of problem when the decision variables can take both discrete and continuous values, the encoding of the objective function into nectar should be carefully implemented, so it can represent the objective effectively. This is still an active area of current research.

14.2.3 Artificial Bee Colony Optimization

The artificial bee colony (ABC) algorithm was developed by D. Karaboga in 2005. Since then, Karaboga and Basturk have systematically studied the performance of the ABC algorithm concerning unstrained optimization problems and its extension to unconstrained optimization.

In the ABC algorithm, the bees in a colony are divided into three groups: employed bees (forager bees), onlooker bees (observer bees) and scouts. For each food source, there is only one employed bee. That is to say, the number of employed bees is equal to the number of food sources. The employed bee of a discarded food site is forced to become a scout for searching new food sources randomly. Employed bees share information with the onlooker bees in a hive so that onlooker bees can choose a food source to forage. Unlike the honey bee algorithm which has two groups of the bees (forager bees and observer bees), bees in ABC are more specialized.

For a given objective function $f(\boldsymbol{x})$, it can be encoded as $F(\boldsymbol{x})$ to represent the amount of nectar at location $\boldsymbol{x}$, thus the probability P_i of an onlooker bee choose to go to the preferred food source at $\boldsymbol{x}_i$ can be defined by

$$P_i = \frac{F(\boldsymbol{x}_i)}{\sum_{j=1}^{S} F(\boldsymbol{x}_j)}, \tag{14.5}$$

where S is the number of food sources. At a particular food source, the intake efficiency is determined by F/τ where F is the amount of nectar and τ is the time spent at the food source. If a food source is tried/foraged at a given number of explorations without improvement, then it is abandoned, and the bee at this location will move on randomly to explore new locations.

Recent studies show that ABC could perform better under the right conditions for various test functions than particle swarm optimization (PSO), differential evolution (DE) and evolutionary algorithms (EA).

14.3 APPLICATIONS

Like the ant colony optimization, bee algorithms are also the right tool for combinatorial and discrete optimization. They have the advantages over other algorithms such as genetic algorithms and simulated annealing in dealing with dynamical network routing and job scheduling problems.

Various applications have been carried out in the last few years, including the combinatorial optimization, job scheduling, web-hosting allocation, engineering design optimization, function optimization, reservoir modeling, and the traveling salesman problem. More applied studies will surely appear in the near future.

EXERCISES

14.1 Write a simple program to implement a version of the bee algorithms with the aim to find the minimum of De Jong's test function $f(\boldsymbol{x}) = \sum_{i=1}^{n} x_i^2$.

14.2 Use your own implementation to solve the following simple job-schedule problem: A job, say, to manufacture the same batch of N car parts by K different work units, and each unit produce a single part requires $\tau_i > 0$ unit time where $i = 1, 2, ..., K$. Try to allocate the work load so that the parts are finished as early as possible.

14.3 For the job scheduling problem in the previous question, if the finish time is not of concern, the overall cost is more important. Suppose each work unit cost p_i for each part per unit time. Modify your program to find the optimal solution for, say, $K = 20$.

REFERENCES

1. A. Afshar, O. B. Haddad, M. A. Marino, B. J. Adams, "Honey-bee mating optimization (HBMO) algorithm for optimal reservoir operation", *J. Franklin Institute*, **344**, 452-462 (2007).
2. B. Basturk and D. Karabogo, "An artificial bee colony (ABC) algorithm for numerical function optimizaton", in: IEEE Swarm Intelligence Symposium 2006, May 12-14, Indianapolis, IN, USA, (2006).
3. C. Chong, M. Y. Low, A. I. Sivakumar, K. L. Gay, "A bee colony optimization algorithm to job shop scheduling", *Proc. of 2006 Winter Simulation Conference*, Eds Perrone L. F. *et al.* , (2006).
4. M. Fathian, B. Amiri, A. Maroosi, "Application of honey-bee mating optimization algorithm on clustering", *Applied Mathematics and Computation*, **190**, 1502-1513 (2007).
5. D. Karaboga, "An idea based on honey bee swarm for numerical optimization", Technical Report TR06, Erciyes University, Turkey, 2005.
6. D. Karaboga and B. Basturk, "On the performance of artificial bee colony (ABC) algorithm", *Applied Soft Computing*,**8**, 687-697 (2008).
7. R. F. Moritz and E. E. Southwick, *Bees as superorganisms*, Springer, 1992.
8. S. Nakrani and C. Tovey, "On honey bees and dynamic server allocation in Internet hosting centers", *Adaptive Behaviour*, **12**, 223-240 (2004).
9. X. S. Yang, "Engineering optimization via nature-inspired virtual bee algorithms", IWINAC 2005, *Lecture Notes in Computer Science*, **3562**, 317-323 (2005).

CHAPTER 15

PARTICLE SWARM OPTIMIZATION

15.1 SWARM INTELLIGENCE

Particle swarm optimization (PSO) was developed by Kennedy and Eberhart in 1995, based on swarm behaviour such as fish and bird schooling in nature. Many algorithms such as ant colony algorithms and virtual ant algorithms use the behaviour of the so-called swarm intelligence. Particle swarm optimization may have some similarities with genetic algorithms and ant algorithms, but it is much simpler because it does not use mutation/crossover operators or pheromone. Instead, it uses the real-number randomness and the global communication among the swarm particles. In this sense, it is also easier to implement as there is no encoding or decoding of the parameters into binary strings as those in genetic algorithms which can also use real-number strings.

This algorithm searches the space of an objective function by adjusting the trajectories of individual agents, called particles, as these trajectories form piecewise paths in a quasi-stochastic manner. The movement of a swarming particle consists of two major components: a stochastic component and a deterministic component. Each particle is attracted toward the position of

Engineering Optimization: An Introduction with Metaheuristic Applications.
By Xin-She Yang
Copyright © 2010 John Wiley & Sons, Inc.

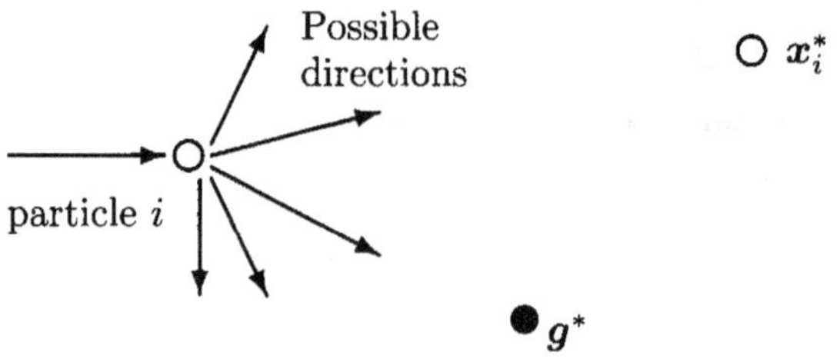

Figure 15.1: Schematic representation of the motion of a particle in PSO, moving towards the global best g^* and the current best x_i^* for each particle i.

the current global best g^* and its own best location x_i^* in history, while at the same time it has a tendency to move randomly.

When a particle finds a location that is better than any previously found locations, then it updates it as the new current best for particle i. There is a current best for all n particles at any time t during iterations. The aim is to find the global best among all the current best solutions until the objective no longer improves or after a certain number of iterations. The movement of particles is schematically represented in Figure 15.1 where x_i^* is the current best for particle i, and $g^* \approx \min\{f(x_i)\}$ for $(i = 1, 2, ..., n)$ is the current global best.

15.2 PSO ALGORITHMS

The essential steps of the particle swarm optimization can be summarized as the pseudo code shown in Figure 15.2.

Let x_i and v_i be the position vector and velocity for particle i, respectively. The new velocity vector is determined by the following formula

$$v_i^{t+1} = v_i^t + \alpha \epsilon_1 \odot [g^* - x_i^t] + \beta \epsilon_2 \odot [x_i^* - x_i^t]. \tag{15.1}$$

where ϵ_1 and ϵ_2 are two random vectors, and each entry taking the values between 0 and 1. The Hadamard product of two matrices $u \odot v$ is defined as the entrywise product, that is $[u \odot v]_{ij} = u_{ij} v_{ij}$. The parameters α and β are the learning parameters or acceleration constants, which can typically be taken as, say, $\alpha \approx \beta \approx 2$.

The initial locations of all particles should distribute relatively uniformly so that they can sample over most regions, which is especially important for multimodal problems. The initial velocity of a particle can be taken as zero, that is, $v_i^{t=0} = 0$. The new position can then be updated by

$$x_i^{t+1} = x_i^t + v_i^{t+1}. \tag{15.2}$$

Although v_i can be any values, it is usually bounded in some range $[0, v_{\max}]$.

Particle Swarm Optimization

Objective function $f(\boldsymbol{x})$, $\boldsymbol{x} = (x_1, ..., x_p)^T$
Initialize locations $\boldsymbol{x}_i$ and velocity $\boldsymbol{v}_i$ of n particles.
Find $\boldsymbol{g}^*$ from $\min\{f(\boldsymbol{x}_1), ..., f(\boldsymbol{x}_n)\}$ (at $t = 0$)
while (criterion)
$t = t + 1$ (pseudo time or iteration counter)
  for loop over all n particles and all p dimensions
  Generate new velocity $\boldsymbol{v}_i^{t+1}$ using equation (15.1)
  Calculate new locations $\boldsymbol{x}_i^{t+1} = \boldsymbol{x}_i^t + \boldsymbol{v}_i^{t+1}$
  Evaluate objective functions at new locations $\boldsymbol{x}_i^{t+1}$
  Find the current best for each particle $\boldsymbol{x}_i^*$
  end for
  Fin the current global best $\boldsymbol{g}^*$
end while
Output the final results $\boldsymbol{x}_i^*$ and $\boldsymbol{g}^*$

Figure 15.2: Pseudo code of particle swarm optimization.

15.3 ACCELERATED PSO

There are many variants which extend the standard PSO algorithm, and the most noticeable improvement is probably to use inertia function $\theta(t)$ so that $\boldsymbol{v}_i^t$ is replaced by $\theta(t)\boldsymbol{v}_i^t$

$$\boldsymbol{v}_i^{t+1} = \theta \boldsymbol{v}_i^t + \alpha \epsilon_1 \odot [\boldsymbol{g}^* - \boldsymbol{x}_i^t] + \beta \epsilon_2 \odot [\boldsymbol{x}_i^* - \boldsymbol{x}_i^t], \tag{15.3}$$

where θ takes the values between 0 and 1. In the simplest case, the inertia function can be taken as a constant, typically $\theta \approx 0.5 \sim 0.9$. This is equivalent to introduce a virtual mass to stabilize the motion of the particles, and thus the algorithm is expected to converge more quickly.

The standard particle swarm optimization uses both the current global best $\boldsymbol{g}^*$ and the individual best $\boldsymbol{x}_i^*$. The reason of using the individual best is primarily to increase the diversity in the quality solutions, however, this diversity can be simulated using some randomness. Subsequently, there is no compelling reason for using the individual best, unless the optimization problem of interest is highly nonlinear and multimodal.

A simplified version which could accelerate the convergence of the algorithm is to use the global best only. Thus, in the accelerated particle swarm optimization, the velocity vector is generated by a simpler formula

$$\boldsymbol{v}_i^{t+1} = \boldsymbol{v}_i^t + \alpha(\boldsymbol{\epsilon} - 1/2) + \beta(\boldsymbol{g}^* - \boldsymbol{x}_i), \tag{15.4}$$

where $\boldsymbol{\epsilon}$ is a random variable with values from 0 to 1. Here the shift 1/2 is purely out of convenience. We can also use a standard normal distribution

$\alpha\epsilon_n$ where ϵ_n is drawn from $N(0, 1)$ to replace the second term. The update of the position is simply

$$\boldsymbol{x}_i^{t+1} = \boldsymbol{x}_i^t + \boldsymbol{v}_i^{t+1}. \tag{15.5}$$

In order to increase the convergence even further, we can also write the update of the location in a single step

$$\boldsymbol{x}_i^{t+1} = (1 - \beta)\boldsymbol{x}_i + \beta\boldsymbol{g}^* + \alpha(\epsilon - 0.5). \tag{15.6}$$

This simpler version will give the same order of convergence. The typical values for this accelerated PSO are $\alpha \approx 0.1 \sim 0.4$ and $\beta \approx 0.1 \sim 0.7$, though $\alpha \approx 0.2$ and $\beta \approx 0.5$ can be taken as the initial values for most unimodal objective functions. It is worth pointing out that the parameters α and β should in general be related to the scales of the independent variables $\boldsymbol{x}_i$ and the search domain.

A further improvement to the accelerated PSO is to reduce the randomness as iterations proceed. This means that we can use a monotonically decreasing function such as

$$\alpha = \alpha_0 e^{-\gamma t}, \tag{15.7}$$

or

$$\alpha = \alpha_0 \gamma^t, \qquad (0 < \gamma < 1), \tag{15.8}$$

where $\alpha_0 \approx 0.5 \sim 1$ is the initial value of the randomness parameter. t is the number of iterations or time steps. $0 < \gamma < 1$ is a control parameter. For example, in our implementation, we will use

$$\alpha = 0.7^t, \tag{15.9}$$

where $t \in [0, 10]$. Obviously, these parameters are fine-tuned to suit the current optimization problems as a demonstration. The implementation of the accelerated PSO in Matlab/Octave is given later in Appendix B.

Theoretically speaking, each particle forms a Markov chain, though this Markov chain is biased towards to the current best, as the transition probability often leads to accept the move towards the current global best. In addition, the multiple Markov chains are interacting in terms of partly deterministic attraction movement. Therefore, the mathematical analysis concerning of the rate of convergence of PSO is very difficult, if not impossible. There is no doubt that any theoretical advance in understanding multiple interacting Markov chains will gain tremendous insightful in understanding how the PSO behaves and may consequently lead to design better or new PSO algorithms. For example, Clerc and Kennedy carried out a convergence analysis using a simplified dynamical system and they obtained

$$\Psi = \frac{2\kappa}{|2 - \phi - \sqrt{\phi(\phi - 4)}|}, \tag{15.10}$$

where

$$\phi = \phi_1 + \phi_2, \qquad \phi_1 = \alpha\epsilon_1, \qquad \phi_2 = \beta\epsilon_2. \tag{15.11}$$

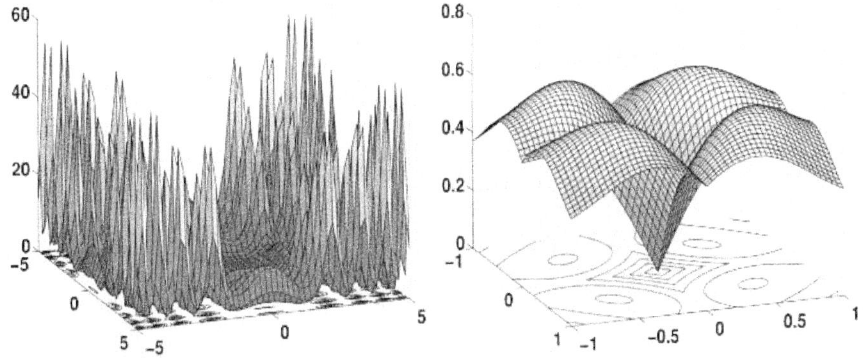

Figure 15.3: A multimodal function with the global minimum $f_* = 0$ at $(0, 0)$, however, it has a singularity at $(0, 0)$ (right).

Here $\kappa \in [0, 1]$ essentially controls the exploration and exploitation in PSO. $\kappa \approx 0$ corresponds to local exploitation and fast convergence, while $\kappa \approx 1$ corresponds to global exploration and slow convergence. It has been proved that the convergence of PSO is guaranteed if $\phi \geq 4$.

Various studies show that PSO algorithms can outperform genetic algorithms and other conventional algorithms for solving many optimization problems. This is partially due to that fact that the broadcasting ability of the current best estimates gives a better and quicker convergence towards the optimality. However, PSO algorithms are almost memoryless since they do not record the movement paths of each particle, and it is expected that it can be further improved using short-term memory in the similar fashion as that in Tabu search. Further development is under active research.

15.4 IMPLEMENTATION

15.4.1 Multimodal Functions

Multiple peak functions are often used to validate new algorithms. For example, the author of this book constructed the following function with multiple peaks,

$$f(\boldsymbol{x}) = \Big(\sum_{i=1}^{d} |x_i|\Big) \exp\Big[-\sum_{i=1}^{d} \sin(x_i^2)\Big], \qquad -2\pi \leq x_i \leq 2\pi, \tag{15.12}$$

which has the global minimum $f_* = 0$ at $\boldsymbol{x}_* = (0, 0, ..., 0)$ for any dimensions $(d = 1, 2, ...)$. This function is highly multimodal, and it has a singularity at the global minimum $(0, 0, ..., 0)$. For example, the landscape of the function

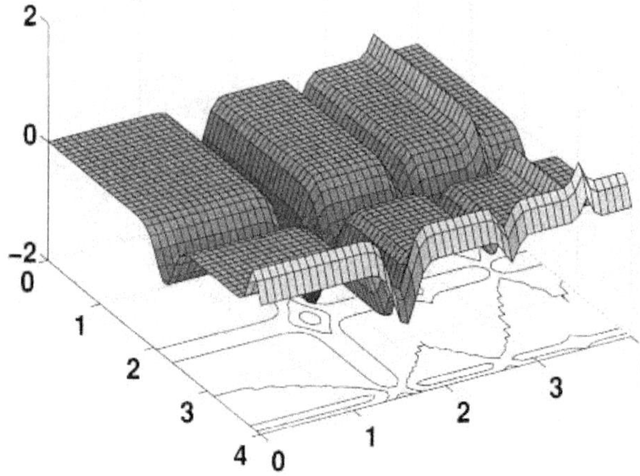

Figure 15.4: Michaelewicz function with a global minimum at about $(2.20319, 1.57049)$.

in 2D is shown in Figure 15.3 and the singularity near the optimum at $(0,0)$ is also shown in the same figure. As we can see that, the multiple peaks look like a forest.

This function can serve as a tough test function, as it is a standard practice to benchmark new algorithms against the test functions in literature. Various studies have already benchmarked the PSO algorithms against standard test functions. Later, we will also use the Michaelewicz function in our implementation as a simple demo.

15.4.2 Validation

The accelerated particle swarm optimization has been implemented using Matlab/Octave, and a simple program is provided in Appendix B. This program can find the global optimal solution of most nonlinear functions in less a minute on most modern personal computers.

■ EXAMPLE 15.1

Now let us look at the 2D Michaelewicz function

$$f(x,y) = -\{\sin(x)[\sin(\frac{x^2}{\pi})]^{2m} + \sin(y)[\sin(\frac{2y^2}{\pi})]^{2m}\},$$

where $m = 10$. The stationary conditions $f_x = f_y = 0$ require that

$$-\frac{4m}{\pi}x\sin(x)\cos(\frac{x^2}{\pi}) - \cos(x)\sin(\frac{x^2}{\pi}) = 0,$$

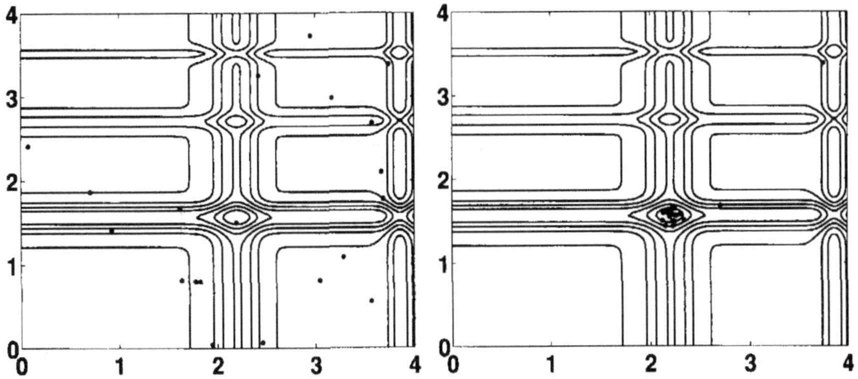

Figure 15.5: Initial locations and final locations of 20 particles after 10 iterations.

and

$$-\frac{8m}{\pi} y \sin(x) \cos(\frac{2y^2}{\pi}) - \cos(y) \sin(\frac{2y^2}{\pi}) = 0.$$

The solution at $(0,0)$ is trivial, and the minimum $f^* \approx -1.801$ occurs at about (2.20319,1.57049) (see Figure 15.4).

If we run the program, we will get the global optimum after about 200 evaluations of the objective function (for 20 particles and 10 iterations). The results and the locations of the particles are shown in Figure 15.5.

15.5 CONSTRAINTS

The implementation we discussed in the previous section is for unstrained problems. For constrained optimization, there are many ways of implementing the constraint equalities and inequalities. However, we will only discussed two approaches: direct implementation and transform to unconstrained optimization.

The simplest direct implementation is to check all the new particle locations to see if they satisfy all the constraints. The new locations are discarded if the constraints are not met, and new locations are replaced by newly generated locations until all the constraints are met. Then, the new solutions are evaluated using the standard PSO procedure. In this way, all the new locations should be in the feasible region, and all infeasible solutions are not selected. For example, in order to maximize $f(\boldsymbol{x})$ subjected to a constraint $g(\boldsymbol{x}) \leq 0$, the standard PSO as discussed earlier is used, however, the new locations $\boldsymbol{x}_i$ of the n particles are checked at each iteration so that they must satisfy $g(\boldsymbol{x}_i) \leq 0$. If any $\boldsymbol{x}_i$ does not satisfy the constraint, it is then replaced by a new (different) location $\tilde{\boldsymbol{x}}_i$ which satisfies the constraint.

Alternatively, we can transform it to an unconstrained problem by using the penalty method or Lagrange multipliers as discussed in Part I. Using a penalty parameter $\lambda \gg 1$ in our simple example here, we have the following penalty function

$$\Psi(\boldsymbol{x}, \nu) = f(\boldsymbol{x}) + \lambda g(x)^2. \tag{15.13}$$

For any fixed value of λ which will determine the accuracy of the corresponding solutions, we can then optimize Ψ as a standard unstrained optimization problem.

There are other variants of particle swarm optimization; in fact, there are about more than 20 different variants. PSO algorithms are often combined with other existing algorithms to produce new hybrid algorithms. In fact, it is still an active area of research with many new studies published each year.

EXERCISES

15.1 To improve the stability and convergence of the standard PSO, a standard technique is to use the inertia function, which essentially replace $\boldsymbol{v}_t$ by $\theta \boldsymbol{v}_t$ where $\theta \in (0, 1]$ is a parameter. Modify the PSO demo program in the book and investigate and observe how θ affect the rate of convergence.

15.2 Write a computer program or modify one of the Matlab programs to find the global solution of

$$f(\boldsymbol{x}) = \sum_{j=1} \Big\{ \sum_{i=1}^{n} (i^j + \beta) \Big[(\frac{x_i}{i})^j - 1 \Big] \Big\}, \quad -n \leq x_i \leq n,$$

where $\beta > 0$. Does the optimal solution depend on β?

15.3 Scalings are important for many optimization problems. If there are some significant difference in scales in different design variables, the rate of convergence of an optimization algorithm can be affected. Using the PSO program, investigate the following modified Easom's function

$$f(x, y) = -\cos(x)\cos(\frac{y}{100}) e^{-(x-\pi)^2 - (\frac{y}{100\pi} - 1)^2},$$

where $0 \leq x \leq 2\pi$ and $0 \leq y \leq 200\pi$.

15.4 Most test functions are deterministic; however, it may become more difficult if the function has some stochastic components. Modify the PSO code to find the global minimum of

$$f(x, y) = -5e^{-\beta[(x-\pi)^2 + (y-\pi)^2]} - \sum_{j=1}^{K} \sum_{i=1}^{K} \epsilon_{ij} e^{-\alpha[(x-i)^2 + (y-j)^2]},$$

where $\alpha, \beta > 0$ are scaling parameters, often we can use $\alpha = \beta = 1$. Here ϵ_{ij} are random variables and can be drawn from a uniform distribution $\epsilon_{ij} \sim$

Unif[0,1]. The domain is $0 \leq x, y \leq K$ and $K = 10$. It is worth pointing that the random numbers, once drawn, should not be changed during each function evaluation (see Appendix A for detail). However, they can vary if you wish.

REFERENCES

1. A. Chatterjee and P. Siarry, "Nonlinear inertia variation for dynamic adapation in particle swarm optimization", *Comp. Oper. Research*, **33**, 859-871 (2006).
2. M. Clerc, J. Kennedy, "The particle swarm - explosion, stability, and convergence in a multidimensional complex space", *IEEE Trans. Evolutionary Computation*, **6**, 58-73 (2002).
3. A. P. Engelbrecht, *Fundamentals of Computational Swarm Intelligence*, John Wiley & Sons, 2005.
4. J. Kennedy and R. C. Eberhart, "Particle swarm optimization", in: *Proc. of IEEE International Conference on Neural Networks*, Piscataway, NJ. pp. 1942-1948 (1995).
5. J. Kennedy and R. C. Eberhart, *Swarm intelligence*, Academic Press, 2001.
6. Swarm intelligence, http://www.swarmintelligence.org

CHAPTER 16

HARMONY SEARCH

16.1 MUSIC-BASED ALGORITHMS

Harmony Search (HS) is a relatively new heuristic optimization algorithm and it was first developed by Z. W. Geem *et al.* in 2001. Since then, it has been applied to solve many optimization problems including function optimization, water distribution network, groundwater modelling, energy-saving dispatch, structural design, vehicle routing, and others. The possibility of combining harmony search with other algorithms such as Particle Swarm Optimization (PSO) has also been investigated.

Harmony search is a music-inspired metaheuristic optimization algorithm. It is inspired by the observation that the aim of music is to search for a perfect state of harmony. This harmony in music is analogous to find the optimality in an optimization process. The search process in optimization can be compared to a musician's improvisation process. This perfectly pleasing harmony is determined by the audio aesthetic standard.

The aesthetic quality of a musical instrument is essentially determined by its pitch (or frequency), timbre (or sound quality), and amplitude (or loudness). Timbre is largely determined by the harmonic content which is

Engineering Optimization: An Introduction with Metaheuristic Applications.
By Xin-She Yang
Copyright © 2010 John Wiley & Sons, Inc.

Figure 16.1: Harmony of two notes with a frequency ratio of 2:3 and their waveform.

in turn determined by the waveforms or modulations of the sound signal. However, the harmonics it can generate will largely depend on the pitch or frequency range of the particular instrument.

Different notes have different frequencies. For example, the note A above middle C (or standard concert A4) has a fundamental frequency of $f_0 = 440$ Hz. As the speed of sound in dry air is about $v = 331 + 0.6T$ m/s where T is the temperature in degrees Celsius near 0°C. So at room temperature $T = 20°$C, the A4 note has a wavelength $\lambda = v/f_0 \approx 0.7795$ m. When we adjust the pitch, we are in fact trying to change the frequency. In music theory, pitch p in MIDI is often represented as a numerical scale (a linear pitch space) using the following formula

$$p = 69 + 12 \log_2 (\frac{f}{440\text{Hz}}), \tag{16.1}$$

or

$$f = 440 \times 2^{(p-69)/12}, \tag{16.2}$$

which means that the A4 notes has a pitch number 69. In this scale, octaves correspond to size 12 and semitone corresponds to size 1. Furthermore, the ratio of frequencies of two notes which are an octave apart is 2:1. Thus, the frequency of a note is doubled (or halved) when it raised (or lowered) by an octave. For example, A2 has a frequency of 110 Hz, while A5 has a frequency of 880 Hz.

The measurement of harmony when different pitches occurring simultaneously, like any aesthetic quality, is somewhat subjective. However, it is possible to use some standard estimation for harmony. The frequency ratio, pioneered by ancient Greek mathematician Pythagoras, is a good way for such estimations. For example, the octave with a ratio of 1:2 sounds pleasant when playing together, so are the notes with a ratio of 2:3 (see Figure 16.1). However, it is unlikely for any random notes such as those shown in 16.2 to produce a pleasant harmony.

Figure 16.2: Random music notes.

16.2 HARMONY SEARCH

Harmony search can be explained in more detail with the aid of the discussion of the improvisation process by a musician. When a musician is improvising, he or she has three possible choices:

- Play any famous piece of music (a series of pitches in harmony) exactly from his or her memory;
- Play something similar to a known piece (this is equivalent to adjusting the pitch slightly);
- Compose new or random notes.

If we formalize these three options for optimization, we have three corresponding components: usage of harmony memory, pitch adjusting, and randomization.

The usage of harmony memory is important as it is similar to choose the best fit individuals in the genetic algorithms. This will ensure the best harmonies will be carried over to the new harmony memory. In order to use this memory more effectively, we cam assign a parameter $r_{\rm accept} \in [0, 1]$, called harmony memory accepting or considering rate. If this rate is too low, only few best harmonies are selected and it may converge too slowly. If this rate is extremely high (near 1), almost all the harmonies are used in the harmony memory, then other harmonies are not explored well, leading to potentially wrong solutions. Therefore, typically, $r_{\rm accept} = 0.7 \sim 0.95$.

To adjust the pitch slightly in the second component, we have to use a method such that it can adjust the frequency efficiently. In theory, the pitch can be adjusted linearly or nonlinearly, but in practice, linear adjustment is used. If $x_{\rm old}$ is the current solution (or pitch), then the new solution (pitch) $\boldsymbol{x}_{\rm new}$ is generated by

$$\boldsymbol{x}_{\rm new} = \boldsymbol{x}_{\rm old} + b_p\,(2\,\texttt{rand} - 1), \tag{16.3}$$

where $\texttt{rand}$ is a random number drawn from a uniform distribution $[0, 1]$. Here b_p is the bandwidth, which controls the local range of pitch adjustment. In fact, we can see that the pitch adjustment (16.3) is a random walk.

Pitch adjustment is similar to the mutation operator in genetic algorithms. We can assign a pitch-adjusting rate ($r_{\rm pa}$) to control the degree of the adjustment. If $r_{\rm pa}$ is too low, then there is rarely any change. If it is too high, then

Harmony Search

Objective function $f(\boldsymbol{x})$, $\boldsymbol{x} = (x_1, ..., x_p)^T$
Generate initial harmonics (real number arrays)
Define pitch adjusting rate ($r_{\rm pa}$) and pitch limits
Define harmony memory accepting rate ($r_{\rm accept}$)
while (t <Max number of iterations)
  Generate new harmonics by accepting best harmonics
  Adjust pitch to get new harmonics (solutions)
  if (rand> $r_{\rm accept}$),
    Choose an existing harmonic randomly
  else if (rand> $r_{\rm pa}$),
    Adjust the pitch randomly within a bandwidth (16.3)
  else
    Generate new harmonics via randomization (16.4)
  end if
  Accept the new harmonics (solutions) if better
end while
Find the current best estimates

Figure 16.3: Pseudo code of Harmony Search.

the algorithm may not converge at all. Thus, we usually use $r_{pa} = 0.1 \sim 0.5$ in most simulations.

The third component is the randomization, which is to increase the diversity of the solutions. Although adjusting pitch has a similar role, it is limited to certain local pitch adjustment and thus corresponds to a local search. The use of randomization can drive the system further to explore various regions with high solution diversity so as to find the global optimality. So we have

$$p_a = p_{\rm lowerlimit} + p_{\rm range} * \texttt{rand}, \tag{16.4}$$

where $p_{\rm range} = p_{\rm upperlimit} - p_{\rm lowerlimit}$. Here **rand** is a random number generator in the range of 0 and 1.

The three components in harmony search can be summarized as the pseudo code shown in Figure 16.3 where we can see that the probability of true randomization is

$$P_{\rm random} = 1 - r_{\rm accept}, \tag{16.5}$$

and the actual probability of pitch adjusting is

$$P_{\rm pitch} = r_{\rm accept} * r_{\rm pa}. \tag{16.6}$$

Furthermore, like genetic algorithms and particle swarm optimization, harmony search is not a gradient-based search, so it avoids most of the pitfalls of any gradient-based search algorithms. Thus, it has fewer mathematical requirements, and subsequently, it can be used to deal with complex objective functions whether continuous or discontinuous, linear or nonlinear, or stochastic with noise.

On the other hand, harmony search could be potentially more efficient than genetic algorithms because harmony search does not use binary encoding and decoding, but it does have multiple solution vectors. Therefore, the implementation of HS algorithm is easier. In addition, there is evidence to suggest that HS is less sensitive to the chosen parameters, which means that we do not have to fine-tune these parameters to get quality solutions.

16.3 IMPLEMENTATION

Using the three components described in above section, we can implement the harmony search algorithm in Matlab/Octave as attached in Appendix B. For

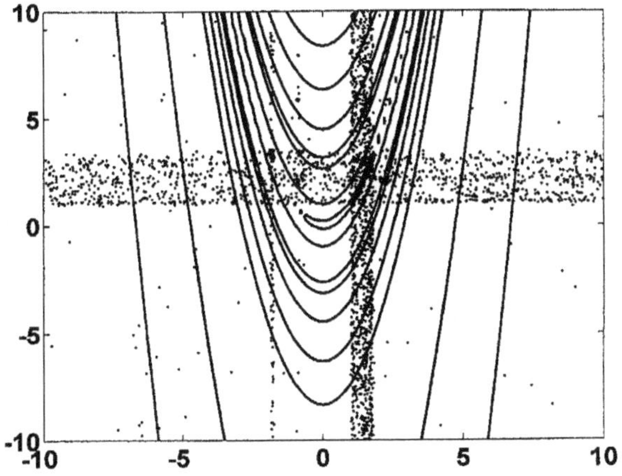

Figure 16.4: The variations of harmonies in harmony search.

Rosenbrock's banana function

$$f(x, y) = (1 - x)^2 + 100(y - x^2)^2, \tag{16.7}$$

where the domain is

$$(x, y) \in [-10, 10] \times [-10, 10], \tag{16.8}$$

it has the global minimum $f_{\min} = 0$ at $(1, 1)$. The best estimate solution $(1.005, 1.0605)$ is obtained after 25000 iterations. On a modern desktop computer, it usually takes about a minute. The variations of these solutions are shown in Figure 16.4.

We have used r_{accept} =HMacceptRate= 0.95, and the pitch adjusting rate r_{pa} =PArate= 0.7. From Figure 16.4, we can see that since the pitch adjustment is more intensive in local regions (two thin strips), it indeed indicates

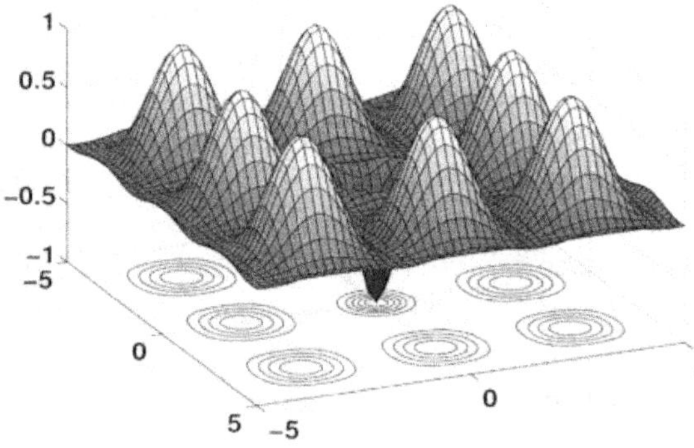

Figure 16.5: Yang's standing wave function with the global minimum at $(0, 0)$.

that the harmony search is more efficient than genetic algorithms. However, such comparisons require more study. Harmony search is emerging as a powerful algorithm, and its relevant literature is expanding. It is still an interesting area of active research.

EXERCISES

16.1 Modify the Matlab code in Appendix B to find the global minimum of the following test function

$$f(\boldsymbol{x}) = \Big\{\exp[-\sum_{i=1}^{n}(x_i/\beta)^{2m}] - 2\exp[-\sum_{i=1}^{n} x_i^2]\Big\} \cdot \prod_{i=1}^{n}\cos^2(x_i),$$

where $m = 5$ and $\beta = 15$. The search domain is $-20 \leq x_i \leq 20$ for $i = 1, 2, ..., n$. This function was designed by the author of this book in 2009 to validate a new search algorithm, Cuckoo Search, developed by Xin-She Yang and Suash Deb.[1] The function is multimodal, similar to a standing wave with a defect at the global optimum (see Figure 16.5).

16.2 Use the standard Harmony Search algorithm to carry out some parameter sensitivity analysis by varying p_a and r_{pa} from 0.1 to 0.99. How will they affect the rate of convergence of the Harmony Search?

16.3 The pitch adjustment formula (16.4) is relatively simple, which ensures that the adjustment is carried out in the neighborhood of the best solutions in the search space. However, it is not necessarily the only best option. In fact, in many variants of Harmony Search, a random-walk style adjustment is

[1] X. S. Yang and S. Deb, "Cuckoo search via Lévy flights", in: *Proc. of World Congress on Nature and Biologically Inspired Computing (NaBIC'09)*, IEEE Publications, USA, pp. 210-214 (2009).

used

$$p_{\text{new}} = p_{\text{old}} + b_p \cdot \mathbf{randn},$$

where b_p is the bandwidth of the pitch adjustment, often $b_p \propto p_{\text{range}}$. Alternatively, you can also use Lévy flights to replace **randn**. Revise the program and see if the rate of convergence improves.

16.4 Try some numerical experiments to see if you could combine Particle Swarm Optimization with Harmony Search and genetic algorithms.

REFERENCES

1. Z. W. Geem, J. H. Kim, and G. V. Loganathan, "A new heuristic optimization algorithm: Harmony search", *Simulation*, **76**, 60-68 (2001).
2. Z. W. Geem, *Music-Inspired Harmony Search Algorithm: Theory and Applications*, Springer, (2009).
3. Z. W. Geem, *Recent Advances in Harmony Search Algorithm*, Studies in Computational Intelligence, Springer, 2010.
4. X. S. Yang, "Harmony search as a metaheuristic algorithm", in: *Music-Inspired Harmony Search: Theory and Applications* (Eds Z. W. Geem), Springer, pp.1-14, 2009.

CHAPTER 17

FIREFLY ALGORITHM

17.1 BEHAVIOUR OF FIREFLIES

The flashing light of fireflies is an amazing sight in the summer sky in the tropical and temperate regions. There are about two thousand firefly species, and most fireflies produce short and rhythmic flashes. The pattern of flashes is often unique for a particular species. The flashing light is produced by a process of bioluminescence, and the true functions of such signaling systems are still being debated. However, two fundamental functions of such flashes are to attract mating partners (communication), and to attract potential prey. In addition, flashing may also serve as a protective warning mechanism to remind potential predators of the bitter taste of fireflies.

The rhythmic flash, the rate of flashing and the amount of time form part of the signal system that brings both sexes together. Females respond to a male's unique pattern of flashing in the same species, while in some species such as *Photuris*, female fireflies can eavesdrop on the bioluminescent courtship signals and even mimic the mating flashing pattern of other species so as to lure and eat the male fireflies who may mistake the flashes as a potential suitable mate.

Engineering Optimization: An Introduction with Metaheuristic Applications.
By Xin-She Yang

Copyright © 2010 John Wiley & Sons, Inc.

Some tropic fireflies can even synchronize their flashes, thus forming emerging biological self-organized behavior.

We know that the light intensity at a particular distance r from the light source obeys the inverse square law. That is to say, the light intensity I decreases as the distance r increases in terms of $I \propto 1/r^2$. Furthermore, the air absorbs light which becomes weaker and weaker as the distance increases. These two combined factors make most fireflies visual to a limit distance, usually several hundred meters at night, which is good enough for fireflies to communicate.

The flashing light can be formulated in such a way that it is associated with the objective function to be optimized, which makes it possible to formulate new optimization algorithms. In the rest of this chapter, we will first outline the basic formulation of the Firefly Algorithm (FA) and then discuss the implementation in detail.

17.2 FIREFLY-INSPIRED ALGORITHM

17.2.1 Firefly Algorithm

Now we can idealize some of the flashing characteristics of fireflies so as to develop firefly-inspired algorithms. For simplicity in describing our new Firefire Algorithm (FA) which was developed by Xin-She Yang at Cambridge University in 2007, we now use the following three idealized rules:

- All fireflies are unisex so that one firefly will be attracted to other fireflies regardless of their sex;
- Attractiveness is proportional to the their brightness, thus for any two flashing fireflies, the less brighter one will move towards the brighter one. The attractiveness is proportional to the brightness and they both decrease as their distance increases. If there is no brighter one than a particular firefly, it will move randomly;
- The brightness of a firefly is affected or determined by the landscape of the objective function.

For a maximization problem, the brightness can simply be proportional to the value of the objective function. Other forms of brightness can be defined in a similar way to the fitness function in genetic algorithms.

Based on these three rules, the basic steps of the firefly algorithm (FA) can be summarized as the pseudo code shown in Figure 17.1.

17.2.2 Light Intensity and Attractiveness

In the firefly algorithm, there are two important issues: the variation of light intensity and formulation of the attractiveness. For simplicity, we can always

Firefly Algorithm

Objective function $f(\boldsymbol{x})$, $\boldsymbol{x} = (x_1, ..., x_d)^T$
Generate initial population of fireflies $\boldsymbol{x}_i$ $(i = 1, 2, ..., n)$
Light intensity I_i at $\boldsymbol{x}_i$ is determined by $f(\boldsymbol{x}_i)$
Define light absorption coefficient γ
while (t <MaxGeneration)
for $i = 1 : n$ all n fireflies
  for $j = 1 : n$ all n fireflies (inner loop)
  if $(I_i < I_j)$, Move firefly i towards j; **end if**
    Vary attractiveness with distance r via $\exp[-\gamma r]$
    Evaluate new solutions and update light intensity
  end for j
end for i
Rank the fireflies and find the current global best $\boldsymbol{g}_*$
end while
Postprocess results and visualization

Figure 17.1: Pseudo code of the firefly algorithm (FA).

assume that the attractiveness of a firefly is determined by its brightness which in turn is associated with the encoded objective function.

In the simplest case for maximum optimization problems, the brightness I of a firefly at a particular location $\boldsymbol{x}$ can be chosen as $I(\boldsymbol{x}) \propto f(\boldsymbol{x})$. However, the attractiveness β is relative, it should be seen in the eyes of the beholder or judged by the other fireflies. Thus, it will vary with the distance r_{ij} between firefly i and firefly j. In addition, light intensity decreases with the distance from its source, and light is also absorbed in the media, so we should allow the attractiveness to vary with the degree of absorption.

In the simplest form, the light intensity $I(r)$ varies according to the inverse square law

$$I(r) = \frac{I_s}{r^2}, \tag{17.1}$$

where I_s is the intensity at the source. For a given medium with a fixed light absorption coefficient γ, the light intensity I varies with the distance r. That is

$$I = I_0 e^{-\gamma r}, \tag{17.2}$$

where I_0 is the original light intensity. In order to avoid the singularity at $r = 0$ in the expression I_s/r^2, the combined effect of both the inverse square law and absorption can be approximated as the following Gaussian form

$$I(r) = I_0 e^{-\gamma r^2}. \tag{17.3}$$

As a firefly's attractiveness is proportional to the light intensity seen by adjacent fireflies, we can now define the attractiveness β of a firefly by

$$\beta = \beta_0 e^{-\gamma r^2}, \tag{17.4}$$

where β_0 is the attractiveness at $r = 0$. As it is often faster to calculate $1/(1+r^2)$ than an exponential function, the above function, if necessary, can conveniently be approximated as

$$\beta = \frac{\beta_0}{1+\gamma r^2}. \tag{17.5}$$

Both (17.4) and (17.5) define a characteristic distance $\Gamma = 1/\sqrt{\gamma}$ over which the attractiveness changes significantly from β_0 to $\beta_0 e^{-1}$ for equation (17.4) or $\beta_0/2$ for equation (17.5).

In the actual implementation, the attractiveness function $\beta(r)$ can be any monotonically decreasing functions such as the following generalized form

$$\beta(r) = \beta_0 e^{-\gamma r^m}, \qquad (m \geq 1). \tag{17.6}$$

For a fixed γ, the characteristic length becomes

$$\Gamma = \gamma^{-1/m} \to 1, \qquad m \to \infty. \tag{17.7}$$

Conversely, for a given length scale Γ in an optimization problem, the parameter γ can be used as a typical initial value. That is

$$\gamma = \frac{1}{\Gamma^m}. \tag{17.8}$$

The distance between any two fireflies i and j at $\boldsymbol{x}_i$ and $\boldsymbol{x}_j$, respectively, is the Cartesian distance

$$r_{ij} = \|\boldsymbol{x}_i - \boldsymbol{x}_j\| = \sqrt{\sum_{k=1}^{d} (x_{i,k} - x_{j,k})^2}, \tag{17.9}$$

where $x_{i,k}$ is the kth component of the spatial coordinate $\boldsymbol{x}_i$ of ith firefly. In 2-D case, we have

$$r_{ij} = \sqrt{(x_i - x_j)^2 + (y_i - y_j)^2}. \tag{17.10}$$

The movement of a firefly i is attracted to another more attractive (brighter) firefly j is determined by

$$\boldsymbol{x}_i = \boldsymbol{x}_i + \beta_0 e^{-\gamma r_{ij}^2} (\boldsymbol{x}_j - \boldsymbol{x}_i) + \alpha\, \boldsymbol{\epsilon}_i, \tag{17.11}$$

where the second term is due to the attraction. The third term is randomization with α being the randomization parameter, and $\boldsymbol{\epsilon}_i$ is a vector of random numbers drawn from a Gaussian distribution or uniform distribution. For example, the simplest form is ϵ_i and can be replaced by rand $- 1/2$ where rand is a random number generator uniformly distributed in $[0, 1]$. For most of our implementation, we can take $\beta_0 = 1$ and $\alpha \in [0, 1]$.

It is worth pointing out that (17.11) is a random walk biased towards the brighter fireflies. If $\beta_0 = 0$, it becomes a simple random walk. Furthermore, the randomization term can easily be extended to other distributions such as Lévy flights.

The parameter γ now characterizes the variation of the attractiveness, and its value is crucially important in determining the speed of the convergence and how the FA algorithm behaves. In theory, $\gamma \in [0, \infty)$, but in practice, $\gamma = O(1)$ is determined by the characteristic length Γ of the system to be optimized. Thus, in most application, it typically varies from 0.1 to 10.

17.2.3 Scaling and Global Optima

It is worth pointing out that the distance r defined above is *not* limited to the Euclidean distance. We can define other distance r in the n-dimensional hyperspace, depending on the type of problem of our interest. For example, for job scheduling problems, r can be defined as the time lag or time interval. For complicated networks such as the Internet and social networks, the distance r can be defined as the combination of the degree of local clustering and the average proximity of vertices. In fact, any measure that can effectively characterize the quantities of interest in the optimization problem can be used as the 'distance' r.

The typical scale Γ should be associated with the scale concerned in our optimization problem. If Γ is the typical scale for a given optimization problem, for a very large number of fireflies $n \gg m$ where m is the number of local optima, then the initial locations of these n fireflies should distribute relatively uniformly over the entire search space. As the iterations proceed, the fireflies would converge into all the local optima (including the global ones). By comparing the best solutions among all these optima, the global optima can easily be achieved. Our recent research suggests that it is possible to prove that the firefly algorithm will approach global optima when $n \to \infty$ and $t \gg 1$. In reality, it converges very quickly and this will be demonstrated later in this chapter.

17.2.4 Two Special Cases

There are two important limiting or asymptotic cases when $\gamma \to 0$ and $\gamma \to \infty$. For $\gamma \to 0$, the attractiveness is constant $\beta = \beta_0$ and $\Gamma \to \infty$, this is equivalent to say that the light intensity does not decrease in an idealized sky. Thus, a flashing firefly can be seen anywhere in the domain. Thus, a single (usually global) optima can easily be reached. If we remove the inner loop for j in Figure 17.1 and replace $\boldsymbol{x}_j$ by the current global best $\boldsymbol{g}_*$, then the Firefly Algorithm becomes the special case of accelerated particle swarm optimization

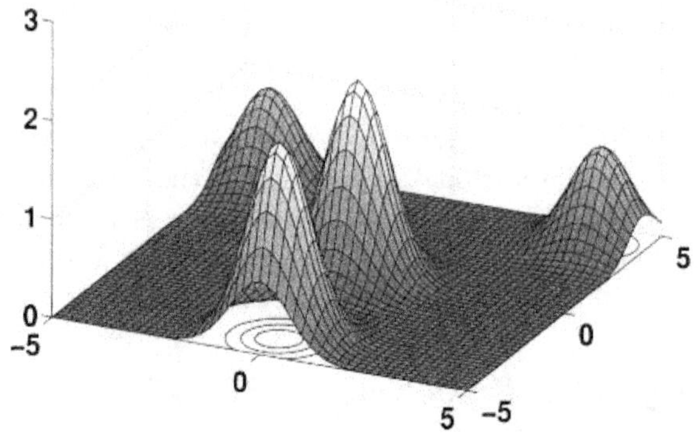

Figure 17.2: Landscape of a function with two equal global maxima.

(PSO) discussed earlier. Subsequently, the efficiency of this special case is the same as that of PSO.

On the other hand, the limiting case $\gamma \to \infty$ leads to $\Gamma \to 0$ and $\beta(r) \to \delta(r)$ which is the Dirac delta function, which means that the attractiveness is almost zero in the sight of other fireflies. This is equivalent to the case where the fireflies randomly roam in a very thick fog region. No other fireflies can be seen, and each firefly roams in a completely random way. Therefore, this corresponds to the completely random search method.

As the firefly algorithm is usually in the case between these two extremes, it is possible to adjust the parameters γ and α so that it can outperform both the random search and PSO. In fact, FA can find the global optima as well as the local optima simultaneously and effectively. This advantage will be demonstrated in detail later in the implementation.

A further advantage of FA is that different fireflies can work almost independently, it is thus particularly suitable for parallel implementation. It is even better than genetic algorithms and PSO because fireflies aggregate more closely around each optimum. It can be expected that the interactions between different subregions are minimal in parallel implementation.

17.3 IMPLEMENTATION

17.3.1 Multiple Global Optima

In order to demonstrate how the firefly algorithm works, we have implemented it in Matlab/Octave. The program is given in Appendix B where we have used

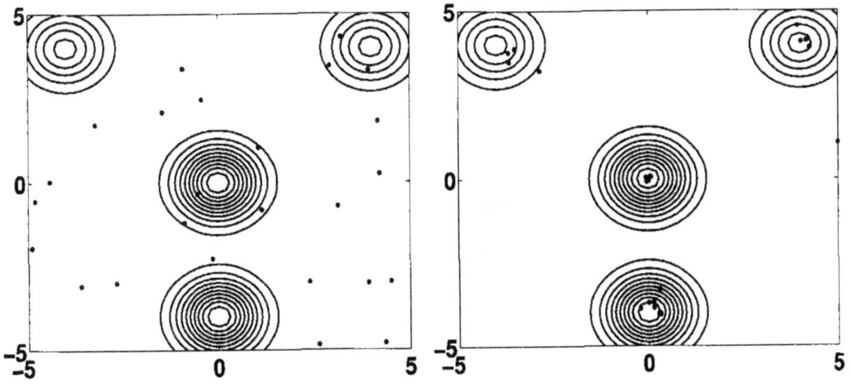

Figure 17.3: The initial locations of 25 fireflies (left) and their final locations after 20 iterations (right).

a simple function

$$f(x, y) = (|x| + |y|) \exp[-0.0625(x^2 + y^2)], \qquad (17.12)$$

which has four equal peaks at $(-2, -2)$, $(-2, 2)$, $(2, -2)$ and $(2, 2)$. This function can easily extended to any higher dimensions.

In order to show that both the global optima and local optima can be found simultaneously, we now use the following four-peak function

$$f(x, y) = e^{-(x-4)^2-(y-4)^2} + e^{-(x+4)^2-(y-4)^2} + 2[e^{-x^2-y^2} + e^{-x^2-(y+4)^2}], \qquad (17.13)$$

where $(x, y) \in [-5, 5] \times [-5, 5]$. This function has four peaks. Two local peaks with $f = 1$ at $(-4, 4)$ and $(4, 4)$, and two global peaks with $f_{\max} = 2$ at $(0, 0)$ and $(0, -4)$, as shown in Figure 17.2. We can see that all these four optima can be found using 25 fireflies in about 20 generations (see Figure 17.3). So the total number of function evaluations is about 500. This is much more efficient than most of existing metaheuristic algorithms.

In the implementation, the values of the parameters are $\alpha = 0.2$, $\gamma = 1$ and $\beta_0 = 1$. Obviously, these parameters can be adjusted to suit for solving various problems with different scales.

17.3.2 Multimodal Functions

Now let us use the FA to find the optima of some tougher test functions such as Ackley's d-dimension function

$$f(\boldsymbol{x}) = -20 \exp\Big[-\frac{1}{5}\sqrt{\frac{1}{d}\sum_{i=1}^{d} x_i^2}\,\Big] - \exp\Big[\frac{1}{d}\sum_{i=1}^{d}\cos(2\pi x_i)\Big] + 20 + e, \qquad (17.14)$$

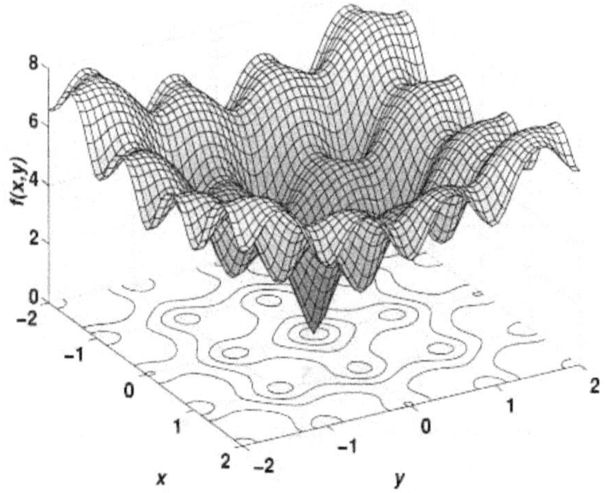

Figure 17.4: Landscape of Ackley's 2D function with the global minimum 0 at $(0, 0)$.

which has the global minimum $f_* = 0$ at $(0, 0, ..., 0)$ in the domain of $-32.768 \leq x_i \leq 32.768$ where $i = 1, 2, ..., d$. For example, the 2D case of Ackley's function is shown in Figure 17.4.

As this is a minimization problem, so we have to slightly modify the Matlab program of the implemented Firefly Algorithm in Appendix B. The simplest change is to use $I_i > I_j$ instead of $I_i < I_j$. Figure 17.5 shows the initial locations of the 25 fireflies and their final locations at the 20th iteration.

17.3.3 FA Variants

The basic firefly algorithm is very efficient, but we can see that the solutions are still changing as the optima are approaching. It is possible to improve the solution quality by reducing the randomness.

A further improvement on the convergence of the algorithm is to vary the randomization parameter α so that it decreases gradually as the optima are approaching. For example, we can use

$$\alpha = \alpha_\infty + (\alpha_0 - \alpha_\infty)e^{-t}, \tag{17.15}$$

where $t \in [0, t_{\max}]$ is the pseudo time for simulations and $t_{\max}$ is the maximum number of generations. α_0 is the initial randomization parameter while α_∞ is the final value. We can also use the similar function to the geometrical annealing schedule. That is

$$\alpha = \alpha_0 \theta^t, \tag{17.16}$$

where $\theta \in (0, 1]$ is the randomness reduction constant.

In addition, in the current version of the FA algorithm, we do not explicitly use the current global best $\boldsymbol{g}_*$, even though we only use it to decode the final

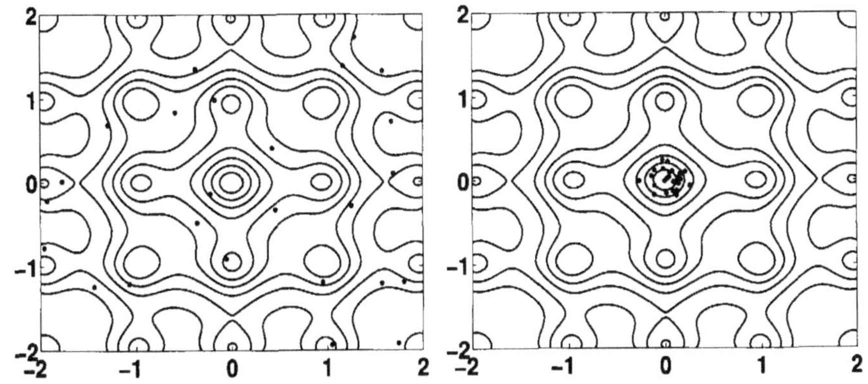

Figure 17.5: The initial locations of the 25 fireflies (left) and their final locations after 20 iterations (right).

best solutions. Our recent studies show that the efficiency may significantly improve if we add an extra term $\lambda \epsilon_i (\boldsymbol{x}_i - \boldsymbol{g}_*)$ to the updating formula (17.11). Here λ is a parameter similar to α and β, and ϵ_i is a set of random numbers. These could form important topics for further research.

EXERCISES

17.1 Schubert's function

$$f(\boldsymbol{x}) = \Big[\sum_{i=1}^{5} i \cos\Big(i + (i+1)x\Big)\Big] \cdot \Big[\sum_{i=1}^{5} i \cos\Big(i + (i+1)y\Big)\Big],$$

in the 2D domain $-10 \leq x, y \leq 10$ is multimodal, and it often serves as a tough test function. Modify the Matlab code in Appendix B or write your own computer program to find all the optima. Can your program find all the optima at the same time? If not, what modification is needed to achieve this? How many fireflies are needed.

17.2 Use the implemented FA code, try to use different methods such as (17.15) and (17.16) to reduce the randomness gradually. Observe if the rate of convergence and the quality of the solution improve?

17.3 The mobility of fireflies in the standard Firefly Algorithm is represented by the Gaussian and/or uniform distribution, which are relatively efficient. For some problems, especially those with significant differences in parameter scales, others probability distributions such as Lévy distribution in terms of Lévy flights may be more efficient. This is true for many metaheuristic algorithms including PSO. Modify the Matlab code to incorporate the Lévy flights or other distributions with a long tail.

REFERENCES

1. S. Lukasik and S. Zak, "Firefly algorithm for continuous constrained optimization tasks", ICCCI 2009, Lecture Notes in Artificial Intelligence (Eds. N. T. Ngugen, R. Kowalczyk and S.-M. Chen), **5796**, 97-106 (2009).
2. S. M. Lewis and C. K. Cratsley, "Flash signal evolution, mate choice, and predation in fireflies", *Annual Review of Entomology*, **53**, 293-321 (2008).
3. C. O'Toole, *Firefly Encyclopedia of Insects and Spiders*, Firefly Books Ltd, 2002.
4. X. S. Yang, *Nature-Inspired Metaheuristic Algorithms*, Luniver Press, 2008.
5. X. S. Yang, "Firefly algorithms for multimodal optimization", in: *Stochastic Algorithms: Foundations and Applications*, SAGA 2009, Lecture Notes in Computer Science, **5792**, 169-178 (2009).
6. X. S. Yang, "Firefly algorithm, Lévy flights and global optimization", in: *Research and Development in Intelligent Systems XXVI*, (Eds M. Bramer *et al.*), Springer, London, pp. 209-218 (2010).
7. X. S. Yang, "Firefly algorithm, stochastic test functions and design optimization", *Int. J. Bio-inspired Computation*, **2**, 78-84 (2010).
8. X. S. Yang and S. Deb, "Engineering optimization by cuckoo search", *Int. J. Math. Modeling Num. Opt.*, **1**, No. 4, (in press) (2010).

PART III

APPLICATIONS

CHAPTER 18

MULTIOBJECTIVE OPTIMIZATION

All the optimization problems we discussed so far have only a single objective. In reality, we often have to optimize multiple objectives simultaneously. For example, we may want to improve the performance of a product while trying to minimize the cost at the same time. In this case, we are dealing with multiobjective optimization problems. Many new concepts are required for solving multiobjective optimization.

18.1 PARETO OPTIMALITY

The optimization problem with a single objective discussed so far can be considered as a scalar optimization problem because the objective function always reaches a single global optimal value or a scalar. For multiobjective optimization, the multiple objective functions form a vector, and thus it is also called vector optimization.

Any multiobjective optimization problem can generally be written as

$$\underset{\boldsymbol{x}\in\Re^n}{\text{minimize}} \ \boldsymbol{f}(\boldsymbol{x}) = [f_1(\boldsymbol{x}), f_2(\boldsymbol{x}), ..., f_p(\boldsymbol{x})],$$
$$\text{subject to} \quad g_j(\boldsymbol{x}) \leq 0, \ \ j = 1, 2, ..., M, \tag{18.1}$$

By Xin-She Yang
Copyright © 2010 John Wiley & Sons, Inc.

$$h_k(\boldsymbol{x}) = 0, \quad k = 1, 2, ..., N, \tag{18.2}$$

where $\boldsymbol{x} = (x_1, x_2, ..., x_n)^T$ is the vector of decision variables. In some formulations used in the optimization literature, inequalities $g_j (j = 1, ..., N)$ can also include any equalities because an equality $\phi(\boldsymbol{x}) = 0$ can be converted into two inequalities $\phi(\boldsymbol{x}) \leq 0$ and $\phi(\boldsymbol{x}) \geq 0$. However, for clarity, we list here the equalities and inequalities separately.

The space $\mathcal{F} = \Re^n$ spanned by the vectors of decision variables $\boldsymbol{x}$ is called the search space. The space $\mathcal{S} = \Re^p$ formed by all the possible values of objective functions is called the solution space or objective space. Comparing with the single objective function whose solution space is (at most) $\Re$, the solution space for multiobjective optimization is considerably much larger. In addition, as we know that we are dealing with multiobjectives $\boldsymbol{f}(\boldsymbol{x}) = [f_i]$, for simplicity, we can write f_i as $f(\boldsymbol{x})$ without causing any confusion.

Multiobjective optimization problems, unlike a single objective optimization problem, do not necessarily have an optimal solution that minimizes all the multiobjective functions simultaneously. Often, different objectives may conflict each other and the optimal parameters of some objectives usually do not lead to optimality of other objectives (sometimes make them worse). For example, we want the first-class quality service on our holidays and at the same time we want to pay as little as possible. The high-quality service (one objective) will inevitably cost much more, and this is in conflict with the other objective (to minimize cost).

Therefore, among these often conflicting objectives, we have to choose some tradeoff or a certain balance of objectives. If none of these are possible, we must choose a list of preferences so that which objectives should be achieved first. More importantly, we have to compare different objectives and make a compromise. This usually requires a formulation of a new evaluation modeling problem, and one of the most popular approaches to such modeling is to find a scalar-valued function that represents a weighted combination or preference order of all objectives. Such a scalar function is often referred to as the preference function or utility function. A simple way to construct this scalar function is to use the weighted sum

$$u(f_1(\boldsymbol{x}), ..., f_p(\boldsymbol{x})) = \sum_{i=1}^{p} \alpha_i f_i(\boldsymbol{x}), \tag{18.3}$$

where α_i are the weighting coefficients. For multiobjective optimization, we have to introduce some new concepts related to Pareto optimality.

A vector $\boldsymbol{u} = (u_1, .., u_n)^T \in \mathcal{F}$, is said to dominate another vector $\boldsymbol{v} = (v_1, ..., v_n)^T$ if and only if $u_i \leq v_i$ for $\forall i \in \{1, ..., n\}$ and $\exists i \in \{1, ..., n\} : u_i < v_i$. This 'partial less' or component-wise relationship is denoted by

$$\boldsymbol{u} \prec \boldsymbol{v}, \tag{18.4}$$

which is equivalent to

$$\forall i \in \{1, ..., n\} : u_i \leq v_i \wedge \exists i \in \{1, ..., n\} : u_i < v_i. \tag{18.5}$$

Here $\wedge$ means the logical 'and'. In other words, no component of $\boldsymbol{u}$ is larger than the corresponding component of $\boldsymbol{v}$, and at least one component is smaller. Similarly, we can define another dominance relationship $\preceq$ by

$$\boldsymbol{u} \preceq \boldsymbol{v} \Longleftrightarrow \boldsymbol{u} \prec \boldsymbol{v} \vee \boldsymbol{u} = \boldsymbol{v}. \tag{18.6}$$

Here $\vee$ means 'or'. It is worth pointing out that for maximization problems, the dominance can be defined by replacing $\prec$ with $\succ$.

A point or a solution $\boldsymbol{x}_* \in \Re^n$ is called a Pareto optimal solution or non-inferior solution to the optimization problem if there is no $\boldsymbol{x} \in \Re^n$ satisfying $f_i(\boldsymbol{x}) \le f_i(\boldsymbol{x}_*), (i = 1, 2, ..., p)$. In other words, $\boldsymbol{x}_*$ is Pareto optimal if there exists no feasible vector (of decision variables in the search space) which would decrease some objectives without causing an increase in at least one other objective simultaneously. That is to say, optimal solutions are solutions which are not dominated by any other solutions. When mapping to objective vectors, they represent different trade-off between multiple objectives.

Furthermore, a point $\boldsymbol{x}_* \in \mathcal{F}$ is called a non-dominated solution if no solution can be found that dominates it. A vector is called ideal if it contains the decision variables that correspond to the optima of objectives when each objective is considered separately.

Unlike the single objective optimization with often a single optimal solution, multiobjective optimization will lead to a set of solutions, called the Pareto optimal set $\mathcal{P}^*$, and the decision vectors $\boldsymbol{x}_*$ for this solution set are thus called non-dominated. That is to say, the set of optimal solutions in the decision space forms the Pareto (optimal) set. The image of this Pareto set in the objective or response space is called the Pareto front. In literature, the set $\boldsymbol{x}_*$ in the decision space that corresponds to the Pareto optimal solutions is also called an efficient set. The set (or plot) of the objective functions of these non-dominated decision vectors in the Pareto optimal set forms the so-called Pareto front $\mathcal{P}$ or Pareto frontier.

Using the above notation, the Pareto front $\mathcal{P}$ can be defined as the set of non-dominated solutions so that

$$\mathcal{P} = \{\boldsymbol{s} \in \mathcal{S} \big| \nexists\, \boldsymbol{s}' \in \mathcal{S} : \boldsymbol{s}' \prec \boldsymbol{s}\}, \tag{18.7}$$

or in term of the Pareto optimal set in the search space

$$\mathcal{P}^* = \{\boldsymbol{x} \in \mathcal{F} \big| \nexists\, \boldsymbol{x}' \in \mathcal{F} : \boldsymbol{f}(\boldsymbol{x}') \prec \boldsymbol{f}(\boldsymbol{x})\}. \tag{18.8}$$

The identification of the Pareto front is not an easy task, and it often requires a parametric analysis, say, by treating all but one objective, say, f_i in a p-objective optimization problem so that f_i is a function of f_1, ..., f_{i-1}, f_{i+1}, ..., and f_p. By maximizing the f_i when varying the values of the other $p - 1$ objectives so that the solutions will trace out the Pareto front.

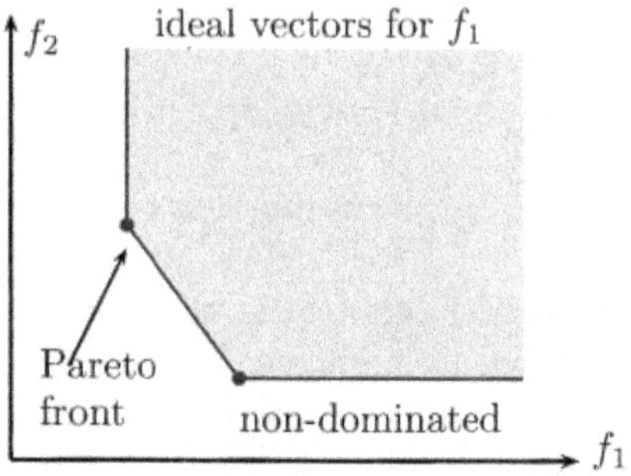

Figure 18.1: Non-dominated set, Pareto front and ideal vectors in a minimization problem with two objectives f_1 and f_2.

■ EXAMPLE 18.1

For example, we have four Internet service providers A, B, C and D. We have two objectives to choose their service 1) as cheap as possible, and 2) higher bandwidth. Their details are listed below:

IP provider	Cost (£/month)	Bandwidth (Mb)
A	20	12
B	25	16
C	30	8
D	40	16

From the table, we know that option C is dominated by A and B because both objectives are improved (low cost and faster). Option D is dominated by B. Thus, solution C is an inferior solution, and so is D. Both solutions A and B are non-inferior solutions or non-dominated solutions. However, which solution (A or B) to choose is not easy, as provider A outperforms B on the first objective (cheaper) while B outperforms A on another objective (faster). In this case, we say these two solutions are incomparable. The set of the non-dominated solutions A and B forms the Pareto front which is a mutually incomparable set.

For a minimization problem with two objectives, the basic concepts of non-dominated set, Pareto front, and ideal vectors are shown in Figure 18.1. Obviously, if we combine these two into a single composite objective, we can compare, for example, the cost per unit Mb. In this case, we essentially reformulate the problem as a scalar optimization problem. For choice A, each Mb costs £1.67, while it costs about £1.56 for choice B. So we should choose B. However, in reality, we usually have many incomparable solutions, and it is often impossible to comprise in some way. In addition, the real choice depends on our preference and emphasis on objectives.

Multiobjective optimization is usually difficult to solve. Loosely speaking, there are three ways to deal with multiobjective problems: direct approach, aggregation or transformation, and Pareto set approximation.

Direct approach is difficult, especially in the case when multiple objectives seem conflicting. Therefore, we often use aggregation or transformation by combining multiple objectives into a single composite objective so that the standard methods for optimization discussed in this book can be used. We will focus on this approach in the rest of the chapter. However, with this approach, the solutions typically depend on the way how we combine the objectives. A third way is to try to approximate the Pareto set so as to obtain a set of mutually non-dominated solutions.

To transform a multiobjective optimization problem into a single objective, we can often use the method of weighted sum, and utility method. We can also choose the most important objective of our interest as the only objective, while rewriting other objectives as constraints with imposed limits.

18.2 WEIGHTED SUM METHOD

Many solution algorithms intend to combine all the multi-objective functions into one scalar objective using the weighted sum

$$F(\boldsymbol{x}) = \alpha_1 f_1(\boldsymbol{x}) + \alpha_2 f_2(\boldsymbol{x}) + ... + \alpha_p f_p(\boldsymbol{x}). \tag{18.9}$$

The important issue arises in assigning the weighting coefficients $(\alpha_1, \alpha_2, ..., \alpha_p)$ because the solution strongly depends on the chosen weighting coefficients. Let us look at an example.

■ EXAMPLE 18.2

The classical three-objective functions are commonly used for testing multi-objective optimization algorithms. These functions are

$$f_1(x, y) = x^2 + (y-1)^2, \tag{18.10}$$

$$f_2(x, y) = (x-1)^2 + y^2 + 2, \tag{18.11}$$

$$f_3(x, y) = x^2 + (y+1)^2 + 1, \tag{18.12}$$

where $(x, y) \in [-2, 2] \times [-2, 2]$.

If we combine all the three functions into a single function $f(x, y)$ using the weighted sum, we have

$$f(x, y) = \alpha f_1 + \beta f_2 + \gamma f_3, \qquad \alpha + \beta + \gamma = 1. \tag{18.13}$$

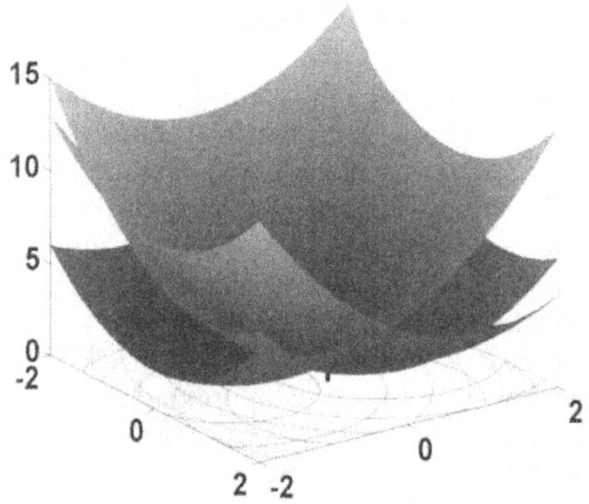

Figure 18.2: Three functions reach the global minimum at $x_* = \beta, y_* = \alpha - \gamma$.

The stationary point is determined by

$$\frac{\partial f}{\partial x} = 0, \qquad \frac{\partial f}{\partial y} = 0, \tag{18.14}$$

which lead to

$$2\alpha + 2\beta(x-1) + 2\gamma = 0, \tag{18.15}$$

and

$$2\alpha(y-1) + 2\beta y + 2\gamma(y+1) = 0. \tag{18.16}$$

The solutions are

$$x_* = \beta, \qquad y_* = \alpha - \gamma. \tag{18.17}$$

This implies that $x_* \in [0,1]$ and $y_* \in [-1,1]$. Consequently, $f_1 \in [0,5], f_2 \in [2,4]$ and $f_3 \in [1,6]$. In addition, the solution or the optimal location varies with the weighting coefficients α, β and γ. In the simplest case $\alpha = \beta = \gamma = 1/3$, we have

$$x_* = \frac{1}{3}, \qquad y_* = 0. \tag{18.18}$$

This location is marked with a short thick line in Figure 18.2.

Now the original multiobjective optimization problem has been transformed into a single objective optimization problem. Thus, the solution methods for solving single objective problems are all valid. For example, we can use the particle swarm to find the optimal for given parameters α, β and γ. Figure 18.3 shows the final locations of 40 particles at $t = 5$ iterations. We can see that the particles converge towards the true optimal location marked with $\circ$. Obviously, the accuracy will improve if we continue the iterations.

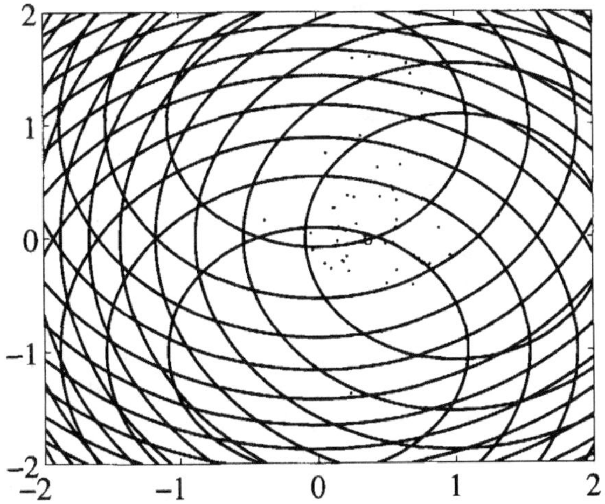

Figure 18.3: Final locations of 40 particles after 5 iterations. The optimal point is at $(1/3, 0)$ marked with $\circ$.

However, there is an important issue here. The combined weighted sum transforms the optimization problem into a single objective, this is not necessarily equivalent to the original multiobjective problem because the extra weighting coefficients could be arbitrary, whilst the final solutions still depend on these coefficients. Furthermore, there are so many ways to construct the weighted sum function and there is not any easy guideline to choose which form is the best for a given problem. When there is no rule to follow, the simplest choice obviously is to use the linear form. But there is no reason why the weighted sum should be linear. In fact, we can use other combinations such as the following quadratic weighted sum

$$\Pi(\boldsymbol{x}) = \sum_{i=1}^{N} \alpha_i f_i^2(\boldsymbol{x}) = \alpha_1 f_1^2(\boldsymbol{x}) + ... + \alpha_p f_p^2(\boldsymbol{x}), \tag{18.19}$$

and the others.

Another important issue is that how to choose the weighting coefficients as the solutions depend on these coefficients. The choice of weighting coefficients is essentially to assign a preference order by the decision maker to the multiobjectives. This leads to a more general concept of utility function (or preference function) which reflects the preference of the decision maker(s).

18.3 UTILITY METHOD

The weighted sum method is essentially a deterministic value method if we consider the weighting coefficients as the ranking coefficients. This implicitly assumes that the consequence of each ranking alternative can be characterized with certainty. This method can be used to explore the implications of alternative value judgement. Utility method, on the other hand, considers uncertainty in the criteria values for each alternative, which is a more realistic method because there is always some degree of uncertainty about the outcome of a particular alternative.

Utility (or preference) function can be associated with risk attitude or preference. For example, if you are offered a choice between a guaranteed £500 and a 50/50 chance of zero and £1000. How much are you willing to pay to take the gamble? The expected payoff of each choice is £500 and thus it is fair to pay $0.5 \times 1000 + (1 - 0.5) \times 0 = £500$ for such a gamble. A risk-seeking decision maker would risk a lower payoff in order to have a chance to win a higher prize, while a risk-averse decision maker would be happy with the safe choice of £500.

For a risk-neutral decision maker, the choice is indifferent between a guaranteed £500 and the 50/50 gamble since both choices have the same expected value of £500. In reality, the risk preference can vary from person to person and may depend on the type of problem. The utility function can have many forms, and one of the simplest is the exponential utility (of representing preference)

$$u(x) = \frac{1 - e^{-(x - x_a)/\rho}}{1 - e^{-(x_b - x_a)/\rho}}, \tag{18.20}$$

where x_a and x_b are the lowest and highest level of x, and ρ is called the risk tolerance of the decision maker.

The utility function defines combinations of objective values $f_1, ..., f_p$ which a decision maker finds equally acceptable or indifference. So the contours of the constant utility are referred to as the indifference curves. The optimization now becomes the maximization of the utility. For a maximization problem with two objectives f_1 and f_2, the idea of the utility contours (indifference curves), Pareto front and the Pareto solution with maximum utility (point A) are shown in Figure 18.4. When the utility function touches the Pareto front in the feasible region, it then provides a maximum utility Pareto solution (marked with A).

For two objectives f_1 and f_2, the utility function can be constructed in different ways. For example, the combined product takes the following form

$$U(f_1, f_2) = k f_1^{\alpha} f_2^{\beta}, \tag{18.21}$$

where α and β are non-negative exponents and k a scaling factor. The aggregated utility function is given by

$$U(f_1, f_2) = \alpha f_1 + \beta f_2 + [1 - (\alpha + \beta)] f_1 f_2. \tag{18.22}$$

There are many other forms. The aim of utility function constructed by the decision maker is to form a mapping $U : \Re^p \mapsto \Re$ so that the total utility function has a monotonic and/or convexity properties for easy analysis. It

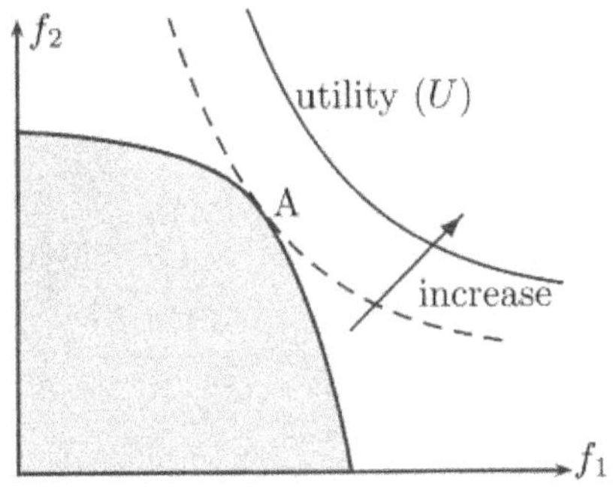

Figure 18.4: Finding the Pareto solution with maximum utility in a maximization problem with two objectives.

will also improve the quality of the Pareto solution(s) with maximum utility. Let us look at a simple example.

■ **EXAMPLE 18.3**

We now try to solve the simple two-objective optimization problem:

$$\underset{(x,y)\in\Re^2}{\text{maximize}} f_1(x, y) = x + y, \qquad f_2(x, y) = x,$$

subject to

$$x + \alpha y \leq 5, \qquad x \geq 0, \; y \geq 0,$$

where $0 < \alpha < 1$. Let us use the simple utility function

$$U = f_1 f_2,$$

which combines the two objectives. The line connecting the two corner points $(5, 0)$ and $(0, 5/\alpha)$ forms the Pareto front (see Figure 18.5). It is easy to check that the Pareto solution with maximum utility is $U = 25$ at $A(5, 0)$ when the utility contours touch the Pareto front with the maximum possible utility.

The complexity of multiobjective optimization makes the construction of the utility function a difficult task as it can be constructed in many ways.

18.4 METAHEURISTIC SEARCH

So far, we have seen that finding solutions of multiobjective optimization is usually difficult, even by the simple weighted sum method and utility function. However, there are other promising methods that work well for multiobjective optimization problems, especially the metaheuristic methods such as simulated annealing and particle swarm optimization.

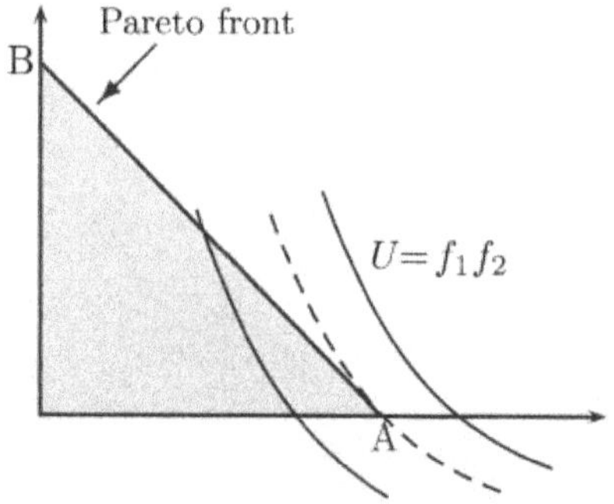

Figure 18.5: Pareto front is the line connecting A(5, 0) and B(0, 5/α). The Pareto solution with maximum utility is $U_* = 25$ at point A.

In 1985, Schaffer was probably the first to use vector evaluated genetic algorithms (VEGA) to solve multiobjective optimization without using any composite aggregation by combining all objectives into a single objective. Since then, many metaheuristic algorithms such as PSO, simulated annealing and ant algorithms have been extended to solve multiobjective optimization problems successfully. Interested readers can refer to the list of more advanced literature at the end of this chapter.

18.5 OTHER ALGORITHMS

There are other powerful algorithms that we have not addressed in this book. Some are widely used such as the Tabu search and neural networks, and others are very specialized but gaining momentum, including photosynthetic algorithm, enzyme algorithm, cuckoo search, bat algorithm, bacteria foraging algorithm, and immune-system based algorithms. As it is not possible to include all the algorithm in an introductory book on metaheuristics, the omission of these algorithms does not mean they are not popular. For example, Tabu search, developed by Fred Glover in the 1970s, is one of most powerful algorithms for optimization. In essence, it uses the memory or search history in terms of Tabu lists so as to record the search moves with the intention to avoid recently visited region or routes and encourage search for optimal solutions in a more efficient way. In fact, more and more metaheuristic algorithms using history and selection are emerging and becoming powerful in various applications.

In addition, we have not covered some of the well-established optimization including random-restart hill climbing, dynamic programming, stochastic optimization, evolution strategy, genetic programming, and many other evolutionary algorithms. Readers interested in these modern techniques can refer to more advanced literature.

The optimization algorithms we have discussed in this book are mainly for the optimization problems with explicit objective functions. However, in reality, it is often difficult to quantify what we want to achieve, but we still try to optimize certain things such as the degree of enjoyment of a quality service. In other cases, it is not possible and/or there is no explicit form of objective function. Whatever the objectives are, we have to evaluate the objectives many times. In most cases, the evaluations of the objective functions consume a lot of computational power (which costs money) and design time. Any efficient algorithm that can reduce the number of objective evaluations will save both time and money. Although, we have mainly focused on the optimization algorithms for objectives which are explicitly known, however, these algorithms will still be applicable to the cases where the objectives are not known explicitly. In most cases, certain modifications are required to suit a particular application. This is an exciting area of active research, and more publications are emerging each year.

An important question is what algorithm to choose for a given problem? This is a difficult question, and we have to try to address it throughout the book. This depends on the many factors such as the type of problem, the solution quality, available computing resource, time limit (before which a problem must be solved), balance of advantages and disadvantages of each algorithm (another optimization problem!), and the expertise of the decision-makers. When we study the algorithms, the efficiency and advantages as well as their disadvantages, to a large extent, essentially determine the type of problem they can solve and their potential applications. In general, for analytical function optimization problems, the nature-inspired algorithms should not be the first choice if analytical methods work well. If the function is simple, we can use the stationary conditions (first derivatives must be zero) and extreme points (boundaries) to find the optimal solution(s). For most optimization problems, this is not possible; then efficient conventional methods such as linear/nonlinear programming, convex optimization, and calculus-based algorithms such as the steepest descent method should be tried. If this again fails, we can try more established evolutionary algorithms such as genetic algorithms and simulated annealing to tackle the problem. If these two options do not provide any satisfactory solutions, then we may try more exotic, nature-inspired metaheuristic algorithms such as particle swarm optimization, ant/bee algorithms, and firefly algorithms.

Another interesting, often asked, question is "To be inspired by nature or not be inspired" when facing new challenging tasks. Nature has evolved over billions of years, she has found almost perfect solutions to almost every problem she has met. Almost all the not-so-good solutions have been discarded via natural selection. The optimal solutions seem (often after a huge number of generations) to appear at the evolutionarily stable equilibrium, even though we may not understand how the perfect solutions are reached. When we try to solve human problems, why not try to be inspired by the nature's great success? The simple answer to the question 'To be inspired or not to be inspired?' is 'why not?'. If we do not have good solutions at hand, it is always a good idea to learn from nature.

EXERCISES

18.1 Modify or extend the simulated annealing for a single objective to multiobjective optimization for m objectives $f_1(\boldsymbol{x}), ..., f_m(\boldsymbol{x})$ without any constraints.

18.2 The method of least-squares works well for most data-fitting problem; however, for polynomials

$$y = f(x) = \alpha_0 + \alpha_1 x + \alpha_2 x^2 + ... + \alpha_p x^p,$$

it tends to result in high-order polynomials to best-fit the given data set (x_i, y_i) where $i = 1, 2, ..., n$, even though the higher-order polynomials are not necessarily the best model. A way to improve this is to use some penalty in coefficients so as to balance the goodness of the fit and the choice of reasonably simple model. Design an optimization problem to achieve this by limiting the sum of the coefficients α_k where $k = 0, 1, ..., p$.

18.3 For multiobjective optimization, the understanding of the Pareto front is very important because the Pareto front can be easily converted to the optimal solutions when needed. Analyze the following optimization problem with two objectives

$$\text{minimize} \quad (f_1(x_1), f_2(\boldsymbol{x})),$$

where

$$\boldsymbol{x} = (x_1, ..., x_n), \qquad n = 30, \qquad x_i \in [0, 1],$$

and

$$f_1(x_1) = x_1, \;\; f_2(\boldsymbol{x}) = g(x_2, ..., x_n) h(f_1(x_1), g(x_2, ..., x_n)),$$

with

$$g(x_2, ..., x_n) = 1 + 9 \sum_{i=2}^{n} \frac{x_i}{(n-1)}, \qquad h(f_1, g) = 1 - \sqrt{f_1/g}.$$

The Pareto optimal front is formed with $g(\boldsymbol{x}) = 1$. Thus, sketch this front.

18.4 Choose a suitable way for formulating a composite objective to solve the following multiple multiobjective optimization designed by the author

$$\text{minimize} \quad \boldsymbol{f}(\boldsymbol{x}) = (f_1(\boldsymbol{x}), ..., f_p(\boldsymbol{x}),$$

$$\text{subject to} \quad \sum_{i=1}^{n} x_i^2 \leq n^2,$$

where

$$f_i(\boldsymbol{x}) = x_i^2 + |x_{i+1}| \;\; (i = 1, 2, ..., p-1), \qquad f_p(\boldsymbol{x}) = x_p^2 + |x_1|,$$

and

$$-n \leq x_i \leq n.$$

If you use the method of the weighted sum, does the optimal solution depend on the weighting coefficients?

18.5 An interesting way of dealing with multiobjective optimization is to write objectives except one as constraints. Try to rewrite the following unconstrained optimization as a single objective constrained optimization problem

$$\text{minimize } f_1(\boldsymbol{x}), f_2(\boldsymbol{x}), ..., f_p(\boldsymbol{x}).$$

18.6 The PSO we discussed in this book is for optimization with a single objective. Modify the standard PSO in a suitable way to solve multiobjective optimization.

REFERENCES

1. C. M. Fonseca and P. J. Fleming, "An overview of evolutionary algorithms in multiobjective optimization", *Evolutionary Computation*, **3** (1), 1-16 (1995).
2. F. Glover and M. Laguna. (1997). *Tabu Search*. Kluwer, Norwell, MA, 1997.
3. F. Glover, "Tabu Search: Part I", *ORSA Journal on Computing*, **1**(3), 190-206 (1989).
4. M. Laumanns, G. Rudolph and H.-P. Schwefel, "A spatial predator-prey approach to multi-objective optimization", in: *Proceedings of Parallel Problem Solving from Nature* (Eds A. E. Eiben, T. Bäck, M. Schoenauer, and H.-P. Schwefel), Springer, Berlin, 1998.
5. M. Mitchell, *An Introduction to Genetic Algorithms*, Cambridge, Mass: MIT Press, 1996.
6. H. Murase, "Finite element analysis using a photosynthetic algorithm", *Computers and Electronics in Agriculture*, **29**, 115-123 (2000).
7. V. Pareto, *Manuale di Economica Politica*, Macmillan, London, 1972.
8. Sawaragi Y., Nakayama H., Tanino T., *Theory of Multiobjective Optimisation*, Academic Press, 1985.
9. D.Schaffer, "Multiple objective optimization with vector evaluated genetic algorithms", in: *Proc. Int. Conf. on Genetic Algorithms and their Applications* (Eds J. Grefenstette), pp. 93-100, (1985).
10. N. Srinivas and K. Deb, "Multipleobjective optimization using nondominated sorting in genetic algorithms", *Evolutionary Computation*, **2** (3), 221-248 (1994).
11. E.-G. Talbi, *Parallel Combinatorial Optimization*, John Wiley & Sons, 2006.
12. E.-G. Talbi, *Metaheuristics: From Design to Implementation*, John Wiley & Sons, 2009.

13. D. H. Wolpert and W. G. Macready, "No free lunch theorems for optimization", *IEEE Transaction on Evolutionary Computation*, **1**, 67-82 (1997).

14. X. S. Yang, "New enzyme algorithm, Tikhonov regularization and inverse parabolic analysis", in: *Advances in Computational Methods in Science and Engineering*, Lecture Series on Computer and Computer Sciences, ICCMSE 2005, Eds T. Simos and G. Maroulis, **4**, 1880-1883 (2005).

15. E. Zitzler, K. Deb, and L. Thiele, "Comparison of multiobjective evoluationary algorithms: empirical results", *Evolutionary Computation*, **8** (2000) (2), 173-195.

16. E. Zitzler, M. Laumanns, and S. Bleuler, "A tutorial on evolutionary multiobjective optimization", in: *Metaheuristics for Multiobjective Optimisation* (Eds X. Gandibleux et al.), Springer, Lecture Notes in Economics and Mathematical Systems, **535**, pp. 3-37 (2004).

CHAPTER 19

ENGINEERING APPLICATIONS

The applications of optimization in engineering are as diverse as the optimization itself. Almost all areas in engineering can use optimization for problem solving. Topics includes pressure vessel design, shape optimization, structural design optimization, building design, energy-efficient design, heat management of electronics, planning, scheduling, and many others. The vast literature and diverse topics make it impossible to include even a decent subset of these applications. In this chapter, we will provide a few examples as a sample, which may provide some insight into the types of problems we are solving in engineering optimization.

19.1 SPRING DESIGN

A very simple design problem in engineering is to design a spring under tension and/or compression for given requirements or constraints including minimum deflection, outer diameter, frequency, and maximum shear stress. For a given problem such as forming a spring using a wire, the adjustable parameters or design variables (see Figure 19.1) are the wire diameter w (thickness), the coil diameter d, and the length L (or equivalently, the number of active coils). For

By Xin-She Yang
Copyright © 2010 John Wiley & Sons, Inc.

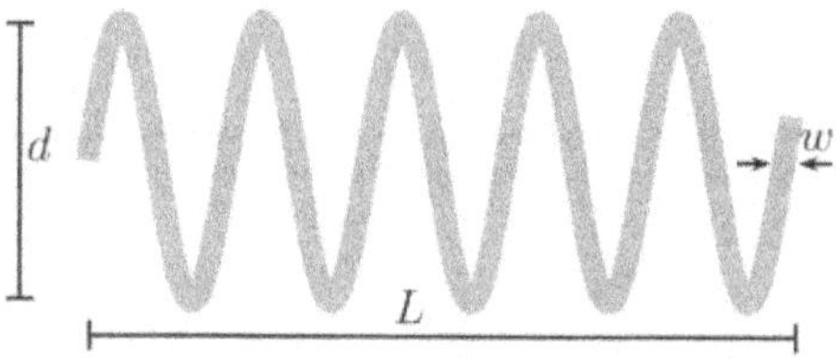

Figure 19.1: The design optimization of a simple spring.

a detailed description, please refer to the work by Arora (1989) and Belegundu (1982).

This design optimization problem can be formulated as follows:

$$\text{minimize} \quad f(\boldsymbol{x}) = (2 + L)dw^2, \tag{19.1}$$

$$\text{subject to} \quad \begin{aligned} g_1(\boldsymbol{x}) &= 1 - \tfrac{d^2 L}{7178w^4} \le 0 \\ g_2(\boldsymbol{x}) &= \tfrac{4d^2 - wd}{12566dw^3 - w^4} + \tfrac{1}{5108w^2} - 1 \le 0 \\ g_3(\boldsymbol{x}) &= 1 - \tfrac{140.45w}{d^2 L} \le 0 \\ g_4(\boldsymbol{x}) &= \tfrac{w+L}{1.5} - 1 \le 0 \end{aligned} \tag{19.2}$$

The simple bounds are

$$0.05 \le w \le 2.0, \qquad 0.25 \le d \le 1.3, \qquad 2.0 \le L \le 15.0. \tag{19.3}$$

If we use the program in Appendix B, we can find the following best solution

$$f_* \approx 0.0075, \tag{19.4}$$

at

$$\boldsymbol{x}_* \approx (0.0500,\ 0.2500,\ 9.9877). \tag{19.5}$$

19.2 PRESSURE VESSEL

Pressure vessels are literarily everywhere such as champagne bottle, bottles of sparkling drink, and gas tanks. For a given volume and working pressure, the basic aim of designing a cylindrical vessel is to minimize the total cost. Typically, the design variables are the thickness d_1 of the head, the thickness d_2 of the body, the inner radius r, and the length L of the cylindrical section (see Figure 19.2).

This is a well-known test problem for optimization (e.g., see Cagnina *et al.* 2008) and it can be written as

$$\text{minimize} \quad f(\boldsymbol{x}) = 0.6224 d_1 r L + 1.7781 d_2 r^2 + 3.1661 d_1^2 L + 19.84 d_1^2 r, \tag{19.6}$$

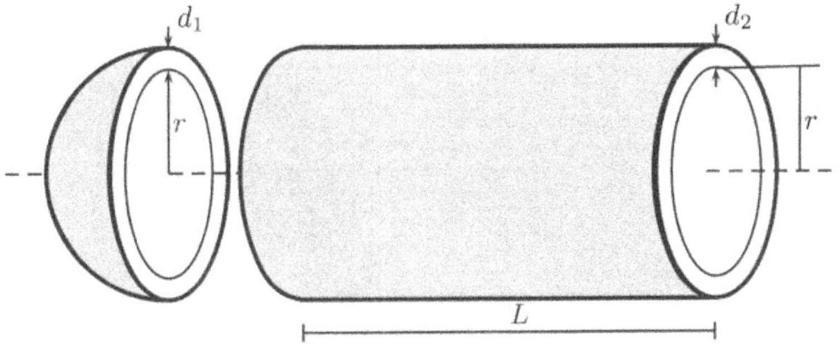

Figure 19.2: Pressure vessel design and optimization.

$$
\text{subject to} \quad \begin{aligned} g_1(\boldsymbol{x}) &= -d_1 + 0.0193r \le 0 \\ g_2(\boldsymbol{x}) &= -d_2 + 0.00954r \le 0 \\ g_3(\boldsymbol{x}) &= -\pi r^2 L - \tfrac{4\pi}{3} r^3 + 1296000 \le 0 \\ g_4(\boldsymbol{x}) &= L - 240 \le 0. \end{aligned} \tag{19.7}
$$

The simple bounds are

$$0.0625 \le d_1, d_2 \le 99 \times 0.0625, \tag{19.8}$$

and

$$10.0 \le r, \qquad L \le 200.0. \tag{19.9}$$

It is worth pointing out that d_1 and d_2 should only take discrete values of integer multiples of 0.0625.

In principle, we can use many algorithms introduced in this book to solve it. For example, a simple Matlab code is attached in Appendix B, which will find the best solution

$$f_* = 6059.714, \tag{19.10}$$

at

$$\boldsymbol{x}_* = (0.8125,\ 0.4375,\ 42.0984,\ 176.6366). \tag{19.11}$$

19.3 SHAPE OPTIMIZATION

Shape optimization is very common in engineering optimization, including design of aerodynamically efficient shape, structural design with minimum materials, and design of special tools. Shape optimization is also called topology optimization as the topology of the structure is changing and evolving with given material properties and design objective and constraints. The literature of topological optimization is vast, and its topics are diverse. Here we can only provide an example of structural design under static loading. For detail, please refer to the excellent tutorial by Sigmund (2001).

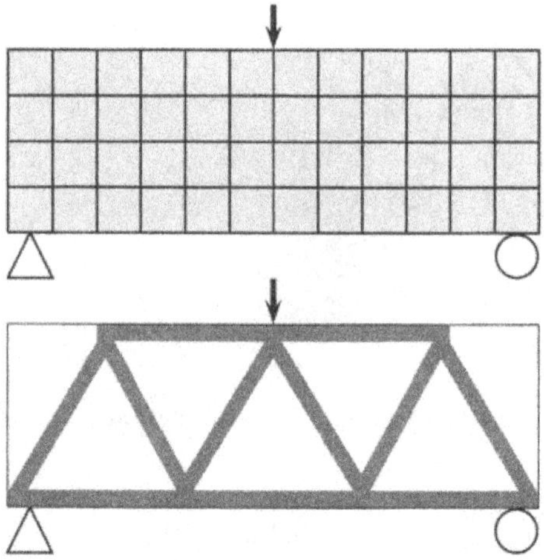

Figure 19.3: The rectangular design domain is divided into N elements. As optimization and material distribution evolve, the shape becomes a truss-style structure (bottom).

The aim is to minimize the overall compliance which is the displacement per unit force (see Figure 19.3). Therefore, we have

$$\text{minimize} \quad f(\boldsymbol{\rho}) = \boldsymbol{U}^T \boldsymbol{K} \boldsymbol{U} = \sum_{e=1}^{N} (\rho_e)^{\beta} \boldsymbol{u}_e^T \boldsymbol{K}_e \boldsymbol{u}_e, \tag{19.12}$$

$$\text{subject to} \quad \frac{V(\boldsymbol{\rho})}{V_0} = \alpha$$

$$\boldsymbol{K}\boldsymbol{U} = \boldsymbol{F} \tag{19.13}$$

$$0 \prec \boldsymbol{\rho}_{\min} \preceq \boldsymbol{\rho} \preceq 1$$

where $\boldsymbol{U} = (u_1, ..., u_{2N})$ are the global displacement vector, while $\boldsymbol{F} = (f_1, ..., F_N)^T$ is the force vector. The parameter $\beta > 0$ is a constant. Here $e = 1, ..., N$ corresponds to the index of finite elements with N elements in total. The design variable vector is the relative densities of each element $\boldsymbol{\rho} = (\rho_1, ..., \rho_N)^T$. The maximum relative density is 1, while the minimum value should be non-negative to avoid any potential singularity. For example, we can use $\boldsymbol{\rho}_{\min} = 0.01$.

The volume fraction ϕ is defined as the ratio of the material volume $V(\boldsymbol{\rho})$ to the design domain volume V_0. If a rectangular domain Ω is divided into N equal quadrilateral elements, each element has the same constant volume

ΔV_e. The material volume can be estimated by

$$V(\boldsymbol{\rho}) = \int_\Omega \rho dV \approx w \sum_{e=1}^{N} \rho_e \Delta V_e, \tag{19.14}$$

where w is the thickness of the domain. The total design domain volume is

$$V_0 = \int_\Omega dV = w \sum_{e=1}^{N} \Delta V_e. \tag{19.15}$$

Since ΔV_e is constant, we have

$$\phi = \frac{w \sum_{e=1}^{N} \rho_e \Delta V_e}{w \sum_{e=1}^{N}} = \frac{1}{N} \sum_{e=1}^{N} \rho_e. \tag{19.16}$$

For a given amount material for such a design problem, we can equivalently impose a fixed ratio $0 < \alpha < 1$. That is

$$\phi = \frac{V(\boldsymbol{\rho})}{V_0} = \alpha. \tag{19.17}$$

In addition, $\boldsymbol{u}_e = (u_{e1}, u_{e2})$ is the element displacement vector which has two degrees of freedom (horizontal and vertical displacements). Its corresponding element stiffness matrix $\boldsymbol{K}_e$ is given by

$$\boldsymbol{K}_e = \frac{E}{(1-\nu^2)} \begin{pmatrix} \boldsymbol{A} & \boldsymbol{B} \\ \boldsymbol{B} & \boldsymbol{A} \end{pmatrix}, \tag{19.18}$$

where

$$\boldsymbol{A} = \begin{pmatrix} \frac{1}{6}(3-\nu) & \frac{1}{8}(1+\nu) & -\frac{1}{12}(3+\nu) & -\frac{1}{8}(1-3\nu) \\ \frac{1}{8}(1+\nu) & \frac{1}{6}(3-\nu) & \frac{1}{8}(1-3\nu) & \frac{\nu}{6} \\ -\frac{1}{12}(3+\nu) & \frac{1}{8}(1-3\nu) & \frac{1}{6}(3-\nu) & -\frac{1}{8}(1+\nu) \\ -\frac{1}{8}(1-3\nu) & \frac{\nu}{6} & -\frac{1}{8}(1+\nu) & \frac{1}{6}(3-\nu) \end{pmatrix}, \tag{19.19}$$

$$\boldsymbol{B} = \begin{pmatrix} -\frac{1}{12}(3-\nu) & -\frac{1}{8}(1+\nu) & \frac{\nu}{6} & \frac{1}{8}(1-3\nu) \\ -\frac{1}{8}(1+\nu) & -\frac{1}{12}(3-\nu) & -\frac{1}{8}(1-3\nu) & -\frac{1}{12}(3+\nu) \\ \frac{\nu}{6} & -\frac{1}{8}(1-3\nu) & -\frac{1}{12}(3-\nu) & \frac{1}{8}(1+\nu) \\ \frac{1}{8}(1-3\nu) & -\frac{1}{12}(3+\nu) & \frac{1}{8}(1+\nu) & -\frac{1}{12}(3-\nu) \end{pmatrix}. \tag{19.20}$$

It is easy to see that all three matrices are symmetric. That is

$$\boldsymbol{K}_e^T = \boldsymbol{K}_e, \qquad \boldsymbol{A}^T = \boldsymbol{A}, \qquad \boldsymbol{B}^T = \boldsymbol{B}. \tag{19.21}$$

We can also see that the stiffness matrix $\boldsymbol{K}_e$ for all elements is the same constant matrix $\boldsymbol{K}_0 = \boldsymbol{K}_e$. The gradient or sensitivity of the objective can be determined by

$$\frac{\partial f}{\partial \rho_e} = -\beta \rho_e^{\beta-1} \boldsymbol{u}_e^T \boldsymbol{K}_0 \boldsymbol{u}_e. \tag{19.22}$$

An excellent Matlab code for this example was developed by Sigmund.[2] Under a static unit load, a rectangular design with initial uniform materials property (see the top part in Figure 19.3) will evolve to the truss-style structure at the bottom as shown in the same figure.

Obviously, this method and the computer program can be extended to multiple loads under various constraints. Interested readers can refer to more advanced literature.

19.4 OPTIMIZATION OF EIGENVALUES AND FREQUENCIES

The design of many engineering structures requires the consideration of frequency design. In general, frequencies of a structure are eigenvalues of a corresponding dynamic system, so the intrinsic frequencies of a structure is often called its eigenfrequencies.

In order to increase the stiffness of structures, we often have to maximize their first eigenfrequency because a higher first frequency means the structure is relatively stiffer for all type loads, both dynamic and static. Thus optimization of eigenfrequencies forms a class of engineering optimization, called dynamic topological optimization. On the other hand, for other purposes, we may wish to limit the highest eigenfrequency, and this becomes an eigenvalue minimization problem.

In general, free oscillations of a mechanical vibration system without damping are governed by the following second-order ordinary differential equation in the matrix form

$$\boldsymbol{M}\ddot{\boldsymbol{u}} + \boldsymbol{K}\boldsymbol{u} = 0, \tag{19.23}$$

where $\boldsymbol{M}$ is called the mass matrix, while $\boldsymbol{K}$ is the stiffness of the structure. $\boldsymbol{u} = (u_1, u_2, ..., u_N)$ is the displacement vector of the vibrations where N is the total degrees of freedom of the system.

Using the complex method for vibration analysis, we can write $\boldsymbol{u}$ as

$$\boldsymbol{u} = \boldsymbol{U} \exp(i\omega t) = \boldsymbol{U}[\cos(\omega t) + i \sin(\omega t)], \tag{19.24}$$

[2] O. Sigmund, "A 99 line topology optimization code written in Matlab", *Struct. Multidisc. Optim.*, **21**, 120-127 (2001). http://www.topopt.dtu.dk/files/matlab.pdf

where ω is the eigenfrequency, and $\boldsymbol{U}$ are the amplitude of the vibration. To extract the solution, we only have to consider the real part of the above expression. Substituting the above expression into (19.23), we have

$$[\boldsymbol{K} - \omega^2 \boldsymbol{M}]\boldsymbol{U} = 0, \tag{19.25}$$

which is an eigenvalue problem whose eigenvalues are the natural frequencies or eigenfrequencies. The system may have multiple eigenfrequencies; the corresponding eigenvector of the jth eigenfrequency ω_j is the jth amplitude $\boldsymbol{U}_j$.

Since we are more concerned with the highest or lowest frequency ω_1, the eigenvalue optimization can now be written as

$$\text{minimize} \quad \omega_1^2 = \frac{\boldsymbol{u}_1^T \boldsymbol{K} \boldsymbol{u}_1}{\boldsymbol{u}_1^T \boldsymbol{M} \boldsymbol{u}_1}, \tag{19.26}$$

$$\text{subject to} \quad [\boldsymbol{K} - \omega_1^2 \boldsymbol{M}]\boldsymbol{u}_1 = 0, \tag{19.27}$$

where we often impose a frequency range $\omega_1 \in [\omega_{\min}, \omega_{\max}]$. Here $\omega_{\min}$ and $\omega_{\max}$ are the minimum and maximum allowed frequencies, respectively, for a design system. Obviously, we have to introduce other constraints such as the amount of material used and geometrical dimensions. Let us demonstrate how to approach such kind of problem through an example.

■ EXAMPLE 19.1

For a relative simple system, we now use three mass blocks attached in by two springs on frictionless surface as shown in Figure 19.4. The masses of the three blocks are m_1, m_2 and m_3, respectively. The two spring constants are k_1 and k_2, respectively. The objective is to choose the mass and spring variables so that the highest eigenfrequency of the system is minimum under a fixed total mass. That is, we have

$$\underset{m_1, m_2, m_3, k}{\text{minimize}} \quad \omega = \max\{w_j\}, \qquad (j = 1, ..., n), \tag{19.28}$$

$$\text{subject to} \quad m_1 + m_2 + m_3 = S, \quad m_1 = m_3, \tag{19.29}$$

$$k_1 = k_2 \geq k_0. \tag{19.30}$$

Here n is the number of eigenfrequencies, and k_0 and S are constants. This is a minimax problem, which is very difficult to solve in general.

Let u_1, u_2, u_3 be the displacement of the three mass blocks m_1, m_2, m_3, respectively. Then, their accelerations will be $\ddot{u}_1$, $\ddot{u}_2$, $\ddot{u}_3$ where $\ddot{u} = d^2u/dt^2$. From the balance of forces and Newton's law, we have

$$m_1 \ddot{u}_1 = k_1(u_2 - u_1), \tag{19.31}$$

$$m_2 \ddot{u}_2 = k_2(u_3 - u_2) - k_1(u_2 - u_1), \tag{19.32}$$

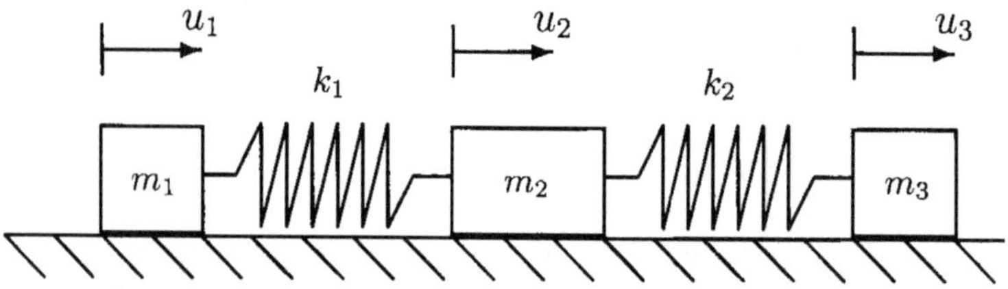

Figure 19.4: Harmonic vibrations.

$$m_3 \ddot{u}_3 = -k_2(u_3 - u_2). \tag{19.33}$$

These equations can be written in a matrix form as

$$\begin{pmatrix} m_1 & 0 & 0 \\ 0 & m_2 & 0 \\ 0 & 0 & m_3 \end{pmatrix} \begin{pmatrix} \ddot{u}_1 \\ \ddot{u}_2 \\ \ddot{u}_3 \end{pmatrix} + \begin{pmatrix} k_1 & -k_1 & 0 \\ -k_1 & k_1 + k_2 & -k_2 \\ 0 & -k_2 & k_2 \end{pmatrix} \begin{pmatrix} u_1 \\ u_2 \\ u_3 \end{pmatrix} = \begin{pmatrix} 0 \\ 0 \\ 0 \end{pmatrix}, \tag{19.34}$$

or

$$\mathbf{M\ddot{u}} + \mathbf{Ku} = \mathbf{0}, \tag{19.35}$$

where $\mathbf{u} = (u_1, u_2, u_3)^T$. The mass matrix is

$$\boldsymbol{M} = \begin{pmatrix} m_1 & 0 & 0 \\ 0 & m_2 & 0 \\ 0 & 0 & m_3 \end{pmatrix}, \tag{19.36}$$

and the stiffness matrix is

$$\boldsymbol{K} = \begin{pmatrix} k_1 & -k_1 & 0 \\ -k_1 & k_1 + k_2 & -k_2 \\ 0 & -k_2 & k_2 \end{pmatrix}. \tag{19.37}$$

From the equality constraints, we know $m_1 = m_3$ and $k_1 = k_2$. So for simplicity, let us use the notation $m_1 = m_2 = m$ and $k_1 = k_2 = k$. We now have

$$\boldsymbol{M} = \begin{pmatrix} m & 0 & 0 \\ 0 & m_2 & 0 \\ 0 & 0 & m \end{pmatrix}, \tag{19.38}$$

and

$$\boldsymbol{K} = \begin{pmatrix} k & -k & 0 \\ -k & 2k & -k \\ 0 & -k & k \end{pmatrix}. \tag{19.39}$$

Equation (19.35) is a second-order ordinary differential equation in terms of matrices. This homogeneous equation can be solved by substituting $u_i = U_i \cos(\omega t)$ where $U_i (i = 1, 2, 3)$ are constants and ω^2 can have several values which correspond to the natural frequencies. Now we have

$$-\omega_i^2 \boldsymbol{M} U_i \cos(\omega t) + \boldsymbol{K} U_i \cos(\omega t) = 0, \tag{19.40}$$

where $i = 1, 2, 3$. Dividing both sides by $\cos(\omega t)$, we have

$$[\boldsymbol{K} - \omega^2 \boldsymbol{M}] U_i = 0, \tag{19.41}$$

which becomes an eigenvalue problem. Any non-trivial solutions for U_i require

$$|\boldsymbol{K} - \omega^2 \boldsymbol{M}| = 0. \tag{19.42}$$

Therefore, the eigenvalues of this equation give the natural eigenfrequencies. Now we have

$$\begin{vmatrix} k - \omega^2 m & -k & 0 \\ -k & 2k - \omega^2 m_2 & -k \\ 0 & -k & k - \omega^2 m \end{vmatrix} = 0,$$

which leads to a cubic equation in terms of ω^2

$$\omega^2(-k + \omega^2 m)(2km + km_2 - mm_2\omega^2) = 0. \tag{19.43}$$

This equation has three solutions. Therefore, the three natural frequencies are

$$\omega_1^2 = 0, \qquad \omega_2^2 = \frac{k}{m}, \qquad \omega_3^2 = \frac{k(2m + m_2)}{mm_2}.$$

The case of $\omega_1^2 = 0$ corresponds the rigid body motion.

In addition, we know that m_2 must be non-negative, we have

$$\omega_3^2 = \frac{k}{m}(\frac{2m}{m_2} + 1) \geq \frac{k}{m} = \omega_2^2. \tag{19.44}$$

Therefore, the highest eigenfrequency is ω_3. Noticing that $m_1 + m_2 + m_3 = 2m + m_2 = S$ is constant, or $m_2 = S - 2m$, we have

$$\omega_3^2 = \frac{2kS}{m(S - 2m)}. \tag{19.45}$$

In order to minimize this frequency, we have to minimize k and maximize the denominator $D = m(S - 2m)$. However, we have another constraint $k \geq k_0$, so the minimum of k is k_0.

To minimize D, we have

$$\frac{dD}{dm} = S - 4m = 0. \tag{19.46}$$

That is

$$m = S/4, \tag{19.47}$$

which gives $m_2 = S - 2(S/4) = S/2 = 2m$. Now we have the minimum

$$\omega_{3,\min}^2 = \frac{4k_0}{S}. \tag{19.48}$$

The three eigenfrequencies are

$$\omega_1^2 = 0, \qquad \omega_2^2 = \frac{4k_0}{S} = \frac{k_0}{m}, \qquad \omega_3^2 = \frac{4k_0}{S} = \frac{2k_0}{m}. \tag{19.49}$$

So the final solutions are

$$m_1 = m_2 = m = \frac{S}{4}, \quad m_2 = 2m = \frac{S}{2}, \quad k_1 = k_2 = k_0. \tag{19.50}$$

It worth pointing out that a naive guess is to use $m_1 = m_2 = m_3 = S/3$, however, this is not the optimal solution because the three eigenfrequencies in this case will be

$$\omega_2^2 = 0, \ \omega_2^2 = \frac{3k_0}{S}, \ \omega_3^2 = \frac{9k_0}{S}. \tag{19.51}$$

It is clear that the highest frequency ω_3^2 is higher than $\omega_{3,\min}^2 = 8k_0/S$ obtained earlier.

19.5 INVERSE FINITE ELEMENT ANALYSIS

Most structural design and finite element analysis are forward modeling with the aim to predict displacement and stress for a given design, load configuration and material settings. However, sometimes inverse problems have to be solved in many applications such as the testing of new materials, design optimization, reverse engineering, and repairing/reinforcing exist structures. For example, for a given set of measurement or observed data such as displacement and load, or temperature distribution, we have to invert or estimate the materials properties such as Young's modulus E, Poisson's ratio ν, and/or thermal conductivities κ. Let us look at a simple example.

■ EXAMPLE 19.2

A rectangular beam with a unit thickness can be divided into N subregions as shown in Figure 19.5 where $N = 10$. We have 10 different pairs of E_i and ν_i. We apply a unit force $\boldsymbol{f} = 1$ at the end of the beam, and try to measure the displacements at the grid points. The aims is to estimate the material properties from the measured displacement.

A test case is to set

$$\boldsymbol{E} = (E_1, ..., E_{10}) = (100, 90, 80, 70, 60, 90, 80, 70, 60, 50) \ \text{GPa},$$

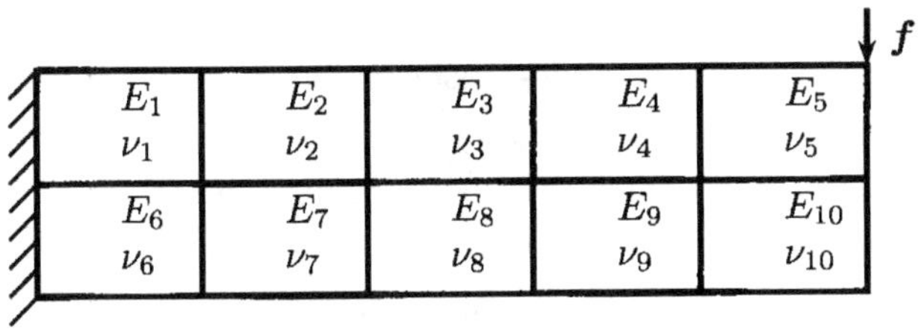

Figure 19.5: A rectangular beam with inhomogeneous materials properties (in 10 different cells).

$$\boldsymbol{\nu} = (\nu_1, ..., \nu_{10}) = (0.2, 0.2, 0.2, 0.2, 0.2, 0.3, 0.3, 0.3, 0.3, 0.3).$$

For a unit loads, we can calculate the displacement at the 18 nodes, under the condition that the three nodes on the left are fixed (zero displacements). For a set of observed values, we can then estimate $\boldsymbol{E}$ and $\boldsymbol{\nu}$ by using iterative forward modeling.

A generic type of such inverse modeling is called the Inverse Initial-Value, Boundary-Value (IVBV) problem.

■ EXAMPLE 19.3

This example was based on the IVBV optimization by Karr *et al.* (2000). On a square plate of unit dimensions, the diffusivity $\kappa(x, y)$ varies with locations (x, y). The initial temperature of the plate is 1, while four sides are held at a constant temperature $u = 0$. The aim is to try to estimate the diffusivity from observed/measured temperatures $u(x, y, t)$.

The heat conduction equation can be written as

$$\frac{\partial u}{\partial t} = \nabla \cdot [\kappa(x, y) \nabla u], \qquad 0 \leq x, y \leq 1,$$

whose appropriate boundary conditions are

$$u(x, y, 0) = 1, \qquad u(x, 0, t) = u(x, 1, t) = u(0, y, t) = u(1, y, t) = 0.$$

The domain is discretized as an $N \times N$ grid, and the measurements are taken at the grid points (x_i, y_j) where $i, j = 1, 2, ..., N$) at three different times t_n with $n = 1, 2, 3$. This means that the data set consists of measured values at N^2 points at three times t_1, t_2, t_3. The objective is now to invert or estimate the N^2 diffusivity values at the N^2 distinct locations by iterative forward modelling using finite element methods (or finite difference methods when appropriate) in combination with an optimization algorithm such as genetic algorithms.

At each iteration, a random $\kappa(i, j)$ matrix is generated, which is then used for forward heat conduction modeling to compute the temperature distribution at all grid points at different times. Good estimates should minimize the error metrics

$$\epsilon_u = A \frac{\sum_{i=1}^{N} \sum_{j=1}^{N} |u_{i,j}^{\text{measured}} - u_{i,j}^{\text{computed}}|}{\sum_{i=1}^{N} \sum_{j=1}^{N} |u_{i,j}^{\text{measured}}|}, \tag{19.52}$$

where $u_{i,j}^{\text{measured}}$ is the measured value at point (x_i, y_j) at a given time, and $u_{i,j}^{\text{computed}}$ is the predicted or computed temperature at the same time t_n. Here A is a scaling constant and can be taken as $A = 100$ for convenience. If there is a set of known diffusivity values for comparison or validation, the optimal estimates should also minimize the relative error between the estimated and the known

$$\epsilon_\kappa = A \frac{\sum_{i=1}^{N} \sum_{j=1} |\kappa_{i,j}^{\text{known}} - \kappa_{i,j}^{\text{esimated}}|}{\sum_{i=1}^{N} \sum_{j=1}^{N} |\kappa_{i,j}^{\text{known}}|}. \tag{19.53}$$

For example, on a grid of 16×16 points, after about 40,000 random $\kappa(i, j)$ matrices were generated using genetic algorithms, the best estimates provide the error metric $\epsilon_u \approx 4.6050$ and $\epsilon_\kappa \approx 1.50 \times 10^{-2}$. The small value of ϵ_κ means the estimated diffusivities are very close to their true values.

EXERCISES

19.1 For the design of a heat exchanger, we often have to maximize the surface area. For a given maximum total cross section area A_0, we can use many, say n, cylindrical pipes of radius r. However, the minimum allowable radius is a. Find the optimal radius r so that the total surface area $A = 2\pi r n$ is maximum.

19.2 Design a lighting pole with a fixed height L such that its natural frequencies are as highest as possible, while its mass as low as possible. Such design will have good resistance to fatigue failure and low cost.

19.3 Calculate the maximum packing factor of identical spheres with a radius a in a large cubic of size L (where $L \gg a$). Here the packing factor can be defined as the ratio of the total volume of the spheres to the volume of the cubic.

REFERENCES

1. J. Arora, *Introduction to Optimum Design*, McGraw-Hill, 1989.
2. A. Belegundu, *A Study of Mathematical Programming Methods for Structural Optimization*, PhD thesis, University of Iowa, Iowa, 1982.
3. M. P. Bendsøe, *Optimization of Structural Topology, Shape and Material*, Springer, Berlin, 1995.

4. M. P. Bendsϕe, O. Sigmund, "Material interpolations in topology optimization", *Arch. Appl. Mech.*, **69**, 635-654 (1999).

5. L. C. Cagnina, S. C. Esquivel, C. A. Coello, "Solving engineering optimization problems with the simple constrained particle swarm optimizer", *Informatica*, **32**, 319-326 (2008).

6. C. A. Coello, "Use of a self-adaptive penalty approach for engineering optimization problems", *Computers in Industry*, **41**, 113-127 (2000).

7. C. L. Kar, I. Yakushin, K. Nicolosi, "Solving inverse initial-value, boundary-value problems via genetic algorithms", *Engineering Applications of Artificial Intelligence*, **13**, 625-633 (2000).

8. E. Sandgren, "Nonlinear integer and discrete programming in mechanical design optimization", *J. Mech. Des.-T. ASME*, **112** (2), 223-229 (1990).

9. O. Sigmund, "A 99 line topology optimization code written in Matlab", *Struct. Multidisc. Optim.*, **21**, 120-127 (2001).

10. X. S. Yang, "Biology-derived algorithms in engineering optimization" (Chapter 32), in: *Handbook of Bioinspired Algorithms and Applications* (Eds. S. Olariu and A. Y. Zomaya), Chapman Hall/CRC Press, (2005).

APPENDIX A

TEST PROBLEMS IN OPTIMIZATION

In order to validate any new optimization algorithm, we have to validate it against standard test functions so as to compare its performance with well-established or existing algorithms. There are many test functions, so there is no standard list or set of test functions one has to follow. However, various test functions do exist, so new algorithms should be tested using at least a subset of functions with diverse properties so as to make sure whether or not the tested algorithm can solve certain type of optimization efficiently.

In this appendix, we will provide a subset of commonly used test functions with simple bounds as constraints, though they are often listed as unconstrained problems in literature. We will list the function form $f(\boldsymbol{x})$, its search domain, optimal solutions $\boldsymbol{x}_*$ and/or optimal objective value f_*. Here, we use $\boldsymbol{x} = (x_1, ..., x_n)^T$ where n is the dimension.

Ackley's function:

$$f(\boldsymbol{x}) = -20 \exp\Big[-\frac{1}{5}\sqrt{\frac{1}{n}\sum_{i=1}^{n} x_i^2}\,\Big] - \exp\Big[\frac{1}{n}\sum_{i=1}^{n}\cos(2\pi x_i)\Big] + 20 + e, \quad \text{(A.1)}$$

Engineering Optimization: An Introduction with Metaheuristic Applications.
By Xin-She Yang
Copyright © 2010 John Wiley & Sons, Inc.

where $n = 1, 2, ..., $ and $-32.768 \leq x_i \leq 32.768$ for $i = 1, 2, ..., n$. This function has the global minimum $f_* = 0$ at $\boldsymbol{x}_* = (0, 0, ..., 0)$.

De Jong's functions: The simplest of De Jong's functions is the so-called sphere function

$$f(\boldsymbol{x}) = \sum_{i=1}^{n} x_i^2, \qquad -5.12 \leq x_i \leq 5.12, \tag{A.2}$$

whose global minimum is obviously $f_* = 0$ at $(0, 0, ..., 0)$. This function is unimodal and convex. A related function is the so-called weighted sphere function or hyper-ellipsoid function

$$f(\boldsymbol{x}) = \sum_{i=1}^{n} i x_i^2, \qquad -5.12 \leq x_i \leq 5.12, \tag{A.3}$$

which is also convex and unimodal with a global minimum $f_* = 0$ at $\boldsymbol{x}_* = (0, 0, ..., 0)$. Another related test function is the sum of different power function

$$f(\boldsymbol{x}) = \sum_{i=1}^{n} |x_i|^{i+1}, \qquad -1 \leq x_i \leq 1, \tag{A.4}$$

which has a global minimum $f_* = 0$ at $(0, 0, ..., 0)$.

Easom's function:

$$f(\boldsymbol{x}) = -\cos(x)\cos(y)\exp\Big[-(x-\pi)^2 + (y-\pi)^2\Big], \tag{A.5}$$

whose global minimum is $f_* = -1$ at $\boldsymbol{x}_* = (\pi, \pi)$ within $-100 \leq x, y \leq 100$. It has many local minima. Xin-She Yang extended in 2008 this function to n dimensions, and we have

$$f(\boldsymbol{x}) = -(-1)^n \Big(\prod_{i=1}^{n} \cos^2(x_i)\Big) \exp\Big[-\sum_{i=1}^{n} (x_i - \pi)^2\Big], \tag{A.6}$$

whose global minimum $f_* = -1$ occurs at $\boldsymbol{x}_* = (\pi, \pi, ..., \pi)$. Here the domain is $-2\pi \leq x_i \leq 2\pi$ where $i = 1, 2, ..., n$.

Equality-Constrained Function:

$$f(\boldsymbol{x}) = -(\sqrt{n})^n \prod_{i=1}^{n} x_i, \tag{A.7}$$

subject to an equality constraint (a hyper-sphere)

$$\sum_{i=1}^{n} x_i^2 = 1. \tag{A.8}$$

The global minimum $f_* = -1$ of $f(\boldsymbol{x})$ occurs at $\boldsymbol{x}_*(1/\sqrt{n}, ..., 1/\sqrt{n})$ within the domain $0 \le x_i \le 1$ for $i = 1, 2, ..., n$.

Griewank's function:

$$f(\boldsymbol{x}) = \frac{1}{4000}\sum_{i=1}^{n} x_i^2 - \prod_{i=1}^{n}\cos(\frac{x_i}{\sqrt{i}}) + 1, \qquad -600 \le x_i \le 600, \tag{A.9}$$

whose global minimum is $f_* = 0$ at $\boldsymbol{x}_* = (0, 0, ..., 0)$. This function is highly multimodal.

Michaelwicz's function:

$$f(\boldsymbol{x}) = -\sum_{i=1}^{n}\sin(x_i)\cdot\Big[\sin(\frac{ix_i^2}{\pi})\Big]^{2m}, \tag{A.10}$$

where $m = 10$, and $0 \le x_i \le \pi$ for $i = 1, 2, ..., n$. In 2D case, we have

$$f(x, y) = -\sin(x)\sin^{20}(\frac{x^2}{\pi}) - \sin(y)\sin^{20}(\frac{2y^2}{\pi}), \tag{A.11}$$

where $(x, y) \in [0, 5] \times [0, 5]$. This function has a global minimum $f_* \approx -1.8013$ at $\boldsymbol{x}_* = (x_*, y_*) = (2.20319, 1.57049)$.

Perm Functions:

$$f(\boldsymbol{x}) = \sum_{j=1}^{n}\Big\{\sum_{i=1}^{n}(i^j + \beta)\Big[(\frac{x_i}{i})^j - 1\Big]\Big\}, \qquad (\beta > 0), \tag{A.12}$$

which has the global minimum $f_* = 0$ at $\boldsymbol{x}_* = (1, 2, ..., n)$ in the search domain $-n \le x_i \le n$ for $i = 1, ..., n$. A related function

$$f(\boldsymbol{x}) = \sum_{j=1}^{n}\Big\{\sum_{i=1}^{n}(i + \beta)\Big[x_i^j - (\frac{1}{i})^j\Big]\Big\}^2, \tag{A.13}$$

has the global minimum $f_* = 0$ at $(1, 1/2, 1/3, ..., 1/n)$ within the bounds $-1 \le x_i \le 1$ for all $i = 1, 2, ..., n$. As $\beta > 0$ becomes smaller, the global minimum becomes almost indistinguishable from their local minima. In fact, in the extreme case $\beta = 0$, every solution is also a global minimum.

Rastrigin's function:

$$f(\boldsymbol{x}) = 10n + \sum_{i=1}^{n}\Big[x_i^2 - 10\cos(2\pi x_i)\Big], \qquad -5.12 \le x_i \le 5.12, \tag{A.14}$$

whose global minimum is $f_* = 0$ at $(0, 0, ..., 0)$. This function is highly multimodal.

Rosenbrock's function:

$$f(\boldsymbol{x}) = \sum_{i=1}^{n-1} \Big[(x_i - 1)^2 + 100(x_{i+1} - x_i^2)^2\Big], \tag{A.15}$$

whose global minimum $f_* = 0$ occurs at $\boldsymbol{x}_* = (1, 1, ..., 1)$ in the domain $-5 \leq x_i \leq 5$ where $i = 1, 2, ..., n$. In the 2D case, it is often written as

$$f(x, y) = (x - 1)^2 + 100(y - x^2)^2, \tag{A.16}$$

which is often referred to as the banana function.

Schwefel's function:

$$f(\boldsymbol{x}) = -\sum_{i=1}^{n} x_i \sin\Big(\sqrt{|x_i|}\Big), \qquad -500 \leq x_i \leq 500, \tag{A.17}$$

whose global minimum $f_* \approx -418.9829n$ occurs at $x_i = 420.9687$ where $i = 1, 2, ..., n$.

Six-hump camel back function:

$$f(x, y) = (4 - 2.1x^2 + \frac{1}{3}x^4)x^2 + xy + 4(y^2 - 1)y^2, \tag{A.18}$$

where $-3 \leq x \leq 3$ and $-2 \leq y \leq 2$. This function has two global minima $f_* \approx -1.0316$ at $(x_*, y_*) = (0.0898, -0.7126)$ and $(-0.0898, 0.7126)$.

Shubert's function:

$$f(\boldsymbol{x}) = \Big[\sum_{i=1}^{n} i \cos\Big(i + (i+1)x\Big)\Big] \cdot \Big[\sum_{i=1}^{n} i \cos\Big(i + (i+1)y\Big)\Big], \tag{A.19}$$

which has 18 global minima $f_* \approx -186.7309$ for $n = 5$ in the search domain $-10 \leq x, y \leq 10$.

Xin-She Yang's functions:

$$f(\boldsymbol{x}) = \Big(\sum_{i=1}^{n} |x_i|\Big) \exp\Big[-\sum_{i=1}^{n} \sin(x_i^2)\Big], \tag{A.20}$$

which has the global minimum $f_* = 0$ at $\boldsymbol{x}_* = (0, 0, ..., 0)$ in the domain $-2\pi \leq x_i \leq 2\pi$ where $i = 1, 2, ..., n$. This function is not smooth, and its derivatives are not well defined at the optimum $(0, 0, ..., 0)$.

A related function is

$$f(\boldsymbol{x}) = -\Big(\sum_{i=1}^{n} |x_i|\Big) \exp\Big(-\sum_{i=1}^{n} x_i^2\Big), \qquad -10 \leq x_i \leq 10, \tag{A.21}$$

which has multiple global minima. For example, for $n = 2$, we have 4 equal minima $f_* = -1/\sqrt{e} \approx -0.6065$ at $(1/2, 1/2)$, $(1/2, -1/2)$, $(-1/2, 1/2)$ and $(-1/2, -1/2)$.

Yang also designed a standing-wave function with a defect

$$f(\boldsymbol{x}) = \Big[e^{-\sum_{i=1}^{n} (x_i/\beta)^{2m}} - 2e^{-\sum_{i=1}^{n} x_i^2} \Big] \cdot \prod_{i=1}^{n} \cos^2 x_i, \qquad m = 5, \tag{A.22}$$

which has many local minima and the unique global minimum $f_* = -1$ at $\boldsymbol{x}_* = (0, 0, ..., 0)$ for $\beta = 15$ within the domain $-20 \le x_i \le 20$ for $i = 1, 2, ..., n$. He also proposed another multimodal function

$$f(\boldsymbol{x}) = \Big\{ [\sum_{i=1}^{n} \sin^2(x_i)] - \exp(-\sum_{i=1}^{n} x_i^2) \Big\} \cdot \exp\Big[-\sum_{i=1}^{n} \sin^2 \sqrt{|x_i|} \Big], \tag{A.23}$$

whose global minimum $f_* = -1$ occurs at $\boldsymbol{x}_* = (0, 0, ..., 0)$ in the domain $-10 \le x_i \le 10$ where $i = 1, 2, ..., n$. In the 2D case, its landscape looks like a wonderful candlestick.

Most test functions are deterministic. Yang designed a test function with stochastic components

$$f(x, y) = -5e^{-\beta[(x-\pi)^2 + (y-\pi)^2]} - \sum_{j=1}^{K} \sum_{i=1}^{K} \epsilon_{ij} e^{-\alpha[(x-i)^2 + (y-j)^2]}, \tag{A.24}$$

where $\alpha, \beta > 0$ are scaling parameters, which can often be taken as $\alpha = \beta = 1$. Here ϵ_{ij} are random variables and can be drawn from a uniform distribution $\epsilon_{ij} \sim$ Unif[0,1]. The domain is $0 \le x, y \le K$ and $K = 10$. This function has K^2 local valleys at grid locations and the fixed global minimum at $\boldsymbol{x}_* = (\pi, \pi)$. It is worth pointing that the minimum $f_{\min}$ is random, rather than a fixed value; it may vary from $-(K^2 + 5)$ to -5, depending α and β as well as the random numbers drawn.

Furthermore, he also designed a stochastic function

$$f(\boldsymbol{x}) = \sum_{i=1}^{n} \epsilon_i \left| x_i - \frac{1}{i} \right|, \qquad -5 \le x_i \le 5, \tag{A.25}$$

where ϵ_i $(i = 1, 2, ..., n)$ are random variables which are uniformly distributed in $[0, 1]$. That is, $\epsilon_i \sim$Unif$[0, 1]$. This function has the unique minimum $f_* = 0$ at $\boldsymbol{x}_* = (1, 1/2, ..., 1/n)$ which is also singular.

Zakharov's functions:

$$f(\boldsymbol{x}) = \sum_{i=1}^{n} x_i^2 + \Big(\frac{1}{2} \sum_{i=1}^{n} i x_i\Big)^2 + \Big(\frac{1}{2} \sum_{i=1}^{n} i x_i\Big)^4, \tag{A.26}$$

whose global minimum $f_* = 0$ occurs at $\boldsymbol{x}_* = (0, 0, ..., 0)$. Obviously, we can generalize this function as

$$f(\boldsymbol{x}) = \sum_{i=1}^{n} x_i^2 + \sum_{k=1}^{K} J_n^{2k}, \tag{A.27}$$

where $K = 1, 2, ..., 20$ and

$$J_n = \frac{1}{2} \sum_{i=1}^{n} i x_i. \tag{A.28}$$

REFERENCES

1. D. H. Ackley, *A Connectionist Machine for Genetic Hillclimbing*, Kluwer Academic Publishers, 1987.
2. C. A. Floudas, P. M., Pardalos, C. S. Adjiman, W. R. Esposito, Z. H. Gumus, S. T. Harding, J. L. Klepeis, C. A., Meyer, C. A. Scheiger, *Handbook of Test Problems in Local and Global Optimization*, Springer, 1999.
3. A. Hedar, Test function web pages, http://www-optima.amp.i.kyoto-u.ac.jp /member/student/hedar/Hedar_files/TestGO_files/Page364.htm
4. M. Molga, C. Smutnicki, "Test functions for optimization needs", http://www.zsd.ict.pwr.wroc.pl/files/docs/functions.pdf
5. X. S. Yang, "Firefly algorithm, Lévy flights and global optimization", in: *Research and Development in Intelligent Systems XXVI*, (Eds M. Bramer *et al.*), Springer, London, pp. 209-218 (2010).
6. X. S. Yang and S. Deb, "Engineering optimization by cuckoo search", *Int. J. Math. Modeling and Numerical Optimization*, **1**, No. 4, (in press) (2010).
7. X. S. Yang, "Firefly algorithm, stochastic test functions and design optimization", *Int. J. Bio-inspired Computation*, **2**, No. 2, 78-84 (2010).

APPENDIX B

MATLAB® PROGRAMS

The following codes intend to demonstrate how each algorithm works, so they are relatively simple and we do not intend to optimize them. In addition, most demonstrative cases are for 2D only, though they can be extended to any higher dimensions in principle. They are not for general-purpose optimization, because there are much better programs out there, both free and commercial. These codes should work using Matlab®[1]. For Octave,[2] slight modifications may be needed.

B.1 GENETIC ALGORITHMS

The following simple demo program of genetic algorithms tries to find the maximum of

$$f(x) = -\cos(x)e^{-(x-\pi)^2}.$$

[1]Matlab, www.mathworks.com
[2]J. W. Eaton, *GNU Octave Manual*, Network Theory Ltd, www.gnu.org/software/octave, 2002.

Copyright © 2010 John Wiley & Sons, Inc.

```
% Genetic Algorithm (Simple Demo) Matlab/Octave Program
% Written by X S Yang (Cambridge University)
% Usage: gasimple     or  gasimple('x*exp(-x)');
function [bestsol, bestfun, count]=gasimple(funstr)
global solnew sol pop popnew fitness fitold f range;
if nargin<1,
  % Easom Function with fmax=1 at x=pi
  funstr='-cos(x)*exp(-(x-3.1415926)^2)';
end
range=[-10 10];        % Range/Domain
% Converting to an inline function
f=vectorize(inline(funstr));
% Initializing the parameters
rand('state',0');      % Reset the random generator
popsize=20;            % Population size
MaxGen=100;            % Max number of generations
count=0;               % counter
nsite=2;               % number of mutation sites
pc=0.95;               % Crossover probability
pm=0.05;               % Mutation probability
nsbit=16;              % String length (bits)
% Generating the initial population
popnew=init_gen(popsize,nsbit);
fitness=zeros(1,popsize);    % fitness array
% Display the shape of the function
x=range(1):0.1:range(2); plot(x,f(x));
% Initialize solution <- initial population
for i=1:popsize,
   solnew(i)=bintodec(popnew(i,:));
end
% Start the evolution loop
for i=1:MaxGen,
   % Record as the history
   fitold=fitness; pop=popnew; sol=solnew;
 for j=1:popsize,
   % Crossover pair
   ii=floor(popsize*rand)+1; jj=floor(popsize*rand)+1;
   % Cross over
   if pc>rand,
      [popnew(ii,:),popnew(jj,:)]=...
                   crossover(pop(ii,:),pop(jj,:));
   % Evaluate the new pairs
   count=count+2;
   evolve(ii); evolve(jj);
   end
```

```
   % Mutation at n sites
   if pm>rand,
    kk=floor(popsize*rand)+1;    count=count+1;
    popnew(kk,:)=mutate(pop(kk,:),nsite);
    evolve(kk);
   end
 end  % end for j
   % Record the current best
   bestfun(i)=max(fitness);
   bestsol(i)=mean(sol(bestfun(i)==fitness));
end
% Display results
subplot(2,1,1); plot(bestsol); title('Best estimates');
subplot(2,1,2); plot(bestfun); title('Fitness');
% ------------- All the sub functions ----------
% generation of the initial population
function pop=init_gen(np,nsbit)
% String length=nsbit+1 with pop(:,1) for the Sign
pop=rand(np,nsbit+1)>0.5;
% Evolving the new generation
function evolve(j)
global solnew popnew fitness fitold pop sol f;
   solnew(j)=bintodec(popnew(j,:));
   fitness(j)=f(solnew(j));
   if fitness(j)>fitold(j),
      pop(j,:)=popnew(j,:);
      sol(j)=solnew(j);
   end
% Convert a binary string into a decimal number
function [dec]=bintodec(bin)
global range;
% Length of the string without sign
nn=length(bin)-1;
num=bin(2:end);   % get the binary
% Sign=+1 if bin(1)=0; Sign=-1 if bin(1)=1.
Sign=1-2*bin(1);
dec=0;
% floating point/decimal place in a binary string
dp=floor(log2(max(abs(range))));
for i=1:nn,
   dec=dec+num(i)*2^(dp-i);
end
dec=dec*Sign;
% Crossover operator
function [c,d]=crossover(a,b)
```

```
nn=length(a)-1;
% generating a random crossover point
cpoint=floor(nn*rand)+1;
c=[a(1:cpoint) b(cpoint+1:end)];
d=[b(1:cpoint) a(cpoint+1:end)];
% Mutatation operator
function anew=mutate(a,nsite)
nn=length(a); anew=a;
for i=1:nsite,
   j=floor(rand*nn)+1;
   anew(j)=mod(a(j)+1,2);
end
```

B.2 SIMULATED ANNEALING

The implemented simulated annealing intends to find the minimum of Rosenbrock's function

$$f(x, y) = (1 - x)^2 + 100(y - x^2)^2.$$

```
% Simulated Annealing (by X-S Yang, Cambridge University)
% Usage: sa_demo
disp('Simulating ... it will take a minute or so!');
% Rosenbrock's function with f*=0 at (1,1)
fstr='(1-x)^2+100*(y-x^2)^2';
% Convert into an inline function
f=vectorize(inline(fstr));
% Show the topography of the objective function
range=[-2 2 -2 2];
xgrid=range(1):0.1:range(2); ygrid=range(3):0.1:range(4);
[x,y]=meshgrid(xgrid,ygrid);
surfc(x,y,f(x,y));
% Initializing parameters and settings
T_init = 1.0;          % Initial temperature
T_min =  1e-10;        % Final stopping temperature
F_min = -1e+100;       % Min value of the function
max_rej=2500;          % Maximum number of rejections
max_run=500;           % Maximum number of runs
max_accept = 15;       % Maximum number of accept
k = 1;                 % Boltzmann constant
alpha=0.95;            % Cooling factor
Enorm=1e-8;            % Energy norm (eg, Enorm=1e-8)
guess=[2 2];           % Initial guess
% Initializing the counters i,j etc
i= 0; j = 0; accept = 0; totaleval = 0;
```

```
% Initializing various values
T = T_init;
E_init = f(guess(1),guess(2));
E_old = E_init; E_new=E_old;
best=guess;  % initially guessed values
% Starting the simulated annealling
while ((T > T_min) & (j <= max_rej) & E_new>F_min)
    i = i+1;
    % Check if max numbers of run/accept are met
    if (i >= max_run) | (accept >= max_accept)
    % Cooling according to a cooling schedule
        T = alpha*T;
        totaleval = totaleval + i;
        % reset the counters
        i = 1;  accept = 1;
    end
    % Function evaluations at new locations
      ns=guess+rand(1,2)*randn;
      E_new = f(ns(1),ns(2));
    % Decide to accept the new solution
    DeltaE=E_new-E_old;
    % Accept if improved
    if (-DeltaE > Enorm)
        best = ns; E_old = E_new;
        accept=accept+1;   j = 0;
    end
    % Accept with a small probability if not improved
    if (DeltaE<=Enorm & exp(-DeltaE/(k*T))>rand );
        best = ns; E_old = E_new;
        accept=accept+1;
    else
        j=j+1;
    end
    % Update the estimated optimal solution
    f_opt=E_old;
end
% Display the final results
disp(strcat('Obj function  :',fstr));
disp(strcat('Evaluations   :', num2str(totaleval)));
disp(strcat('Best  solution:', num2str(best)));
disp(strcat('Best objective:', num2str(f_opt)));
```

B.3 PARTICLE SWARM OPTIMIZATION

The following PSO program tries to find the global minimum of Michaelwicz's 2D function

$$f(x,y) = -\sin(x)\sin^{20}(\frac{x^2}{\pi}) - \sin(y)\sin^{20}(\frac{2y^2}{\pi}).$$

```
% Particle Swarm Optimization (by X-S Yang, Cambridge University)
% Usage: pso_demo(number_of_particles,Num_iterations)
%  eg:   best=pso_demo(20,15);
% where best=[xbest ybest zbest]  %an n by 3 matrix

function [best]=pso_demo(n,Num_iterations)
% n=number of particles; Num_iterations=number of iterations
if nargin<2,   Num_iterations=15;  end
if nargin<1,   n=20;            end
% Michaelwicz Function f*=-1.801 at [2.20319,1.57049]
fstr='-sin(x)*(sin(x^2/pi))^20-sin(y)*(sin(2*y^2/pi))^20';
% Converting to an inline function and vectorization
f=vectorize(inline(fstr));
% range=[xmin xmax ymin ymax];
range=[0 4 0 4];
% ------------------------------------------------------
% Setting the parameters: alpha, beta
alpha=0.2; beta=0.5;
% ------------------------------------------------------
% Grid values of the objective for visualization only
Ndiv=100;
dx=(range(2)-range(1))/Ndiv;dy=(range(4)-range(3))/Ndiv;
xgrid=range(1):dx:range(2); ygrid=range(3):dy:range(4);
[x,y]=meshgrid(xgrid,ygrid);
z=f(x,y);
% Display the shape of the function to be optimized
figure(1); surfc(x,y,z);
% -----------------------------------------------------
best=zeros(Num_iterations,3);   % initialize history
% ----- Start Particle Swarm Optimization -----------
% generating the initial locations of n particles
[xn,yn]=init_pso(n,range);
% Display the particle paths and contour of the function
 figure(2);
% Start iterations
for i=1:Num_iterations,
% Show the contour of the function
  contour(x,y,z,15); hold on;
```

```
% Find the current best location (xo,yo)
zn=f(xn,yn);
zn_min=min(zn);
xo=min(xn(zn==zn_min));
yo=min(yn(zn==zn_min));
zo=min(zn(zn==zn_min));
% Trace the paths of all roaming particles
% Display these roaming particles
plot(xn,yn,'.',xo,yo,'*'); axis(range);
% Move all the particles to new locations
[xn,yn]=pso_move(xn,yn,xo,yo,alpha,beta,range);
drawnow;
% Use "hold on" to display paths of particles
hold off;
% History
best(i,1)=xo; best(i,2)=yo; best(i,3)=zo;
end   %%%%% end of iterations
% ----- All subfunctions are listed here -----
% Intial locations of n particles
function [xn,yn]=init_pso(n,range)
xrange=range(2)-range(1); yrange=range(4)-range(3);
xn=rand(1,n)*xrange+range(1);
yn=rand(1,n)*yrange+range(3);
% Move all the particles toward (xo,yo)
function [xn,yn]=pso_move(xn,yn,xo,yo,a,b,range)
nn=size(yn,2);  %a=alpha, b=beta
xn=xn.*(1-b)+xo.*b+a.*(rand(1,nn)-0.5);
yn=yn.*(1-b)+yo.*b+a.*(rand(1,nn)-0.5);
[xn,yn]=findrange(xn,yn,range);
% Make sure the particles are within the range
function [xn,yn]=findrange(xn,yn,range)
nn=length(yn);
for i=1:nn,
   if xn(i)<=range(1), xn(i)=range(1); end
   if xn(i)>=range(2), xn(i)=range(2); end
   if yn(i)<=range(3), yn(i)=range(3); end
   if yn(i)>=range(4), yn(i)=range(4); end
end
```

B.4 HARMONY SEARCH

This simple program for Harmony Search intends to find the global minimum of Rosenbrock's function $f(x, y) = (1 - x)^2 + 100(y - x^2)^2$, which can easily be extended to any dimension.

```
% Harmony Search (Simple Demo) by X-S Yang (Cambridge University)
% Usage: hs_demo
% or     hs_demo('(x-1)^2+100*(y-x^2)^2',25000);
function [solution,fbest]=hs_simple(funstr,MaxAttempt)
disp('It may take a few minutes ...');
% MaxAttempt=25000;  % Max number of Attempt
if nargin<2, MaxAttempt=25000; end
if nargin<1,
% Rosenbrock's Banana function with the global f*=0 at (1,1).
funstr = '(1-x1)^2+100*(x2-x1^2)^2';
end
% Converting to an inline function
f=vectorize(inline(funstr));
ndim=2;  %Number of independent variables
% The range of the objective function
range(1,:)=[-5 5]; range(2,:)=[-5 5];
% Pitch range for pitch adjusting
pa_range=[100 100];
% Initial parameter setting
HS_size=20;          %Length of solution vector
HMacceptRate=0.95; %HM Accepting Rate
PArate=0.7;          %Pitch Adjusting rate
% Generating Initial Solution Vector
for i=1:HS_size,
   for j=1:ndim,
   x(j)=range(j,1)+(range(j,2)-range(j,1))*rand;
   end
   HM(i, :) = x;
   HMbest(i) = f(x(1), x(2));
end %% for i
% Starting the Harmony Search
for count = 1:MaxAttempt,
  for j = 1:ndim,
    if (rand >= HMacceptRate)
      % New Search via Randomization
      x(j)=range(j,1)+(range(j,2)-range(j,1))*rand;
    else
      % Harmony Memory Accepting Rate
      x(j) = HM(fix(HS_size*rand)+1,j);
      if (rand <= PArate)
      % Pitch Adjusting in a given range
      pa=(range(j,2)-range(j,1))/pa_range(j);
      x(j)= x(j)+pa*(rand-0.5);
      end
    end
```

```
  end %% for j
  % Evaluate the new solution
   fbest = f(x(1), x(2));
  % Find the best in the HS solution vector
   HSmaxNum = 1; HSminNum=1;
   HSmax = HMbest(1); HSmin=HMbest(1);
   for i = 2:HS_size,
      if HMbest(i) > HSmax,
        HSmaxNum = i;
        HSmax = HMbest(i);
      end
      if HMbest(i)<HSmin,
         HSminNum=i;
         HSmin=HMbest(i);
      end
   end
   % Updating the current solution if better
   if fbest < HSmax,
       HM(HSmaxNum, :) = x;
       HMbest(HSmaxNum) = fbest;
   end
   solution=x;    % Record the solution
end %% for count (harmony search)
```

B.5 FIREFLY ALGORITHM

Here we use a non-smooth multi-peak function that has 2^d peaks in the d-dimensional case

$$f(\boldsymbol{x}) = \Big(\sum_{i=1}^{d} |x_i|\Big) \exp\Big(-\alpha \sum_{i=1}^{d} x_i^2\Big). \tag{B.1}$$

In the simplest 2-D case, we have

$$f(x,y) = (|x| + |y|)e^{-\alpha(x^2+y^2)}, \tag{B.2}$$

which has four equal peaks. In order to obtain the analytical solutions, we have to differentiate it, but this function does not have derivative at $(0,0)$. However, the function is symmetric in x and y. That is, the function remains the same when x and y are interchanged. So we only have to find the solution in the first quadrant $x > 0$ and $y > 0$. In the first quadrant, we have

$$\frac{\partial f}{\partial x} = [1 - 2\alpha x(x+y)]e^{-\alpha(x^2+y^2)} = 0, \tag{B.3}$$

and

$$\frac{\partial f}{\partial y} = [1 - 2\alpha y(x+y)]e^{-\alpha(x^2+y^2)} = 0. \tag{B.4}$$

The symmetry in x and y implies that $x_* = y_*$ at the stationary points, so we have

$$1 - 2\alpha x_*(x_* + x_*) = 1 - 4\alpha x_*^2 = 0, \tag{B.5}$$

which gives

$$x_* = \frac{1}{2\sqrt{\alpha}}. \tag{B.6}$$

For $\alpha = 0.0625$, we have $x_* = 2$. So the four peaks are at

$$(x_*, y_*) = (2,2),\ (2,-2),\ (-2,2),\ (-2,-2). \tag{B.7}$$

The maximum at the peaks is $f_* = 4\exp(-1/2) \approx 2.4261226$. A demonstrative code for the Firefly Algorithm is as follows:

```
% Firefly Algorithm by X-S Yang (Cambridge University)
% Usage: firefly_demo([number_of_fireflies,MaxGeneration])
%  eg:   firefly_demo([25,20]);
function [best]=firefly_demo(instr)
% n=number of fireflies
% MaxGeneration=number of pseudo time steps
if nargin<1,    instr=[25 20];      end
n=instr(1);  MaxGeneration=instr(2);
rand('state',0);  % Reset the random generator
% ------ Four peak functions ---------------------
funstr='(abs(x)+abs(y))*exp(-0.0625*(x^2+y^2))';
% Converting to an inline function
f=vectorize(inline(funstr));
% range=[xmin xmax ymin ymax];
range=[-5 5 -5 5];
% ------------------------------------------------
alpha=0.2;       % Randomness 0--1 (highly random)
gamma=1.0;       % Absorption coefficient
% ------------------------------------------------
% Grid values are used for display only
Ndiv=100;
dx=(range(2)-range(1))/Ndiv; dy=(range(4)-range(3))/Ndiv;
[x,y]=meshgrid(range(1):dx:range(2),range(3):dy:range(4));
z=f(x,y);
% Display the shape of the objective function
figure(1);    surfc(x,y,z);
% ------------------------------------------------
% generating the initial locations of n fireflies
[xn,yn,Lightn]=init_ffa(n,range);
```

```
% Display the paths of fireflies in a figure with
% contours of the function to be optimized
 figure(2);
% Iterations or pseudo time marching
for i=1:MaxGeneration,     %%%%% start iterations
% Show the contours of the function
 contour(x,y,z,15); hold on;
% Evaluate new solutions
zn=f(xn,yn);
% Ranking the fireflies by their light intensity
[Lightn,Index]=sort(zn);
xn=xn(Index); yn=yn(Index);
xo=xn;   yo=yn;    Lighto=Lightn;
% Trace the paths of all roaming  fireflies
plot(xn,yn,'.','markersize',10,'markerfacecolor','g');
% Move all fireflies to the better locations
[xn,yn]=ffa_move(xn,yn,Lightn,xo,yo,...
        Lighto,alpha,gamma,range);
drawnow;
% Use "hold on" to show the paths of fireflies
    hold off;
end   %%%%% end of iterations
best(:,1)=xo'; best(:,2)=yo'; best(:,3)=Lighto';
% ----- All subfunctions are listed here ---------
% The initial locations of n fireflies
function [xn,yn,Lightn]=init_ffa(n,range)
xrange=range(2)-range(1);
yrange=range(4)-range(3);
xn=rand(1,n)*xrange+range(1);
yn=rand(1,n)*yrange+range(3);
Lightn=zeros(size(yn));
% Move all fireflies toward brighter ones
function [xn,yn]=ffa_move(xn,yn,Lightn,xo,yo,...
    Lighto,alpha,gamma,range)
ni=size(yn,2); nj=size(yo,2);
for i=1:ni,
% The attractiveness parameter beta=exp(-gamma*r)
    for j=1:nj,
r=sqrt((xn(i)-xo(j))^2+(yn(i)-yo(j))^2);
if Lightn(i)<Lighto(j), % Brighter and more attractive
beta0=1;     beta=beta0*exp(-gamma*r.^2);
xn(i)=xn(i).*(1-beta)+xo(j).*beta+alpha.*(rand-0.5);
yn(i)=yn(i).*(1-beta)+yo(j).*beta+alpha.*(rand-0.5);
end
    end % end for j
```

```
end % end for i
[xn,yn]=findrange(xn,yn,range);
% Make sure the fireflies are within the range
function [xn,yn]=findrange(xn,yn,range)
for i=1:length(yn),
   if xn(i)<=range(1), xn(i)=range(1); end
   if xn(i)>=range(2), xn(i)=range(2); end
   if yn(i)<=range(3), yn(i)=range(3); end
   if yn(i)>=range(4), yn(i)=range(4); end
end
```

If you run the Matlab program with 25 fireflies, all four peaks can be found simultaneously.

B.6 LARGE SPARSE LINEAR SYSTEMS

To solve a large-scale sparse linear system, Krylov subspace methods are most efficient in most cases. For example, the conjugate gradient method is an efficient solver for

$$\boldsymbol{A}\boldsymbol{u} = \boldsymbol{b},$$

where $\boldsymbol{A}$ is a normal matrix. In our simple implementation, it requires that $\boldsymbol{A}$ is real and symmetric.

```
% Simple implementation of the Conjugate Gradient Method
% by X. S. Yang (Cambridge University)
% This part forms a known system Av=b for demo only
n=50;                % try to use n=50, 250, 500, 5000
A=randn(n,n);        % generate a random square matrix
A=(A+A')/2;          % make sure A is symmetric
v=(1:n)';            % the target solution
b=A*v;               % form a linear system

% Solve the system by the conjugate gradient method
u=rand(n,1);         % initial guess u0
r=b-A*u;             % intial r0
d=r;                 % intial d0=r0
delta=10^(-5);       % accuracy

while max(abs(r))>delta
   alpha=r'*r/(d'*A*d);  % alpha_n
   K=r'*r;               % temporary value
   u=u+alpha*d;          % u_{n+1}
   r=r-alpha*A*d;        % r_{n+1}
   beta=r'*r/K;          % beta_n
```

```
    d=r+beta*d;                % d_{n+1}
    u'                         % disp u on the screen
end
```

Initially, you can run the program using a small $n = 50$, which will show almost exact answers. Then, you can try $n = 500$ and 5000 as further numeric experiments.

B.7 NONLINEAR OPTIMIZATION

To write our own codes is often time-consuming. In many cases, we can simply use existing well-tested programs. For example, the optimizer `fmincon` of the Matlab optimization toolbox can be used to solve many optimization problems. Let us use two examples to demonstrate how to turn a nonlinear optimization problem into simple Matlab codes.

B.7.1 Spring Design

The original spring design optimization is

$$\text{minimize} \quad f(\boldsymbol{x}) = (2+L)dw^2, \tag{B.8}$$

$$\text{subject to} \quad \begin{aligned} g_1(\boldsymbol{x}) &= 1 - \tfrac{d^2 L}{7178w^4} \le 0 \\ g_2(\boldsymbol{x}) &= \tfrac{4d^2 - wd}{12566dw^3 - w^4} + \tfrac{1}{5108w^2} - 1 \le 0 \\ g_3(\boldsymbol{x}) &= 1 - \tfrac{140.45w}{d^2 L} \le 0 \\ g_4(\boldsymbol{x}) &= \tfrac{w+L}{1.5} - 1 \le 0 \end{aligned} \tag{B.9}$$

The simple bounds are

$$0.05 \le w \le 2.0, \qquad 0.25 \le d \le 1.3, \qquad 2.0 \le L \le 15.0. \tag{B.10}$$

In order to use the Matlab optimization toolbox/function `fmincon`, we can write it using $\boldsymbol{x} = (w, d, L)^T = (x_1, x_2, x_3)^T$, and we have

$$\text{minimize} \quad f(\boldsymbol{x}) = (2+x_3)x_2x_1^2, \tag{B.11}$$

$$\text{subject to} \quad \begin{aligned} g_1(\boldsymbol{x}) &= 1 - \tfrac{x_2^3 x_3}{7178x_1^4} \le 0 \\ g_2(\boldsymbol{x}) &= \tfrac{4x_2^2 - x_1x_2}{12566x_2x_1^3 - x_1^4} + \tfrac{1}{5108x_1^2} - 1 \le 0 \\ g_3(\boldsymbol{x}) &= 1 - \tfrac{140.45x_1}{x_2^2 x_3} \le 0 \\ g_4(\boldsymbol{x}) &= \tfrac{x_1+x_2}{1.5} - 1 \le 0 \end{aligned} \tag{B.12}$$

with simple bounds

$$0.05 \le x_1 \le 2.0, \quad 0.25 \le x_2 \le 1.3, \quad 2.0 \le x_3 \le 15.0. \tag{B.13}$$

These bounds can be written as

$$Lb \le \boldsymbol{x} \le Ub, \tag{B.14}$$

with the lower bound

$$\texttt{Lb} = [0.05; 0.25; 2.0]; \tag{B.15}$$

and the upper bound

$$\texttt{Ub} = [2.0; 1.3; 15.0]; \tag{B.16}$$

Starting from an educated guess

$$\texttt{x0} = [0.1; 0.5; 10]; \tag{B.17}$$

we can write the program as the following matlab code

```
% Spring Design Optimization using Matlab fmincon
function spring
  x0=[0.1; 0.5; 10];
  Lb=[0.05; 0.25; 2.0];  Ub=[2.0; 1.3; 15.0];
  % call matlab optimization toolbox
  [x,fval]=fmincon(@objfun,x0,[],[],[],[],Lb,Ub,@nonfun)

% Objective function
function f=objfun(x)
f=(2+x(3))*x(1)^2*x(2);

% Nonlinear constraints
function [g,geq]=nonfun(x)
% Inequality constraints
  g(1)=1-x(2)^3*x(3)/(7178*x(1)^4);
  gtmp=(4*x(2)^2-x(1)*x(2))/(12566*x(2)*x(1)^3-x(1)^4);
  g(2)=gtmp+1/(5108*x(1)^2)-1;
  g(3)=1-140.45*x(1)/(x(2)^2*x(3));
  g(4)=x(1)+x(2)-1.5;
% Equality constraints [none]
geq=[];
```

If we run this simple Matlab program, we can obtain the optimal solution

$$\boldsymbol{x}_* \approx (0.0500, 0.2500, 9.9877), \tag{B.18}$$

which gives $f_{\min} \approx 0.0075$. This solution is better than the best solution obtained by Cagnina *et al.*

$$f_* \approx 0.012665, \tag{B.19}$$

at

$$\boldsymbol{x}_* \approx (0.051690,\ 0.356750,\ 11.287126). \tag{B.20}$$

B.7.2 Pressure Vessel

The standard pressure vessel optimization can be written as

$$\text{minimize} \quad f(\boldsymbol{x}) = 0.6224 d_1 r L + 1.7781 d_2 r^2 + 3.1661 d_1^2 L + 19.84 d_1^2 r, \tag{B.21}$$

$$\begin{aligned} \text{subject to} \quad g_1(\boldsymbol{x}) &= -d_1 + 0.0193 r \le 0 \\ g_2(\boldsymbol{x}) &= -d_2 + 0.00954 r \le 0 \\ g_3(\boldsymbol{x}) &= -\pi r^2 L - \frac{4\pi}{3} r^3 + 1296000 \le 0 \\ g_4(\boldsymbol{x}) &= L - 240 \le 0 \end{aligned} \tag{B.22}$$

with simple bounds $0.0625 \le d_1, d_2 \le 99 \times 0.0625$, and $10.0 \le r, L \le 200.0$.

To transform into the Matlab code, we now use $\boldsymbol{x} = (x_1, x_2, x_3, x_4)^T = (d_1, d_2, r, L)^T$ to rewrite the above optimization as

$$\text{minimize} \quad f(\boldsymbol{x}) = 0.6224 x_1 x_3 x_4 + 1.7781 x_2 x_3^2 + 3.1661 x_1^2 x_4 + 19.84 x_1^2 x_3,$$

$$\begin{aligned} \text{subject to} \quad g_1(\boldsymbol{x}) &= -x_1 + 0.0193 x_3 \le 0 \\ g_2(\boldsymbol{x}) &= -x_2 + 0.00954 x_3 \le 0 \\ g_3(\boldsymbol{x}) &= x_4 - 240 \le 0 \\ g_4(\boldsymbol{x}) &= -\pi x_3^2 x_4 - \frac{4\pi}{3} x_3^3 + 1296000 \le 0. \end{aligned}$$

Here we have interchanged the third and forth inequalities so the the first three are simple inequalities, while the last inequality is nonlinear. The first three inequalities g_1, g_2, g_3 can be written as

$$\boldsymbol{A}\boldsymbol{x} \le \boldsymbol{b}, \tag{B.23}$$

where

$$\boldsymbol{A} = \begin{pmatrix} 1 & 0 & 0.0193 & 0 \\ 0 & -1 & 0.00954 & 0 \\ 0 & 0 & 0 & 1 \end{pmatrix}, \qquad \boldsymbol{b} = \begin{pmatrix} 0 \\ 0 \\ 240 \end{pmatrix}. \tag{B.24}$$

The simple bounds can be rewritten as

$$\texttt{Lb} = [\texttt{d}; \texttt{d}; 10; 10]; \qquad \texttt{Ub} = [99 * \texttt{d}; 99 * \texttt{d}; 200; 200], \tag{B.25}$$

where $d = 0.0625$. With the above reformulation, we can program it as the following Matlab code

```
% Design Optimization of a Pressure Vessel
function pressure_vessel
  x0=[1; 1; 20; 50];
  d=0.0625;
% Linear inequality constraints
  A=[-1 0 0.0193 0;0 -1 0.00954 0;0 0 0 1];
```

```
  b=[0; 0; 240];
% Simple bounds
  Lb=[d; d; 10; 10];  Ub=[99*d; 99*d; 200; 200];
  options=optimset('Display','iter','TolFun',1e-08);
  [x,fval]=fmincon(@objfun,x0,A,b,[],[],Lb,Ub,@nonfun,options)
% The objective function
function f=objfun(x)
f=0.6224*x(1)*x(3)*x(4)+1.7781*x(2)*x(3)^2 ...
  +3.1661*x(1)^2*x(4)+19.84*x(1)^2*x(3)
% Nonlinear constraints
function [g,geq]=nonfun(x)
% Nonlinear inequality
  g=-pi*x(3)^2*x(4)-4*pi/3*x(3)^3+1296000;
% Equality constraint [none]
  geq=[];
```

In the above code, '...' means the line continues. That is, the current line and the next line should form a single line in Matlab. This '...' is only added for typesetting the code to avoid an ugly long line.

If we run the above program, we have

$$\boldsymbol{x}_* \approx (0.7782, 0.3846, 40.3196, 200.0000)^T, \qquad \text{(B.26)}$$

with $f_{\min} \approx 5885.33$, which is lower that the best solution $f_* \approx 6059.714$ at $\boldsymbol{x}_* \approx (0.8125,\ 0.4375,\ 42.0984,\ 176.6366)$, obtained by Cagnina *et al.* It is worth pointing out that we have relaxed the constraints of d_1 and d_2 in the above implementation, as d_1 and d_2 can only take discrete values of integer multiples of 0.0625 in the original design problem. If we impose these constraints, our program will produce exactly the same results as the best solution $f_* \approx 6059.714$ obtained by Cagnina *et al.* Here it is left as an exercise for the readers to modify the program to achieve this. You can either add extra constraints or change the formulation slightly.

From the above two examples (spring and pressure vessel design), we can see that the optimizer `fmincon` in Matlab is a very good optimizer and suitable for many applications.

REFERENCES

1. L. C. Cagnina, S. C. Esquivel, C. A. Coello, "Solving engineering optimization problems with the simple constrained particle swarm optimizer", *Informatica*, **32**, 319-326 (2008).
2. J. W. Eaton, "Gnu Octave Manual", Network Theory Ltd, 2002.
3. O. Sigmund, "A 99 line topology optimization code written in Matlab", *Struct. Multidisc. Optim.*, **21**, 120-127 (2001).
4. Matlab's optimization toolbox, www.mathworks.com

APPENDIX C

GLOSSARY

Agent-based modeling: A category of search methods using multiple agents such as particles and populations so as to increase the overall search efficiency. Many popular metaheuristic algorithms such as ant algorithms and bee algorithms are agent-based. Also called population-based, though agent-based is more often used in robotics and machine learning.

Algorithm: A step-by-step description of a solution procedure to a problem, though it means numerical algorithms in computing.

Algorithm complexity: Also called the time complexity of an algorithm, which is the number of steps needed to complete the execution of an algorithm. It is a measure of how efficient an algorithm is. For example, the inverse of an $n \times n$ matrix often has $O(n^3)$ complexity.

Annealing schedule: A procedure of lowering or varying the temperature parameter in simulated annealing algorithms so as to lower the energy of the system, leading to a good rate of convergence of the algorithm.

Ant algorithms: A class of search algorithms using the behavior of an ant colony. One of most popular ant algorithms is the ant colony optimization (ACO).

By Xin-She Yang
Copyright © 2010 John Wiley & Sons, Inc.

Ant colony optimization: An optimization algorithm, developed by Marco Dorigo in 1992, uses the so-called swarm intelligence of social ants. The routes in combinatorial optimization are marked by pheromone concentration deposited by the ants/agents. An ant will preferably choose a route with its probability proportional to some function of pheromone concentration. The aim is to evolve the system to find the best routes according to certain criteria.

Artificial bee colony: An optimization algorithm based on intelligent foraging behavior of honey bees, first developed by D. Karaboga in 2005. Bees communicate by waggle dances so that the colony intends to maximize their honey intake for various flower patches.

Artificial immune system: A computational system or paradigm based on the immune systems of vertebrates, which uses memory and learning characteristics to problem-solving such as machine learning, artificial intelligence and optimization.

Artificial intelligence: A branch of computer sciences concerning the intelligence of machines and machine learning, pioneered by Alan Turing. Now artificial intelligence (AI) has a wide range of applications.

Banana function: See Rosenbrock's banana function.

Bee algorithms: A class of algorithms based on the foraging behavior of bees, including bee colony optimization (BCO), artificial bee colony (ABC), honeybee algorithm, virtual bee algorithm and other variants.

Bee colony optimization: A special class of optimization using bee algorithms based on honeybee behavior. Many real-world problems such as routing and scheduling can be solved by bee colony optimization.

Bioinspired algorithm: Also biology-inspired algorithms. A special class of metaheuristic algorithm for optimization, which was developed by using certain characteristics of evolution in biological systems, including reproduction, crossover or recombination, mutation and selection of the fittest, chemical messenger, and swarm intelligence. Examples are genetic algorithms and the honeybee algorithm.

Black-box optimization: A special type of optimization whose objective function has no explicit form of dependence on the design variables. For example, minimization of engine noise by changing the geometry of a car engine is a black-box optimization problem.

Calculus of variations: A branch of mathematics dealing with the functional of unknown functions, often in terms of integral and the Euler-Lagrange equation. For example, finding the shape of a soap bubble is an optimization problem in calculus of variations.

Characteristic equation: A equation associated with eigenvalues of a matrix $\boldsymbol{A}$. The characteristic equation is written as $\det(\boldsymbol{A} - \lambda \boldsymbol{I}) = |\boldsymbol{A} - \lambda \boldsymbol{I}| = 0$. It also means the auxiliary equation for ordinary differential equations.

Class C^n functions: A function $f(x)$ is said to belong to class C^n if it is differentiable and all its derivatives (up to order n) exist and are continuous. For example, $|x|$ is class C^0 because $|x|$ is not differentiable at $x = 0$.

Combinatorial optimization: A special class of optimization whose set of all possible and feasible solutions is discrete. For example, the set of all possible routes in the traveling salesman problem is discrete, and the number of such routes is at most $n!$.

Complexity: Also related to computational complexity and algorithm complexity. It measures the degree of the inherent difficulty in computations.

Computational complexity: A branch of computation science dealing with the complexity of computational problems, resources (e.g., memory) and algorithms. It uses order notation to represent complexity. For example, the quick-sort algorithm has the complexity of $O(n \log n)$. In most cases, this often means the complexity of the algorithm. See also algorithm complexity.

Conjugate gradient method: An iterative method for solving large linear systems and unconstrained optimization for the case when the matrix is symmetric and positive-definite. This method uses conjugate directions, instead of direct downhill, and subsequently it could be more efficient than the steepest descent method for the case of local valley is narrow and distorted.

Constrained optimization: An optimization problem with at least one constraint. If there is no constraint, it becomes a case of unconstrained optimization.

Constraint: A condition which poses an extra rule on the values of design variables. For example, certain design variables such as physical length should be non-negative. If the condition is an equality, it is called equality constraint. If it is an inequality, it is called inequality constraint.

Continuous optimization: Optimization whose design variables take real values, rather than discrete values such as integers.

Control variable: Also called decision or design variable.

Convergence: A limiting behavior of a sequence such as an iterative procedure, which approaches a fixed value, point, or set. In optimization, this often means that algorithm produces the desired solutions as the iterations proceeds.

Convex: An important concept in optimization. An object such as a sphere is called convex if for any two points in the object, then every point on the straight line joining these two points is also within the object. So a hollow shell is not convex.

Convex function: A function $f(x)$ on Ω is said to be convex if and only if $f(\alpha x_1 + \beta x_2) \leq \alpha f(x_1) + \beta f(x_2)$ for all $x_1, x_2 \in \Omega$, $\alpha + \beta = 1$ and $\alpha, \beta \geq 0$.

Convex optimization: A special but very important class of optimization concerning convex functions and ways to transform problems into one with convexity.

Cost function: An objective function.

Critical point: A point x of a function $f(x)$ is called critical if $f'(x) = 0$ (stationary condition) or $f(x)$ is not differentiable. So all stationary points are critical points. In addition, $|x|$ is not differential at $x = 0$, so $x = 0$ is a critical point of $|x|$. Similarly, $x = 0$ is a critical point of $x^{2/3}$ which is a cusp. Depending on context, this is also called a singular point of the curve.

Cuckoo search: A search strategy, developed by Xin-She Yang and Suash Deb in 2009, based on the breeding behavior of some cuckoo species. The idea is to represents the solutions using eggs/nests of hosting birds, and new solutions are generated via Lévy flights and represented by cuckoos which intend to replace existing solutions by selecting the best.

Decision/Design variable: The parameters or variables that can be adjusted so that certain objective functions can increase or decrease. Sometimes, design variables are also interchangeably called parameters, independent variables.

Decision/Design space: All the possible, allowable, combinations forms a space, called design space. Also called search space.

Derivative-free methods: Also called gradient-free methods.

Determinant: A quantity associated with a square matrix. For example,
$$\begin{vmatrix} a & b \\ c & d \end{vmatrix} = ad - bc, \begin{vmatrix} 3 & 1 \\ 4 & -2 \end{vmatrix} = 3 \times (-2) - 1 \times 4 = -10.$$

Deterministic algorithm: An algorithm is said to be deterministic if its procedure will produce exactly the same results when starting from the same initial condition(s). There is no random component in the algorithm. For example, hill-climbing is a deterministic local search algorithm.

Differential equation: A mathematical equation that contains derivatives. It includes both ordinary differential equation and partial differential equations.

Discrete optimization: A class of optimization whose design variables can only take discrete values such as integers.

Dot product: The dot product is also called the inner product. For any two vectors $\boldsymbol{u} = (a \quad b)^T$ and $\boldsymbol{v} = (c \quad d)^T$, their dot product is $\boldsymbol{u} \cdot \boldsymbol{v} = ac + bd$.

Eagle strategy: A two-stage search algorithm for stochastic optimization problems, recently developed by Xin-She Yang and Suash Deb in 2009, based on the foraging behavior of eagles. The initial solutions are generated by Lévy flights, and they are followed by more intensive local search algorithms such as the steepest descent method and firefly algorithms.

Efficiency: Also called algorithmic efficiency. It is a very vague term commonly used in optimization literature. Loosely speaking, it means an indicator to measure the performance of an algorithm, often in terms of the consumption of computing resources (time and memory). It is typically cited by speed, computing time, algorithm complexity, and number of steps to find the desired solutions.

Elitism: A form of exploiting existing best solutions in a search algorithm. For example, in genetic algorithms, the simplest form of elitism is to allow the best solution to pass on to the next generation without any modification.

Engineering optimization: A type of optimization with the emphasis of solving the common optimization problems in engineering design and applications. Structural design optimization and shape optimization are two examples.

Equality: A condition in the form an equation. For example, $x^2 - y^2 = 2$ is an equality.

Equality constraint: A constraint in the form of an equation.

Equation: A mathematical statement using an equal '=' sign. For example, $x^2 - 1 = 5$ and $a^2 + b^2 = c^2$ are equations.

Entropy: An important concept or measure of the order or information of a disordered system in multiple states. If p_i is the probability of the system in state i, then the entropy can be defined by $-k \sum_i p_i \log p_i$ where k is constant. In thermodynamics, $k = k_B$ the Boltzmann constant. In information theory, $k = \ln(2)$ is often used.

Evolutionary algorithms: A subset of optimization algorithms in evolutionary computation, mainly uses genetic operators including reproduction, crossover or recombination, mutation, and selection. Genetic algorithms belong to evolutionary algorithms. Four major evolutionary algorithms are: genetic algorithm, evolutionary programming, genetic programming, and evolutionary strategy.

Evolutionary computation: Also evolutionary computing. A branch of artificial intelligence often use iterative procedure and evolutionary algorithms to find the desired solutions. For example, evolutionary programming developed by L. J. Fogel *et al.* in the 1960s is an essence part of evolutionary computation.

Evolutionary programming: An evolutionary algorithm, developed by L. J. Fogel in the 1960s, to produce artificial intelligence using finite-state machine with mutation being the main genetic operator.

Evolutionary strategy: An evolutionary algorithm, developed by Ingo Rechenberg and Hans-Paul Schwefel in the 1960s, for finding solutions automatically to optimization problems using mutation and selection. No crossover operator was used at that time.

Exploitation: An intensified local search, often exploiting existing solutions such as by selecting some of among the best, and passing on to the next generation by elitism or some form of memory.

Exploration: A search process in the search space, intending to search for new solution by visit mostly (new) regions of the whole space, though not necessarily evey part.

Extremum: Either a minimum or maximum, though it is more often mean the endpoints of a region.

Factorial: A factorial is a product defined by $n! = 1 \times 2 \times ... \times n$. For example, $4! = 1 \times 2 \times 3 \times 4 = 24$.

Feasible region: A candidate solution to an optimization problem is called feasible if it satisfies all the constraints. All the feasible solution form the feasible set, or solution space. Also called the search space.

Fibonacci sequence: A sequence of numbers, named after Leonardo of Pisa, generated by a simple recurrence relationship $F_n = F_{n-1} + F_{n-2}$ starting from $F_0 = F_1 = 1$.

Firefly algorithm: A metaheuristic optimization algorithm, developed by Xin-She Yang at Cambridge University in 2008. This algorithm was inspired by the flashing light and behavior of fireflies. Under special cases, it can reduce either to a random search or particle swarm optimization. Preliminary studies show that it is more powerful than PSO.

Frobenius norm: Also called Hilbert-Schmidt norm, an extension of 2-norm to p-norm of vectors to a matrix $\boldsymbol{A} = [a_{ij}]$. That is $\|\boldsymbol{A}\|_F = \|\boldsymbol{A}\|_2 = (\sum_{i=1}^m \sum_{j=1}^n a_{ij}^2)^{1/2}$.

Function: A mathematical relationship between an independent variable such as x and a dependent or response variable such as y, often written as $y = f(x)$. For example, $y = x^2 - 2x$ is a function.

Functional: A mathematical map which transforms a vector space into a scalar field. Integrals as used in calculus of variations are functionals. For example, the arc length $L(s)$ of curve $f(x)$ defined by $L(s) = \int_{x_0}^{s} \sqrt{1 + [f'(x)]^2} dx$ is a functional of $f(x)$.

Gaussian distribution: A very widely used probability distribution, also called the normal distribution $p(x) = \frac{1}{\sigma\sqrt{2\pi}} \exp[-\frac{(x-\mu)^2}{2\sigma^2}]$ with a mean of μ and a standard deviation of σ.

Genetic algorithms: A class of optimization algorithms, developed by John Holland in 1970s, which was inspired by Darwin's evolution theory and natural selection. There are so many different variants which are now collectively' called genetic algorithms (GA). Typically, genetic algorithms uses crossover or recombination, mutation, selection and certain elitism.

Genetic programming: An evolutionary method, developed by J. R. Koza, for producing computer programs or codes using genetic algorithms, while each representation is a computer program, rather than a simple solution. So it is machine learning strategy.

Gibbs sampler: A special case of Metropolis-Hastings sampling method with an acceptance probability of 1. For the case of multivariate distribution, at each step a sample is drawn from a univariate distribution while other variables remain fixed. It is thus simple to implement, but not so efficient.

Gradient: The rate of change of a curve or a function $f(x)$, often written as the first derivative df/dx.

Gradient-based method: A search method which uses the gradient or derivatives as the information to guide the search direction and determine the step size. A classic example is Newton's method for finding the root of a function $f(x)$, or the Newton-Raphson method for finding the minimum of a function.

Gradient-free method: Also called non-gradient based or derivative-free search method. It does not use any derivatives, only the value of the function itself are used. Subsequently, it works well for functions even with discontinuity where gradient-based methods will fail. For example, the bisection method for finding roots and the Nelder-Mead method for optimization are gradient-free methods.

Global optimization: The optimization with the emphasis of finding the global optimality, rather than satisfying with any local optima. Global optimization problems are most difficult to solve.

Global optimum: A solution is said to be a global optimum if it is the best (either minimum or maximum) among all possible solutions in all the feasible region.

Global search: Also a global search algorithm. A search algorithm with the aim to find the global optimality. Most modern algorithms intend to suit for global search, though not always successful. The important issue of designing a global search method is to introduce certain ability to jump out of local optima, often by randomization.

Greedy method: Also called greedy algorithm. A greedy algorithm use a heuristic style local choice at every stage of the algorithm execution, though with an intention to find the global optimal solutions. It is a short-sighted approach, and do not work well in most cases. For example, the traveling salesman problem, going to the nearest city at each step usually do not produce the shortest route globally. However, if the global convergence is proved for some very specialized cases, then it is one of the quickest and most efficient algorithms for such cases. Typically, it works well for small problems or only good enough solutions are needed.

Harmony search: A music-inspired optimization algorithm, developed by Z. W. Geem *et al.* in 2001, which uses harmony set to represent solutions. New solutions are generated by pitch adjustment, randomization and high-qaulity solutions are more likely to be kept for next generations.

Hessian matrix: Or simply Hessian. The matrix formed by second-derivatives of a function $f(\boldsymbol{x})$, often the objective function. $\boldsymbol{H} = [H_{ij}] = [\frac{\partial^2 f}{\partial x_i \partial x_j}]$.

Heuristics: A class of algorithms for optimization by trial-and-error. The aim is to find good feasible solutions in practically acceptable time. In literature, heuristic algorithms are simply referred to as heuristics.

Heuristic algorithm: An optimization algorithm which works in a heuristic manner. Quality solutions to an optimization problem can be found easily, but the global best solutions cannot be guaranteed.

Hill-climbing: A search optimization intends to improve an existing sub-optimal solution until no further improvement is gained. The search direction or move tends to move up the hill along the steepest direction (or against the maximum gradient locally). The aim is to reach the (local) peak. It is a local search method. The Newton-Raphson method belongs to the hill-climbing category. If the movement is downhill, then it becomes the steepest descent method.

Honeybee algorithm: One of the earliest variants of bee algorithm, developed by C. Tovey and S. Nakrani in 2004. It was first used for dynamic server allocation in Internet hosting centers and optimization.

Hooke-Jeeves pattern search: A classic search algorithm for multidimensional nonlinear optimization. The basic idea to start from a base point, and an exploratory move to a new point by changing one variable while other variables remains fixed. If the solution at the new point is improved (lower for a minimization problem), then a pattern move is carried out so that the new point will be the new base. If the new point is not an improvement, then abandon this point, and try again. This is a local search method and is gradient free.

Identity matrix: A square matrix with all diagonal elements being 1s and off diagonal elements being 0s. For example, a 2×2 identity matrix is $\boldsymbol{I} = \begin{pmatrix} 1 & 0 \\ 0 & 1 \end{pmatrix}$.

Importance sampling: A sampling technique which emphasizes the more important samples. In a distribution, some samples may be more important, in terms of weights, in estimating certain statistic measures, thus the algorithm intends to ensure these important samples are drawn so as to obtain more accurate estimates.

Inequality: A relationship with $<$ or > 0. For example, $x + y \geq 0$ is an inequality.

Inequality constraint: A constraint in a form of inequality.

Integer: whole numbers such as $-2, -1, 0, +1, +5$, and $+56789$.

Integer programming: A special class of discrete programming or optimization where all design variables are integers. In the case of linear programming with design variables being integers, it is called integer linear programming. Most real-world integer programming problems are NP-hard.

Intensification: An intensive local search around the neighborhood of a current best solution.

Interior point method: A powerful search method for linear and nonlinear convex programming. It intends to find the global solution by trespassing the interior points, rather than going along the boundaries which are typically used in the simplex method. It is also referred to as the barrier method.

Inverse: An inverse of a function $y = f(x)$ is to find $x = g(y)$. For example, $y = \sin x$, we have $x = \sin^{-1} y$. For a square matrix $\boldsymbol{A}$, its inverse is $\boldsymbol{A}^{-1}$ if $\boldsymbol{A}$ is not singular or $\det(\boldsymbol{A}) \neq 0$.

Iteration: A procedure to obtain a solution numerically.

Karush-Kuhn-Tucker condition: A requirement for a solution to be optimal in nonlinear optimization, and it is an extension of the Lagrange multiplier for the nonlinear optimization with inequality constraints.

Kuhn-Tucker condition: Also Karush-Kuhn-Tucker condition.

ℓ_2-norm: Also the two-norm. For a vector $\boldsymbol{u}$, it becomes $|\boldsymbol{u}| = \|\boldsymbol{u}\| = \sqrt{u_1^2 + ... + u_n^2}$.

Lagrange multiplier: Also the method of Lagrange multiplier. A method of dealing with equality constraints in an optimization problem by converting an constrained problem to an unconstrained one. For example,

$$\min x^2 \sin(y^3), \text{ subject to } x^2 - y^2 = 1,$$

can be written as

$$\min x^2 \sin(y^3) + \lambda(x^2 - y^2 - 1).$$

Here λ is the Lagrange multiplier. Then, the aim is to find a point (x, y, λ) which is stationary.

Landscape: A not-well-defined term used in optimization. Loosely speaking, it means the structure or topological variations of the objective function. So it is also called response surface.

Least squares: Also called method of least squares. See method of least squares.

Lévy flight: A random process or random walk whose step size distribution obeys the Lévy distribution $L(s) = s^{-n}$ where $1 \leq n \leq 3$. Some species of birds and fruitflies appear to have the Lévy flight behaviour in their flight paths.

Line search: A simple search algorithm for finding the minimum of a univariate function $f(x)$ by repetitively using the direction of a chosen step and reducing its step size.

Linear: A function is called linear if its value will double if the independent variables doubles. A linear function can be written as $y = ax + b$ where a, b are constants. Here x, y can be scalars, vectors or matrices, and b should be the same type as x. In general, a function or map is called linear if $f(x + y) = f(x) + f(y)$ and $f(\alpha x) = \alpha f(x)$ where α is a constant.

Linear congruential generator: A deterministic algorithm for generating pseudo-random numbers using a recurrence relationship, often in the form of $u_{n+1} = (au_n + c) \bmod m$ where a, c and m are positive integers.

Linear search: A sequential search method in computer science by going through each element in a sequence of data. For example, matching (to a known value) is a simple linear search with $O(n)$ complexity.

Linear programming: A special class of mathematical programming whose objectives and constraints are all linear function. Simplex method, developed by George Dantzig in 1947 is one of the most powerful algorithm for linear programming.

Linear system: A set of multiple linear equations. For example, $x - 2y = 1$ and $x + 2y = 5$ form a simple linear system which can be written as the matrix form $\begin{pmatrix} 1 & -2 \\ 1 & 2 \end{pmatrix} \begin{pmatrix} x \\ y \end{pmatrix} = \begin{pmatrix} 1 \\ 5 \end{pmatrix}$.

Local optimum: A solution is called a local optimum if it is the best (either minimum or maximum) in a neighborhood of feasible solutions. This is in contract with the global optimum which is the best among all possible solutions.

Local search: A search algorithm focus on the search in a neighborhood of solutions, and the optimum it can reach is often not the global optimum. For example, the steepest descent method is an efficient local search method.

Low discrepancy sequence: A deterministic sequence to generate quasi-random numbers or low-discrepancy numbers. For example, the Sobol sequence developed by I. M. Sobol in 1967 is such a sequence, which intends to generate numbers as far apart as possible in an interval.

Machine learning: A branch of artificial intelligence, concerning the algorithms for machine to learning from data, recognize complex systems, adapt to changing environment and make intelligent decisions or actions.

Markov chain: A stochastic process which obeys certain Markov property. That is, the next state of the process only depends on the current state and some transition probability jumping from the current state to the next state. For example, the Brownian motion, which is also a random walk, of a molecule is a Markov chain.

Markov chain Monte Carlo: A Monte Carlo sampling method using Markov chain as a generated process. A good Markov chain Monte Carlo (MCMC) should mix quickly, and converge quickly. Modern Bayesian statistics and high-dimension sampling all uses MCMC as a powerful tool.

Matlab: A popular commercial computer package for manipulating matrices very easily, developed by Mathworks. It serves a powerful integrated environment for numerical computing, using the fourth-generation programming language.

Matrix: An rectangular array of numbers. For example, $\begin{pmatrix} 2 & -3 & 4 \\ -5 & 0 & 2 \end{pmatrix}$ is a 2×3 matrix. A vector is a special case of a matrix.

Metaheuristics: A class of optimization algorithms which are higher level than heuristics. It is also the generic name for metaheuristic algorithms.

Metaheuristic algorithms: A class of stochastic algorithms using a combination of randomization and local search. They are often based on learning from nature or biological systems. Popularly algorithms includes genetic algorithms, particle swarm optimization, ant algorithms, and bee algorithms. Metaheuristic algorithms are usually designed for global optimization.

Method of least squares: A very popular method for data analysis and curve fitting. It intends to best-fit data by minimizing the sum of the residual squares. For example, for a set of observations (x_i, y_i). In order to best fit the simple mathematical model $y = f(x, \boldsymbol{\alpha})$ where $\boldsymbol{\alpha}$ are the parameters, we have to minimize $\sum_i (y_i - f(x_i, \boldsymbol{\alpha}))^2$.

Metropolis algorithm: A sampling algorithm for Monte Carlo simulations using a proposal move and acceptance probability. The sample drawn from a simpler distribution, called proposal distribution is accepted or rejected by an acceptance criterion. It was first proposed by Metropolis *et al.* in 1953, and it belongs to the general Markov chain Monte Carlo method.

Metropolis-Hastings algorithm: A more efficient sampling method of Markov chain Monte Carlo, extended by W. K. Hastings in 1970, which uses asymmetric proposal distribution, in contrast with the Metropolis algorithm where symmetric proposal is used.

Monte Carlo: See Monte Carlo method.

Monte Carlo method: Also simply called Monte Carlo. A computational method which uses random sampling repetitively so as to estimate certain quantities in terms of statistical measure. For example, the area of complex geometry or integral can be estimated by sampling a region using the Monte Carlo method. The large number theroem guarantees that as the sampling increases, the approximation will approach the true value eventually.

Mutlimodal: A function is called multimodal if it has more than one local mode (valley and peak).

Multimodal optimization: The optimization deals with multimodal objective functions. Most practical optimization problems are nonlinear and multimodal. Algorithms work for unimodal optimization usually do not work well for multimodal cases.

Multiobjective: The multiple objective functions in a multiobjective optimization problem. Multiobjectives are also called multi-criteria.

Multi-criteria: See multiobjective.

Multiobjective optimization: An optimization problem with more than one objective function. The objectives are often conflicting. For example, finding the values x so that both x^2 and $(x-2)^2$ are minimal.

Multivariate: More than one independent variable. For example, $f(x, y)$ is a bivariate function.

Natural computing: An emerging branch of computer science using nature-inspired algorithms for computation, optimization, hardware design and system-level modeling.

Nature-inspired algorithm: An algorithm has been developed by learning from nature. Almost all metaheuristic algorithms are nature-inspired such as bioinspired algorithms.

Nelder-Mead method: Also called Nelder-Mead downhill simplex, pioneered by J. Nelder and R. Mead in 1965. It is a nonlinear optimization technique for finding the minimum using downhill simplex in terms of a polytope with $n+1$ vertices in an n-dimensional search space. As it can converge at non-stationary points (so not the actual optima), it is essentially a heuristic algorithm. There have been various improved variants of this method. This method is also a gradient-free method.

Neural networks: Also artificial neural networks. A classic and well-studied type of method for artificial intelligence and machine learning. It uses multilayer (often three) of neurons and interconnections forming a network to simulate system output from inputs or stimuli. It has many applications such as adaptive control and speech recognition.

Newton-Raphson method: A numerical method for solving or finding the root(s) of a nonlinear function.

No-free-lunch theorems: A few theorems, proved by D. H. Wolpert and W. G. Macready in 1997, concern the algorithms for optimization. The main conclusion is that if algorithm A is better than algorithm B for some problems, then B will outperform A for other problems. That is to say, both algorithms are statistically equivalent over all possible problems. In other words, it is theoretically impossible to design a universally efficient all-purpose optimization algorithm.

Noise: The stochastic component of a quantity with uncertainty. In optimization, this often means the stochastic variations of a parameter or function values.

Nonlinear: A nonlinear function is a function which is not linear. Mathematically, a function is called nonlinear if $f(x+y) \neq f(x) + f(y)$ and $f(\alpha x) \neq \alpha f(x)$ where α is a constant. Most functions such as $\sin(x)$ and $\exp(x)$ are not linear.

Nonlinear programming: A very general class of optimization whose objective and/or constraints are nonlinear functions. Most practical problems are nonlinear. Krush-Kuhn-Tucker conditions are linked with the optimality of nonlinear optimization, though there is no universality in nonlinear optimization. Global optimality can only be guaranteed for a few special class of optimization, such as convex optimization.

Norm: For a real number, its norm is the absolute value. For a vector $\boldsymbol{u} = (u_1, ..., u_n)^T$, its p-norm is defined as $\|\boldsymbol{u}\|_p = (\sum_{i=1}^n |u_i|^p)^{1/p}$ where $p = 1, 2, ...$ is an integer.

Normal distribution: A probability distribution. See Gaussian distribution.

NP: Nondeterministic polynomial (NP).

NP-complete: A problem is called NP-complete if it is an NP-hard problem and all other problems are reducible to it via certain reduction algorithms and such reduction only takes a polynomial time.

NP-hard: A problem is called hard if the solution time is an exponential function of its problem size n. An NP-hard problem has no efficient algorithms to find its solutions, and the solutions can only be obtained either by guess or heuristically.

Numerical integration: A numerical method to approximate or evaluate the value of a complex integral.

Objective function: Also simply objective, or cost function, sometimes also called a criterion. An objective function is a function to be optimized for an optimization problem.

Octave: A matlab-like open source high-level environment for numerical computing under Gnu public license. It uses Gnuplot as a graphic engine. See www.gnu.org/software/octave

ODE: An ordinary differential equation.

Optima: The minima or maxima of an objective function are referred to as optima. An optimum can be local or global. If a minimum is the smallest among all minima, then it is a global minimum. If a maximum is the highest among all maxima, then it is a global maximum.

Optimal: A solution is said to be optimal if it is the best for a given criterion in a neighborhood of feasible solutions. This best (either minimal or maximal) can be local, or global.

Optimality: Also called Optimality criterion.

Optimality criteria: The conditions that the objective functions reach their maxima or minima. For example, $\nabla f(\boldsymbol{x}) = 0$ is the necessary condition for optimality. That is, $f(x)$ can be only optimal when $\nabla f(\boldsymbol{x}) = 0$ is true. But $\nabla f(\boldsymbol{x}) = 0$ is not a sufficient condition.

Optimization: A problem or solution procedure which intends to find the optimal solutions according to a single or multiple objectives under given constraints or conditions.

Order: The degree of an equation, the sequence of a list, and the estimate of magnitude of a quantity, often used in complexity of algorithm. For example, a linear algorithm often has $O(n)$ complexity.

Ordinary differential equations: A relationship or equation including derivatives. For example, $dy/dx = -\exp(-x)$ is an ordinary differential equation.

Pareto optimality: This is an important concept in multiobjective optimization. A point $\boldsymbol{x}_*$ is called Pareto optimal if there exist no feasible solution vector which would decrease some objectives without causing an increase in at least one of other objectives simultaneously.

Partial derivative: For a function of two or more independent variables, its partial derivative is the derivative with respect to one of those variables. For example, $f(x, y) = xy + y^2$, we have $\frac{\partial f}{\partial x} = y$ and $\frac{\partial f}{\partial y} = x + 2y$.

Particle swarm optimization: A metaheuristic optimization technique, first developed by James Kennedy and Russell Eberhart in 1995, which uses a set of particles or agents and each particle represents a solution to the optimization problem. The particle motion consists of a random component, and a deterministic component attracting towards the current best solutions.

Partial differential equation: A relationship or equation which contains partial derivatives such as $\frac{\partial u}{\partial x}$ and/or derivatives. For example, $\frac{\partial u}{\partial t} = D\frac{\partial^2 u}{\partial x^2}$ is a partial differential equation.

Pheromone: A chemical messenger using by many social insects such as ants.

Piecewise continuous: A function over an interval $[a, b]$ is called piercewise continuous if it is continuous in each subinterval of $[a, b]$ with at most a finite number of discontinuity.

Piecewise differentiable: A function is called piecewise differentiable if it is differentiable in each subdomain except for a finite number of points. For example, $|x|$ is not differentiable at $x = 0$, but is piecewise differentiable in $\Re$ because it is differentiable in both $(-\infty, 0)$ and $(0, \infty)$.

Piecewise smooth: A curve is called piecewise smooth if it can be partitioned into smaller pieces and on each piece both the continuity and smoothness condition hold.

p-Norm: For a vector $\boldsymbol{u} = (u_1, ..., u_n)^T$, its p-norm is defined as $\|\boldsymbol{u}\|_p = (\sum_{i=1}^n |u_i|^p)^{1/p}$ where $p = 1, 2, ...$ is an integer. For example, the two-norm is $\|\boldsymbol{u}\|_2 = \sqrt{u_1^2 + ... + u_n^2}$ which is often called ℓ_2-norm, also Euclidean norm. In addition, $\|\boldsymbol{u}\|_\infty = \max |u_i|$. The norm of a matrix can be defined in a similar manner, though slightly more complicated.

Poisson's distribution: A probability distribution for discrete events. $p(k) = \lambda^k e^{-\lambda}/k!$ where $k = 0, 1, 2, ...$ and $\lambda > 0$.

Polynomial: An expression with a finite number of terms involving the powers of x. For example, $x^2 - 2x + 3$ a quadratic polynomial while $x^5 - x^3 - x$ is a polynomial of degree 5.

Polynomial time: A problem if called polynomial-time or P-problem if the number of steps needed to find the solution is bounded by a polynomial of problem size n. If an algorithm has a polynomial solution time, we say it is efficiently in the context of computational complexity, though in practice it is not necessarily quick enough.

Population-based: An algorithm is called population-based if it uses multiple representations and/or agents. For example, genetic algorithms use multiple gene strings to represent solutions, while particle swarm optimization uses many particles for solutions.

Programming: A mathematical term means planning and/or optimization. So loosely speaking, nonlinear programming is nonlinear optimization, though in mathematical literature, many prefer to use the term 'programming'.

Principle of maximum entropy: A general principle that the true distribution of a probabilistic system tends to be in the state with maximum entropy. It is also an important concept related to convex optimization.

Probability: A number or expected frequency of an event occurring. It is always in the range of $[0, 1]$.

Problem functions: A collective term means both the objective functions and the constraints of an optimization problem.

Pseudo code: A schematic representation or procedure to explain the steps of an algorithm so as to guide its implementation in a programming language such as C and Matlab. It is a higher level code or meta-code, rather than an actual computer code or program.

Pseudo-random numbers: A pseudo-random number is a random-like number generated by a deterministic computer algorithm. They can pass certain statistical tests, and can effectively be used as random numbers.

Pseudo-random number generator: A deterministic algorithm, often in terms of a congruential linear sequence, to generate pseudo-random numbers for Monte Carlo simulations. The uniform distribution generator in also almost all computer languages is a good example.

PSO: see particle swarm optimization.

Quadratic form: A mathematical form or expression written as homogeneous quadratic polynomials. For example, for the bivariate case, we have $Q(x, y) = \alpha x^2 + \beta xy + \gamma y^2$ where $\beta = \alpha + \gamma$.

Quadratic programming: A special case of mathematical optimization where the objective function takes the quadratic form: $f(\boldsymbol{u}) = \boldsymbol{u}^T \boldsymbol{A} \boldsymbol{u} + \boldsymbol{b}^T \boldsymbol{u} + \boldsymbol{c}$.

Quasi-Monte Carlo method: Also simply quasi-Monte Carlo. A variant of Monte Carlo methods that uses a more 'regular' distribution of sampling points, rather the pseudo-random sampling. The convergence of standard Monte Carlo is $O(1/\sqrt{N})$ where N is the sample size. In quasi-Monte Carlo, the convergence rate is $O(1/N)$, and thus more efficient.

Quasi-random: A random-style number used for quasi-Monte Carlo simulations. See low discrepancy sequence.

Random number: A number generated by a realization of a random process. For example, the values of noise level on the street are random numbers. However, the random numbers generated by a computer algorithm is not really random, though they may pass certain statistical tests. So they are called pseudo-random numbers.

Random walk: A stochastic process which consists of a series of random steps, forming a trajectory of consecutive random steps. For example, a drunkard's walk pattern, a gas molecule and the Brownian motion are all random walks. Diffusion process is also a random walk. A random walk is a Markov chain.

Random variable: A variable to represent the outcome of an event such as the noise level or the number of earthquakes in a period.

Randomization: A stochastic operation for almost all stochastic algorithms. New solutions are often generated by adding some random component to an existing solution, or purely randomly sample the search space. The aim is to generate as diverse solutions as possible.

Random restart: Most local search algorithm will need to start from an initial random guess. In the case of multimodal optimization, the final results will depend on the initial guess. In order to explore different modes or regions of the search space, some iterations with different starting points are needed. This is called random restart.

Random restart hill-climbing: A strategy of hill-climbing where the objective function is multimodal. In order to avoid being stuck in a local optimum (peak), an iterative procedure combined with random restart is used so that the system can climb to the global peak if randomization is good enough.

Random search: A stochastic search which tries to find solution by randomly explore the search space. There are two main types: blind random search and localized random search. The blind random search does not use any historical informaton of previous search, and it searches by sampling the search space randomly. It is the simplest but not very efficient. A better way is the localized random search which randomly search the neighborhood of a previous or existing solution. This is similar to a random walk around a previously known solution.

Response: The objective function. The structure or the variations of the objective is often called the response surface, or landscape. The space spanned by its values is called response space.

Robust: A vague term. In optimization, it usually means that an algorithm still works despite there is uncertainty or noise in the inputs or a wide range of changes in parameters.

Robust optimization: A special class of optimization deal with how the optimal solutions may be affected by the uncertainty in the design variables. For example, many engineering design of product, the actual design may be affected by the real properties of available materials.

Rosenbrock's banana function: Also called, Rosenbrock's function, or banana function. A classic test function $f(x, y) = (1-x)^2 + 100(y-x^2)^2$ which has a global minimum $f(1, 1) = 0$ at $(1, 1)$. The valley of the function in 2D looks like a curved banana.

Robust optimization: A special class of optimization, involving uncertainty. For deterministic optimization problems, a global solution is exact. If there is uncertainty in design variables, a small change or uncertainty will usually render its global optima meaningless. This type of optimization focuses on how uncertainty may affect the optimal solutions.

Sampling: A way of drawing a sample or generating random values or numbers for a given probability distribution. It also means the way of distributing sampling points stochastically in the search space.

Scilab: A clone of matlab and free available package for computing. See www.scilab.org

Search space: Also design space. The space spanned by the design variables is called the search space. All optimization algorithms intend to search through this space in certain way, not necessarily efficiently.

Sensitivity analysis: The analysis of how the variations of the response variable(s) of a mathematical model can be attributed to design variables or the inputs. The aim is to identify the most important variables or parameters for a given problem, and to answer what-if type of questions.

Shape optimization: Also topology optimization. A branch of optimization aiming to find the optimal shape such as the airfoil so as to *minimize*

certain objective functionals (rather than simple function) subjected to multiple constraints. Now it is widely used in optimal control and structural optimization in engineering.

Simplex method: A powerful method in linear programming, pioneered by George Dantzig in 1947, which intends to search along the boundaries linked by extreme points determined by the constraints. Another simplex method for nonlinear optimization, which has nothing to with the simplex method for linear programming, is a search method in n-dimensional polytope using a simplex with $n+1$ vertices (or a polytope or convex hull) and by flipping over the landscape.

Simulated annealing: A metaheuristic algorithm for optimization, inspired by the annealing process of metals and developed by Kirkpatrick *et al.* in 1983. It is a trajectory-based algorithm, and is essentially a Markov chain by random walk. A search move for better solutions is accepted with a probability if there is no improvement in the quality, while a better solution is always accepted. A global convergence is almost guaranteed due to its link with the stationary property of Markov chains.

Singular point: A point of function or curve is called a singular point if the function blows up or significant change in property. For complex functions, that is the point at which the function is no longer analytic. For a real function, it usually the function itself is finite, or its derivative is not defined. For example, $x = 0$ is a singular point of $1/x$, and it is also a singular point of $x^{2/3}$ because it forms a cusp and it is not differentiable at $x = 0$.

Slack variable: A non-negative auxiliary variable used to convert an inequality constraint to an equality in linear programming. For example, the inequality $5x + 6y \leq 20$ can be written as $5x + 6y + s = 20$ where $s \geq 0$ is a slack variable.

Smooth: A function is called smooth if it belongs to C^{∞}. That is to say, the function is continuous and has derivatives of all orders (up to infinity). For example, $\exp(-x)$ is smooth.

Soft computing: A branch of computer science dealing with methods of finding inexact solutions to hard computation problems. For example, neural networks and fuzzy systems are considered as part of soft computing, though this definition is not well accepted.

Standard deviation: A number which is the square root of the variance of a random variable.

Stationary condition: A mathematical condition which the first derivative or gradient of a function is zero. For example, $df(x)/dx = 0$ for a univariate function.

Stationary point: A point at which the stationary condition holds. For a univariate function $f(x)$, it is the solution of $df(x)/dx = 0$.

Statistics: The studies of data collection, interpretation, analysis and characterisation of numerical data and sampling.

Steepest descent method: A classic method for finding the minimum of a function. It starts with an initial solution and descent downhill in the maximum or steepest gradient direction so as to reach the local minimum as quickly as possible. A related method is the hill-climbing where the algorithm intends to move up hill to a local peak.

Stochastic optimization: A special class of optimization problems where some probabilistic components or uncertainty are intrinsically presented in objective functions, constraints and/or design variables. Some forms of averaging or statistic measures should be used. In some literature, this is confused with stochastic search algorithms. Alternatively, stochastic programming, and/or optimization with uncertainty or noise is used in some literature. If design variables are not exact, or known with uncertainty, the optimization becomes robust optimization.

Stochastic programming: A class of algorithm or a framework for optimization problems with uncertainty or parameters with probabilistic distributions. For example, robust optimization belongs to such a framework.

Stochastic search: A class of search methods employing a strong component of randomization with the guided aim. The random search is a stochastic search algorithm, however, most stochastic search methods are rarely purely random. Selection of the best is often used.

Stochastic tunneling: An optimization algorithm which uses the tunneling idea in quantum mechanics to transform the landscape of an objective function so as to suppress certain modes and retain the modes of unvisited regions. For example, the tunneling function for an objective f is defined as $\Psi = 1 - \exp[-\gamma(f - f_{\min})]$ where $f_{\min}$ is the current minimum found so far, and $\gamma > 0$ is a scaling parameter which can be adjusted dynamically.

Strong maximum: A point $\boldsymbol{x}_*$ is called a strong local maximum in a neighborhood of $N(\boldsymbol{x}_*, \epsilon)$ if $f(\boldsymbol{x}_*) > f(\boldsymbol{u})$ for all $\forall \boldsymbol{u} \in N(\boldsymbol{x}_*, \epsilon)$ with $\epsilon > 0$ and $\boldsymbol{u} \neq \boldsymbol{x}_*$. The inclusion of $=$ will lead to a weak maximum. The strong minimum can be defined similarly.

Support vector machine: A kernel-based technique for supervised machine learning, data classification and regression. For example, the construction of a high-dimensional hyperplane with maximum separation distance can be used for efficient classification of data sets.

Swarm intelligence: A collective name for a special class of artificial intelligence and agent-based system. It uses multiple agents in a decentralized, often rule-based manner. The emerging self-organized behavior of these swarming agent can mimic certain characteristics of intelligence and thus powerful to problem solving. For example, ant colony optimization and particle swarm optimization all belong to the category of swarm intelligence.

Tabu search: A powerful search optimization, developed by F. Glover in later 1980s and 1990s, which uses memory via a Tabu list to avoid revisiting recently visited neighborhood. This may substantially increase the efficiency for solving some problems.

Test function: A standard function in literature, specially designed to test and compare the performance of optimization algorithms. All new algorithms should be tested against these functions. For example, Rosenbrock's banana function $f(x, y) = (1-x)^2 + 100(y-x^2)^2$ is a simple 2D test function.

Trajectory-based: A search algorithm updating its solution through time or iteration, and its search path forms a piecewise trajectory in the search space. For example, simulated annealing is a trajectory-based search algorithm.

Traveling salesman problem: A combinatorial optimization problem of finding the shortest route to visit each city exactly once of n known cities. This is an NP-hard problem and no efficient algorithm exists.

Trust region: It is a mathematical term in optimization. The idea is to carry out some approximations over a selected region, so-called, trust region, so as to evaluate the objective functions more efficiently and the algorithm can converge more quickly. The trust region is often circular or a hyper-sphere.

Trust region method: A search method using the idea of trust region. It is a local search method, and typically represent the function by some quadratic form, and the size of the trust region is adjusted according to how well the approximation agree with the actual objective function.

Uncertainty: A measure of the variations of the values of a random variable of a stochastic process. In most cases, the uncertainty is measured by the standard derivation, or sometimes variance.

Unconstrained optimization: A special class of optimization which has no constraint at all. For example, the minimization of x^2 is a simple unconstrained optimization problem.

Uniform distribution: A simple probability distribution where each small interval of the same length in a large interval $[a, b]$ will have an equal probability. The probability density in $[a, b]$ is $1/(b-a)$ if $a \leq x \leq b$, otherwise 0.

Unimodal: A response surface is called unimodal if it has a single mode. In this case, the local optimum is the same as the global one. For example, a parabola $f(x) = x^2$ is unimodal because $f_{\min} = 0$. A special case of unimodal functions is convex. A convex function is unimodal, but unimodal does not necessarily mean convex.

Univarite: A univariate function is a function which only depends on a single independent variable. For $f(x)$ is a univariate function.

Utility function: A utility function is also called a preference function. It is a function associated with the risk preference of multiobjective optimization, and it is a way of representing preference and combining the multiple objective functions so that the optimization becomes optimizing the utility.

Variance: A number or a measure of statistical dispersion of a random variable, often denoted by σ^2. The square root of the variance is called the standard deviation σ.

Vector: A vector is a quantity such as force with both a magnitude or length and a direction.

Virtual ant algorithm: A variant of ant algorithms, developed by Xin-She Yang *et al.* at Cambridge university in 2006, for black-box type optimization in combination with finite element analysis, and it was first applied to continuous optimization.

Virtual bee algorithm: A special variant of bee algorithms, developed by Xin-She Yang at Cambridge University in 2005, for engineering optimization and continuous optimization problems.

Weak maximum: A point $\boldsymbol{x}_*$ is called a weak maximum in a neighborhood $N(\boldsymbol{x}_*, \epsilon)$ if $f(\boldsymbol{x}_*) \geq f(\boldsymbol{u})$ for all $\forall \boldsymbol{u} \in N(\boldsymbol{x}_*, \epsilon)$ for $\epsilon > 0$ and $\boldsymbol{u} \neq \boldsymbol{x}_*$. Similarly, we can define the weak minimum accordingly.

Weighted sum: A sum or linear combination of n functions $f_i(\boldsymbol{x})$ so as to form a single composite function. This is often used in least squares and multiobjective optimization. For example, for n objective functions, we can combine them into $F(\boldsymbol{x}) = \alpha_1 f_1(\boldsymbol{x}) + \alpha_2 f_2(\boldsymbol{x}) + ... + \alpha_n f_n(\boldsymbol{x})$ where $\alpha_i \geq 0$ and $\sum_{i=1}^{n} \alpha_i = 1$.

APPENDIX D

PROBLEM SOLUTIONS

SOLUTIONS FOR CHAPTER 1

1.1 From basic mathematics, we know that $f(x) = x^2 - x - 6 = (x+2)(x-3) = 0$ has two solutions $x_1 = -2$ and $x_2 = 3$. Thus, the optimum occurs at $x_* = (x_1 + x_2)/2 = 1/2$, and this minimum is $f_{\min} = f(x_*) = f(1/2) = (1/2)^2 - 1/2 - 6 = -25/4$.

1.2 From simple calculus, we know that an optimum occurs when $f'(x) = 0$. That is, $f'(x) = 2x - 1 = 0$, or $x_* = 1/2$. In addition, the second derivative $f''(x) = 2 > 0$, this optimum is a minimum and we have $f_*(1/2) = -25/4$.

1.3 For a fixed total length L of a rectangle, let x and y be the lengths of two adjacent sides, respectively. We have $2x + 2y = L$ or $y = L/2 - x$. So the area is $A = xy = x(L/2 - x)$. A reaches its maximum if $dA/dx = 0$ or $L/2 - 2x = 0$. This means $x = L/4$ and $y = L/2 - x = L/4 = x$. The rectangle becomes a square. As the second derivative $d^2A/dx^2 = -2 < 0$, $A = L^2/16$ is indeed the maximum.

Engineering Optimization: An Introduction with Metaheuristic Applications.
By Xin-She Yang
Copyright © 2010 John Wiley & Sons, Inc.

1.4 To design a cylindrical water tank with a base radius r and a height h, the volume is $V = \pi r^2 h$. The total material used is proportional to the total surface area $S = \pi r^2$ (top) + $2\pi r h$ (side) + πr^2 (bottom). Without the loss of generality, we can assume that S is constant, and the objective now is to maximize V for a fixed S. From $S = 2\pi r^2 + 2\pi r h$, we have $h = (S - 2\pi r^2)/(2\pi r)$. This means that

$$V = \frac{r}{2}(S - 2\pi r^2).$$

From $dV/dr = \frac{1}{2}(S - 2\pi r^2) - 2\pi r^2 = \frac{S}{2} - 3\pi r^2 = 0$, we have $S = 6\pi r^2$. Then, we have

$$h = (S - 2\pi r^2)/(2\pi r^2) = 2r,$$

which leads to $V_{\max} = \pi r^2 h = 2\pi r^3 = S^{3/2}/(3\sqrt{6\pi})$. Interestingly, if there is no top cover, then $S = \pi r^2 + 2\pi r h$. Following the same process again, we have $h = r$ and $V_{\text{new}} = \pi r^3 = S^{3/2}/(3\sqrt{3\pi}) = \sqrt{2} V_{\max}$. Obviously, this means that for the same fixed amount of material, the volume is higher. It is worth pointing that most water bottles are not designed in an optimal way according to this simple criterion discussed here. The main reason is that there are other important criteria to consider for such a product including the shape of the bottle, the manufacturability of the bottle top, and pressure design rules (for champagne bottles, and sparkling water bottles). In addition, they should also look right.

1.5 Heron's proof in about 100BC is very interesting, which uses the inequality of a triangle where the total length of any two sides is always greater than the length of the third side. With reference to Figure D.1, we assume that the path PAQ is the right path for light reflection such that $\alpha = \beta$. The only thing we have to prove that any other path such as PBQ is longer than PAQ. From simple geometry, we know that AQ=AS, and the total length of PAQ is equal to PAS where S is the mirror image of Q. For any other point B, which is different from A, we know that BQ=BS, this implies that the total length of PBQ is the same as the length of PBS. For the triangle PSB, the total length of PB and BS is always greater than the third side PS=PAQ. Therefore, we have the length of PBS=PBQ which is greater than PAQ, which means that PAQ is the shortest.

SOLUTIONS FOR CHAPTER 2

2.1 As there is no constraint at all, so this is an unconstrained optimization problem. By observation, since both x^2 and y^2 are non-negative when x and y are real variables, then the minimum $f = 0$ occurs at $x = y = 0$.

2.2 (a) constrained nonlinear optimization with a single objective; (b) unconstrained multiobjective optimization; (c) linear optimization with inequality

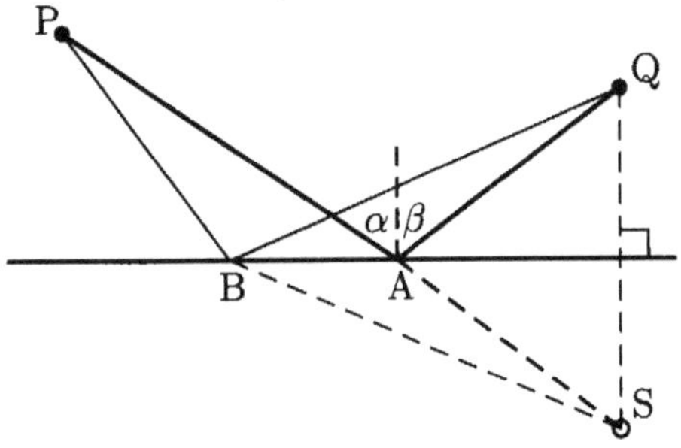

Figure D.1: Heron's proof of the shortest path.

constraints; (d) nonlinear optimization with an equality constraint; (e) multiobjective optimization with multiple constraints; (f) stochastic, black-box, multiobjective optimization with multiple constraints;

2.3 (a) The addition of n numbers, usually takes n floating-point operations, so the computational complexity is $O(n)$. (b) Each new entry of the product is the sum of n terms (row by column), and the product of the two matrix has a size of $n \times n$, so we have n^2 entries, and the complexity is $O(n^2 \times n) = O(n^3)$. (c) To find the maximum value of n unsorted random numbers, we can search through each value once, so $O(n)$ is the complexity. (d) Evaluation of a polynomial of degree n leads to the addition of $n+1$ terms. That is $O(n)$. (e) For simple sorting of a set of n number using simple pair comparison, we need to use two loops as there are n^2 pairs. So the complexity is simply $O(n^2)$. Of course, many sorting algorithms intend to do this more efficiently. For example, quicksort has a complexity $O(n \log n)$.

2.4 (a) $O(n)$, (b) $O(n^2)$, (c) $O(n^3)$, (d) $O(n^5)$, (e) $O(n \log n)$.

2.5 From the constraints, we know that x, y are non-negative, so the objective is to increase both as high as possible. However, the maximum allowable x and y are 5 and 2, respectively. From $2x + y = 8$, we have $y = 8 - 2x$. In order for y to remain non-negative while increasing x, we have $x = 4$ as the maximum, which corresponds to point A at $(4, 0)$ in Figure D.2. However, the maximum possible y is 2, this leads to $x = 3$ (point B). As both x and y increase, the objective function f reaches the highest possible value at $B(3, 2)$. That is $f = x + y = 3 + 2 = 5$.

SOLUTIONS FOR CHAPTER 3

3.1 The stationary condition is simply $d\cos(x)/dx = -\sin(x) = 0$. This means that $x = n\pi$ (where $n = 0, \pm 1, \pm 2, ...$). The infection points are deter-

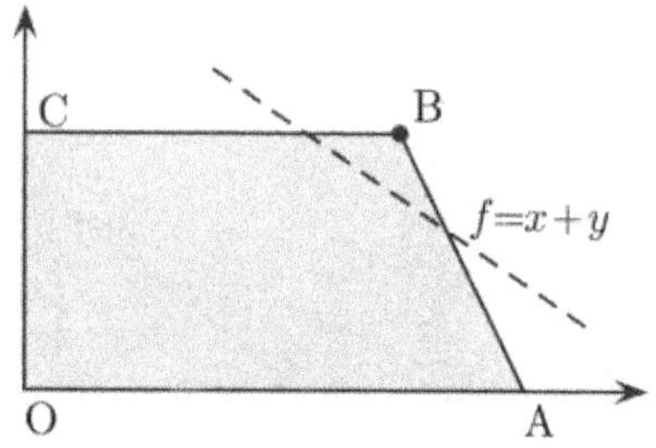

Figure D.2: The feasible region of design variables of a simple linear programming problem.

mined by $\cos''(x) = -\cos(x) = 0$ which leads to $x = n\pi + \pi/2 = (n + 1/2)\pi$, $(n = 0, \pm 1, \pm 2, ...)$.

3.2 From the stationary condition $f'(x) = -2x\exp(-x^2) = 0$, we have $x = 0$. As $f''(x) = 4x^2\exp(-x^2) - 2\exp(-x^2)$ and $f''(0) < 0$, then $f(0) = \exp(-0^2) = 1$ is the maximum. The inflection points are given by

$$f''(x) = 2(2x^2 - 1)\exp(-x^2) = 0.$$

As $\exp(-x^2) > 0$, we have $2x^2 - 1 = 0$ or $x = \pm\sqrt{2}/2$.

3.3 To find the maximum, we have

$$f'(x) = \frac{\cos(x)}{x} - \frac{\sin(x)}{x^2} = \frac{1}{x^2}[x\cos(x) - \sin(x)] = 0.$$

There are two cases $x = 0$ and $x \neq 0$. If $x = 0$, then we have a solution, and $f(0) = \sin(0)/0 \to 1$. We have to use limits, otherwise, in real-number computing, this may cause the division-by-zero error.

For the case of $x \neq 0$, then we have

$$x\cos(x) - \sin(x) = 0.$$

We have to solve it numerically using a root-finding algorithm. The non-zero solutions or critical points are $x \approx 4.49, 7.73, 10.9, 14.1, ...$ As x increases, $f(x)$ will decrease since $|\sin(x)| \leq 1$, so the global optimum occurs at $f_{\max} = 1$ (see Figure D.3).

3.4 The maximum can only occur at $f'(x) = -\exp(-x^x)x^x(1 + \ln x) = 0$. Since $x > 0$, both x^x and $\exp(-x^x) > 0$, then $1 + \ln x = 0$. That is $x = 1/e$. So the maximum at $x = 1/e$ is $\exp[-(1/e)^{1/e}]$.

3.5 Even though $k > 0$ and each term of $x^2 + k\cos^2(x)$ is always non-negative, the minimization of each term does not lead to the minimization of the whole expression as we will see below.

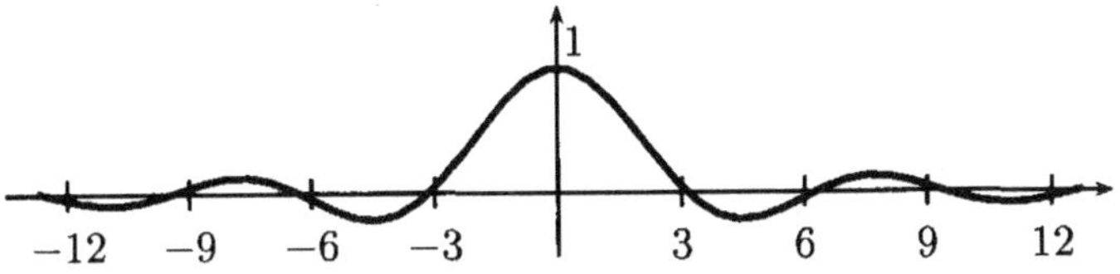

Figure D.3: The plot of $\text{sinc}(x) = \sin(x)/x$.

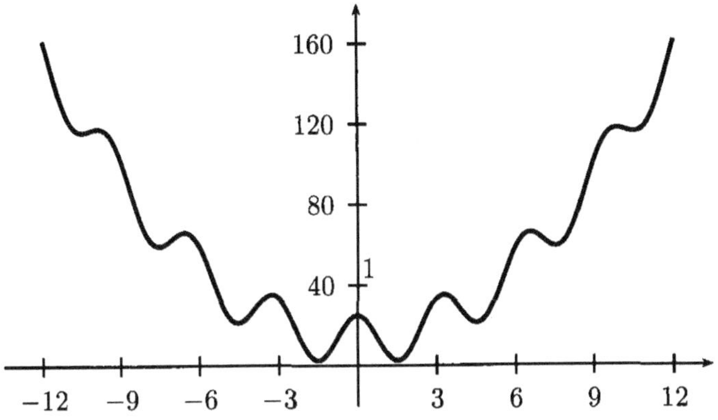

Figure D.4: The plot of $f(x) = x^2 + 25\cos^2(x)$.

From $f'(x) = 0$, we have $2x - 2\cos(x)\sin(x) = 0$, which becomes $k\sin(2x) = 2x$. We can see that $x = 0$ is a solution whatever the value of k is. If we increase k up to $k = 1$, we still get the same solution. So there is at least one minimum $f_{\min} = 0$ which at the moment occurs at $x = 0$ for $0 < k \leq 1$.

The situation starts to change once $k > 1$. There are now two solutions and even more. For example, for the case of $k = 2$, we have $\sin(2x) = x$, and we can get $x_* \approx \pm 0.9477...$ numerically. For example, the case of $k = 25$ is shown in Figure D.4 where the global two minima are obvious.

3.7 The main difficulty is that the derivative of $|\tan x|$ is not well defined, especially at $x = \pm\pi/2$. If we take the derivative in each appropriate interval and carefully make sure the continuity, we have

$$f'(x) = -\text{sign}[\tan x](1 + \tan^2 x)/(1 + |\tan x|)^2,$$

where $\text{sign}(x)$ is the sign function, which is $+1$ if $x > 0$, -1 if $x < 0$, and 0 if $x = 0$. From the expression, we know that if $x \to \pm\pi/2$, $\tan(x) \to \pm\infty$, and $(1+\tan^2 x)/(1+|\tan(x)|)^2 \to 1$, which means that $f'(\pi/2) \to 1$ from the right $\frac{\pi}{2}+$ and -1 from the left $\frac{\pi}{2}-$. So $f'(x)$ is not defined at $x = \pm\pi/2$. If we plot out the graph of $f(x)$, we know its maxima $f_{\max} = 1$ occur at $x = 0$ and $x = \pm\pi$, while its minima $f_{\min} = -1$ occur at $x = \pm\pi/2$. This example has demonstrated that optimization is not straightforward even for seemingly simple cases, especially when the objective function has singularity. In fact, here $f(x)$ is piecewise smooth, and is thus a piecewise function. In convex analysis, the derivative concept should be replaced by more appropriate subderivatives.

3.8 (a) It is positive definite. The principal minors of $\boldsymbol{A}$ are $\boldsymbol{A}_1 = (2)$ and $\boldsymbol{A}_2 = \begin{pmatrix} 2 & 3 \\ 3 & 5 \end{pmatrix}$, and their determinants 2 and 1 are both positive, respectively. (b) It is neither positive definite or negative definite because the determinants of its principal minors -2 and -10 are all negative. In addition, one eigenvalue is positive and the other is negative. (c) It is negative definite because its two eigenvalues are simply -6 and -1. Since two eigenvalues are all negative, $\boldsymbol{B}$ is negative definite. If we use its two principal minors $\boldsymbol{A}_1 = (-5)$ and $\boldsymbol{A}_2 = \begin{pmatrix} -5 & 2 \\ 2 & -2 \end{pmatrix}$, their determinants are -5 and 6, respectively. They alternative in sign starting $-5 < 0$, so the matrix is negative definite.

3.9 (a) The Hessian of $f(x, y) = x^2 + e^y$ is

$$\boldsymbol{H} = \begin{pmatrix} 2 & 0 \\ 0 & e^y \end{pmatrix},$$

which is positive definite because its two eigenvalues 2 and e^y are always positive. So $f(x, y)$ is convex.

(b) The Hessian of $f(x, y) = 1 - x^2 - y^4$ is $\boldsymbol{H} = \begin{pmatrix} -2 & 0 \\ 0 & -12y^2 \end{pmatrix}$ which is negative definite (as its two eigenvalues -2 and $-12y^2$ are always negative). Therefore, $1 - x^2 - y^4$ is concave in the domain $\Re^2$.

(c) The Hessian of $f(x, y) = xy$ is simply $\boldsymbol{H} = \begin{pmatrix} 0 & 1 \\ 1 & 0 \end{pmatrix}$, and it is indefinite. So $f(x, y)$ is not convex or concave.

(d) The Hessian of $f(x, y) = x^2/y$ for $y > 0$ is $\boldsymbol{H} = \frac{2}{y^2}\begin{pmatrix} y & -x \\ -x & x^2/y \end{pmatrix}$ which is positive semidefinite. So $f(x, y) = x^2/y$ is convex for $y > 0$.

SOLUTIONS FOR CHAPTER 4

4.1 We can define a penalty function $\Pi(x)$ using a penalty parameter $\mu \gg 1$. We have

$$\Pi(x, \mu) = f(x) + \frac{\mu}{2} g(x)^T g(x) = 100(x - b)^2 + \pi + \frac{\mu}{2}(x - a)^2,$$

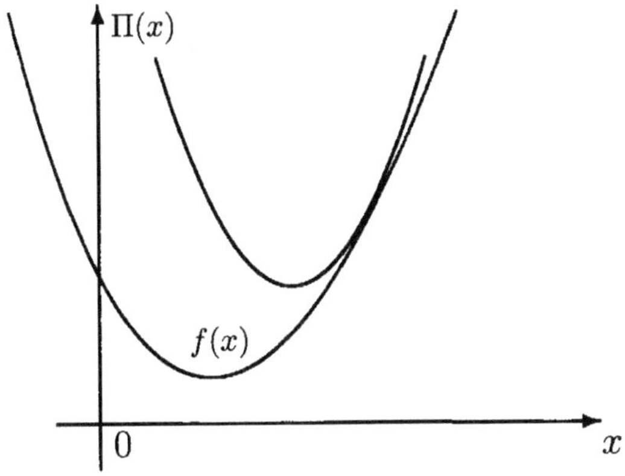

Figure D.5: Quadratic penalty function $\Pi(x,\mu) = 100(x-1)^2 + \pi + \frac{\mu}{2}(x-a)^2$ and $\mu = 2000$.

where the typical value for μ is $2000 \sim 10000$.

This essentially transforms the original constrained optimization into an unconstrained problem. From the stationary condition $\Pi'(x) = 0$, we have

$$200(x_* - b) - \mu(x_* - a) = 0,$$

which gives

$$x_* = \frac{200b + \mu a}{200 + \mu}.$$

For $\mu \to \infty$, we have $x_* \to a$. For $\mu = 2000$, $a = 2$ and $b = 1$, we have $x_* \approx 1.9090$. This means the solution depends on the value of μ, and it is difficult to use extremely large values without causing extra computational difficulties.

4.2 It is easy to verify that the optimal solution is $f_* = 0$ at $\boldsymbol{x}_* = (0, 0, ..., 0)$. A good program should produce an approximate solution with a certain accuracy, say 10^{-5}.

4.3 As you know, this is the well-known travelling salesman problem and there is no efficient way to find the optimal solution. The idea here is to let you try various methods or even invent your own method. Often, a naive strategy is to go to the nearest next city, which does not produce an optimal solution in general. To demonstrate this, let us see a 4-city case located at A(0,0), B(0,50), C(50,50) and D(25,25) (see Figure D.6). Here it is easy to check among all $n! = 4! = 24$ combinations, the route ABCD is the shortest. However, if we start at city B, the shortest neighbor is city D, not city A or C. If we use the naive nearest neighbor strategy, we reach D from B. Then,

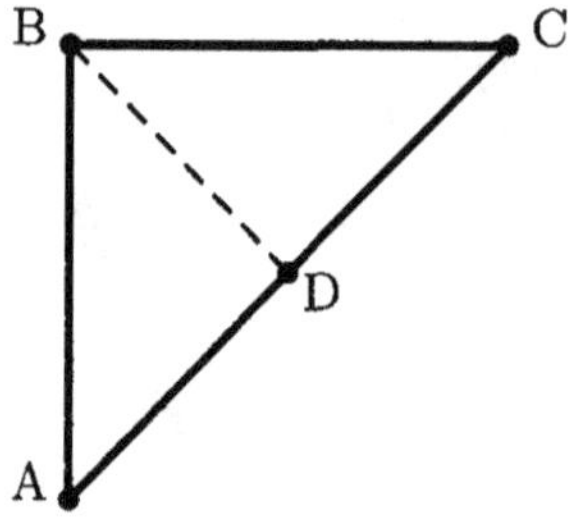

Figure D.6: A simple route to tour 4 cities.

we can either go to C or A from D. In either case, the overall distance is not the shortest.

4.4 This function is not differentiable at $(0,0)$, and this singularity corresponds to a kink point. Thus, it may pose some difficulty using gradient-based method. Fortunately, the kink point at $(0,0)$ corresponds the minimum point $f_{\min} = 0$, as $|x| + |y| > 0$ and $e^{-x^2-y^2} > 0$ if $x \neq 0$ and $y \neq 0$.

As we intend to find the maximum value of f, we can simply exclude the only singularity at $(0,0)$, and the function is differentiable in the rest of the domain. In addition, f is symmetric with respect to x and y, so we can simply only consider one quadrant $x > 0, y > 0$. The stationary conditions lead to

$$\frac{\partial f}{\partial x} = [1 - 2x(x+y)]e^{-x^2-y^2} = 0,$$

$$\frac{\partial f}{\partial y} = [1 - 2y(x+y)]e^{-x^2-y^2} = 0.$$

The symmetry implies that the optimality requires $x_* = y_*$, or

$$1 - 4x_*^2 = 0,$$

or $x_* = \pm 1/2$. Therefore, the maximum $f_{\max} = e^{-1/2}$ occurs at $(1/2, 1/2)$, $(1/2, -1/2)$, $(-1/2, 1/2)$ and $(-1/2, -1/2)$.

4.5 It seems an awkward way to solve a linear system $\boldsymbol{Au} = \boldsymbol{b}$ using linear programming. The idea here is to see the potential of converting some optimization problems to a linear programming problem as there are many efficient solvers for linear programming. To do this, we often use an epigraph approach using an extra variable t, so that we have an equivalent problem

$$\begin{array}{ll} \text{minimize} & t \\ \text{subject to} & \boldsymbol{Au} - \boldsymbol{b} - t\mathbf{1} \leq 0, \\ & -\boldsymbol{Au} + \boldsymbol{b} - t\mathbf{1} \leq 0 \end{array} ,$$

where **1** is the vector of the same size as $\boldsymbol{b}$ with each entry being one.

4.6 There are many ways to implement this method. A simple version is provided in Appendix B.

SOLUTIONS FOR CHAPTER 5

5.1 The global minimum $f_* = -1$ occurs at $x_* = \pi$. You may notice, starting from $x_0 = 5$ may obtain your optimal solution much quicker than starting from $x_0 = -5$.

5.2 If your implementation is proper, your program should be able to find the global minimum $\boldsymbol{x}_* = (0, 1/2, 3 - 1/3, ..., n - 1/n)$. If you have access to the Matlab optimization toolbox, you can use `fmincon` to verify your solutions.

5.3 This tunneling idea is based on the quantum tunneling phenomenon. It intends to mobilize the systematical search towards the optimum so that it can explore even the most less accessible regions. The optimal solution $\boldsymbol{x}_*$ is preserved by this transform.

SOLUTIONS FOR CHAPTER 6

6.1 The conversion is related to the epigraph method by introducing an extra variable t such that

$$\underset{\boldsymbol{x},t}{\text{minimize}} \; t,$$

$$\text{subject to} \quad f(\boldsymbol{x}) - t \le 0, \quad g(\boldsymbol{x}) \le 0, \quad h(\boldsymbol{x}) = 0.$$

It is straightforward to verify that the new problem with a linear objective is indeed equivalent to the original problem

$$\text{minimize} \quad f(\boldsymbol{x}),$$

$$\text{subject to} \quad g(\boldsymbol{x}) \le 0, \qquad h(\boldsymbol{x}) = 0.$$

6.2 As x^2 is convex, so is ix^2. Using the simple addition rule of convex function, the sum $\sum_{i=1}^{n} ix_i^2$ is also convex.

6.3 The global minimum $f_{\min}$ occurs at $\boldsymbol{x}_* = (0, 1, 2, ..., n-1)$, which is independent of α and β. The objective value $f_{\min}$ could vary from $-(K+1)$ to -1, which is essentially random.

The main problem of this stochastic function is that its derivatives also have some randomness, thus most deterministic methods do not work well. Special care is needed in implementation.

SOLUTIONS FOR CHAPTER 7

7.1 In order to find the extremal of the functional J, we can use the Euler-Lagrange equation

$$\frac{\partial \psi}{\partial x} - \frac{d}{dt}\Big(\frac{\partial \psi}{\partial \dot{x}}\Big) = 0,$$

where ψ is the integrand $\psi = \dot{x}^2(1+t)$. Substituting it into the above equation, we have

$$0 - \frac{d}{dt}[2\dot{x}(1+t)] = 0.$$

Integrating it with respect to t, we have

$$2\dot{x}(1+t) = A,$$

where A is an integration constant. This means that

$$\dot{x} = \frac{A}{2(1+t)}.$$

Integrating it again, we have

$$x(t) = \frac{A}{2}\ln(1+t) + B,$$

where B is another integration constant. Using the condition $x(0) = 0$, we have

$$0 = \frac{A}{2}\ln 1 + B,$$

which gives $B = 0$. From $x(1) = 1$, we have

$$1 = \frac{A}{2}\ln 2,$$

or $A = 2/\ln 2$. Therefore, we have

$$x(t) = \frac{\ln(1+t)}{\ln 2}.$$

7.2 This is a case when $\psi(x, ;t) = \dot{x}^2/x^2$ is independent of t. This means that the Euler-Lagrange equation becomes a simpler form

$$\psi - \dot{x}\frac{\partial \psi}{\partial \dot{x}} = \text{const}.$$

Without the loss of generality, we can assume this constant is $-K$. Thus, we have

$$\dot{x}^2/x^2 - \dot{x}(2\dot{x})/x^2 = -K,$$

or

$$\dot{x}^2/x^2 = K,$$

or

$$\dot{x}/x = \sqrt{K}.$$

This means that

$$\frac{d}{dt}(\ln x) = \sqrt{K}.$$

Integrating it with respect to t, we have

$$\ln x = \sqrt{K}t + B.$$

From $x(0) = 1$, we have $B = 0$. From $x(1) = e$, we have

$$\ln e = \sqrt{K}\tau,$$

which means that $K = 1/\tau^2$. Finally, we have

$$x(t) = e^{t/\tau}.$$

This leads to the optimum $J = \int_0^\tau \dot{x}^2/x^2 dt = 1/\tau$.

7.3 Let $h(t)$, $V(t)$ and $M(t)$ are the height, velocity, and mass of the spacecraft at time t, respectively. From Newton's second law, we know that

$$M(t)\ddot{h}(t) = p(t) - M(t)g_m,$$

where g_m is the acceleration due to gravity on the moon. We can write this second-order differential equation as a system

$$\dot{h} = V(t),$$

$$\dot{V} = p(t)/M(t) - g_m.$$

In addition, the fuel consumption to produce the thrust will reduce the mass of the spacecraft. That is

$$\dot{M} = -\beta p(t),$$

where $\beta > 0$. We can write the above three equations in a vector form using

$$\boldsymbol{u}(t) = \begin{pmatrix} h(t) & V(t) & M(t) \end{pmatrix}^T.$$

The objective is to minimize the fuel consumption, that is to maximize the mass $M(\tau)$ at the landing time τ. Now we have the optimal control problem

$$\text{maximize} \;\; M(\tau),$$

$$\text{subject to} \;\; \dot{\boldsymbol{u}} = \boldsymbol{F}(\boldsymbol{u}(t), p(t), t),$$

and $h > 0, M(t) > 0$ and $p(t) > 0$ for all time $t > 0$. The only problem here is that we do not know τ which should be determined by the final landing condition $V(\tau) = h(\tau) = 0$. Therefore, this is a free-time optimal control problem.

7.4 Let A_0 be the fixed area of a small hole on the base, and u be the leakage velocity. This problem is an optimization with optimal final time t_f

$$\text{maximize} \quad t_f \quad \text{with} \quad \int_0^{t_f} A_0 u dt = V_0,$$

$$\text{subject to} \quad 2\pi r h_0 + \pi r^2 = S_0,$$

where $V_0 = \pi r^2 h_0$ is the total water volume.

At any time t, the leakage velocity is $u = \sqrt{2gh}$ where $h(t)$ is the current height of the water column and g is the acceleration due to gravity. The rate of height reduction is related to u by

$$A_w \frac{dh}{dt} = -uA_0 = -A_0\sqrt{2gh},$$

where $A_w = \pi r^2$ is the area of the cross section of the water tank. The above equation becomes

$$\frac{dh}{dt} = -\alpha\sqrt{h}, \qquad \alpha = \frac{A_0\sqrt{2g}}{A_w},$$

or

$$\frac{1}{\sqrt{h}}\frac{dh}{dt} = -\alpha.$$

Integrating from $t = 0$ (for h_0) to t_f (for $h = 0$), we have

$$\int_{h_0}^{0} \frac{1}{\sqrt{h}} dh = -\int_0^{t_f} \alpha dt,$$

which leads to

$$t_f = \frac{2\sqrt{h_0}}{\alpha} = \frac{2\sqrt{h_0}A_w}{A_0\sqrt{2g}} = \frac{2\sqrt{h_0}\;\pi r^2}{A_0\sqrt{2g}}.$$

Here we try to maximize t_f. However, we have to incorporate the constraint $S_0 = 2\pi r h_0 + \pi r^2$ which gives

$$r = \sqrt{h_0^2 + S_0/\pi} - h_0,$$

where we have used the fact that $r > 0$. Now we finally have

$$t_f = \frac{2\pi}{A_0\sqrt{2g}}[(\sqrt{h_0^2 + S/\pi} - h_0)\sqrt{h_0}].$$

The only variable here is h_0. Since $h_0 \geq 0$, the maximum of t_f is determined by

$$\frac{dt_f}{dh_0} = 0.$$

This condition leads to an equation involving surds; however, its solution is straightforward. After some lengthy algebra, we have the optimal solution

$$h_0^* = (\frac{S_0}{3\pi})^{3/4}.$$

Therefore, the optimal height for a water tank with fixed total surface area S_0 is $(S_0/3\pi)^{3/4}$, which will result in a maximum leakage time if there is any hole on the base.

7.5 The Hamiltonian for this optimal control problem can be written as

$$H = \frac{1}{n}u^n + \lambda_1 x + \lambda_2 u,$$

which leads to the following optimal conditions

$$\dot{\lambda}_1 = -\frac{\partial H}{\partial s} = 0,$$

and

$$\dot{\lambda}_2 = -\frac{\partial H}{\partial x} = -\lambda_1,$$

which lead to

$$\lambda_1 = k_1,$$

and

$$\lambda_2 = -k_1 t + k_2,$$

where k_1, k_2 are undetermined integration constants. In addition, another optimality condition

$$\frac{\partial H}{\partial u} = u^{n-1} + \lambda_2 = 0,$$

gives

$$u^{n-1} = -\lambda_2 = k_1 t - k_2,$$

or

$$u = (k_1 t - k_2)^{1/(n-1)}.$$

From $\dot{x} = u$, we have

$$\dot{x} = (k_1 t - k_2)^{1/(n-1)}.$$

Integrating once, we get

$$x = \frac{(n-1)}{k_1 n}(k_1 t - k_2)^{n/(n-1)} + k_3,$$

where k_3 is a constant. Again from $\dot{s} = x$, we have after integration

$$s = \frac{(n-1)^2}{k_1^2 n(2n-1)}(k_1 t - k_2)^{(2n-1)/(n-1)} + k_3 t + k_4,$$

where k_4 is a constant. We have four conditions $x(0) = 1, x(1) = 0$, $s(0) = 0$ and $s(1) = 1$, and thus the four constants k_1, k_2, k_3, k_4 can be determined. However, the expressions are lengthy

For simplicity, we assume $n = 2$. Using the four conditions, we have the final expressions as

$$u = 6t - 4, \qquad x = 3t^2 - 4t + 1,$$

and

$$s = t^3 - 2t + t.$$

This gives the optimal cost or objective

$$J_* = \frac{1}{2}\int_0^1 (6t-4)^2 dt = 2.$$

7.6 The Hamiltonian can be written as

$$H(\boldsymbol{x}, \boldsymbol{v}) = \frac{1}{2}(\boldsymbol{x}^T \boldsymbol{P}\boldsymbol{x} + \boldsymbol{v}^T \boldsymbol{Q}\boldsymbol{v}) + \boldsymbol{\lambda}^T(\boldsymbol{A}\boldsymbol{x} + \boldsymbol{B}\boldsymbol{v}).$$

This extension is straightforward as we can essentially consider the multivariate vector as a variable. The Pontryagin's maximum principle leads to the following optimality conditions

$$\frac{\partial H}{\partial \boldsymbol{v}} = \boldsymbol{Q}\boldsymbol{v} + \boldsymbol{B}^T\boldsymbol{\lambda} = 0,$$

$$\dot{\boldsymbol{\lambda}} = -\frac{\partial H}{\partial \boldsymbol{x}} = -\boldsymbol{P}\boldsymbol{x} - \boldsymbol{A}^T\boldsymbol{\lambda},$$

and

$$\boldsymbol{\lambda}_*(\tau) = \frac{\partial \boldsymbol{G}}{\partial \boldsymbol{x}}\Big|_\tau = \boldsymbol{K}\boldsymbol{x}(\tau),$$

where $\boldsymbol{G} = \frac{1}{2}\boldsymbol{x}^T\boldsymbol{K}\boldsymbol{x}$ is the terminal cost. We can write the above equations as a matrix form

$$\begin{pmatrix} \dot{\boldsymbol{x}} \\ \dot{\boldsymbol{\lambda}} \end{pmatrix} = -\begin{pmatrix} -\boldsymbol{A} & \boldsymbol{B}\boldsymbol{Q}^{-1}\boldsymbol{B}^T \\ \boldsymbol{P} & \boldsymbol{A}^T \end{pmatrix}\begin{pmatrix} \boldsymbol{x} \\ \boldsymbol{\lambda} \end{pmatrix},$$

where we have used the first condition $\boldsymbol{v} = -\boldsymbol{Q}^{-1}\boldsymbol{B}^T\boldsymbol{\lambda}$.

SOLUTIONS FOR CHAPTER 8

8.1 The cumulative probability function is

$$\phi(x) = \int_0^x p(v)dv = 1 - \exp[-x^2/2\sigma^2].$$

The inverse transform from a uniform distribution u becomes

$$x = \sigma\sqrt{2}\sqrt{-\ln(1-u)}.$$

As $u \in (0,1)$, we can simply replace $1-u$ by u, we have

$$x = \sigma\sqrt{-\ln u^2}.$$

As most pseudo-random generators cannot generate exactly 0 and 1, so the logarithm does not have a problem at $u = 0$. For example, in Matlab, you can use the following code to generate $n = 25000$ numbers.

```
n=25000; sigma=3; u=rand(1,n);
v=sigma*sqrt(-log(u.^2));
hist(v,40);
```

8.2 The cumulative probability function is

$$\phi(x) = \int_{-\infty}^{x} p(v)dv = \begin{cases} 1 - \frac{1}{2}e^{-(x-\mu)/\beta} & \text{if } x \geq \mu \\ \frac{1}{2}e^{(x-\mu)/\beta} & \text{if } x < \mu \end{cases},$$

which can be written as

$$\phi(x) = \frac{1}{2}[1 + \text{sign}(x-\mu)(1 - e^{-|x-\mu|/\beta})],$$

where $\text{sign}(v)$ is the sign function which is 1 if $v > 0$, 0 if $v = 0$ and -1 if $v < 0$. So for the inverse transform from a uniform distribution $u \sim$Unif$[0,1]$, we have

$$x = \mu - \beta\ln(1 - 2|u - 1/2|)\text{sign}(u - 1/2).$$

8.3 It is straightforward to obtain the cumulative probability function by integration

$$\phi(x) = \int_{-\infty}^{x} p(v)dv = \frac{1}{\pi}\tan^{-1}(\frac{x-\mu}{\beta}) + \frac{1}{2}.$$

Its inverse is

$$x = \mu + \beta\tan[\pi(u - \frac{1}{2})],$$

where u is drawn from a uniform distribution $[0,1]$. If you write a program and run the simulation, you will have a very sharp peak. The main difficulty is that when $v \to \pm\pi/2$, $x \to \pm\infty$. However, such extreme values are rare (almost zero probability). To be numerically sensible so as to construct a histogram, you can replace π by $\pi\gamma$ where $\gamma = 0.99$ is a scaling parameter. To start with, you can try $\gamma = 0.9$ and then increase to 1 and see what happens. Such phenomena near singularity are common for monotonically decrease probability distributions including Pareto distribution and Lévy distribution.

SOLUTIONS FOR CHAPTER 9

9.1 If u has a distribution $f(x)$ and v has a distribution $g(x)$, the the sum distribution $h(x)$ of $u+v$ is given by the convolution theorem

$$h = (f * g)(x) = \int_{-\infty}^{\infty} f(x-\xi)g(\xi)d\xi = \int_{-\infty}^{\infty} g(x-\xi)f(\xi)d\xi.$$

If u and v are both a uniform distribution in [0,1], it is straightforward to verify that

$$h(x) = \begin{cases} x, & \text{if } 0 \le x \le 1 \\ 2-x, & \text{if } 1 < x \le 2 \\ 0, & \text{elsewhere.} \end{cases}$$

Similarly, $u-v$ obeys a distribution $h(x) = |1-x|$ for $|x| \le 1$ and zero elsewhere. However, there is no simple analytical solution for the case of uv.

Using the following simple matlab code, you can simply try a few combinations

```
n=50000;
u=rand(1,n); v=rand(1,n);
hist(u+v,40);
```

9.2 The sum distribution of u and v can be obtained using the convolution theorem[1]

$$h(x) = \frac{1}{2\pi\sigma^2}\int_{-\infty}^{\infty} e^{-(x-\xi)^2/2\sigma^2} e^{-\xi^2/2\sigma^2} d\xi = \frac{e^{-x^2/4\sigma^2}}{2\pi\sigma^2} \cdot \int_{-\infty}^{\infty} e^{-(\xi-x/2)^2/\sigma^2} d\xi$$

$$= \frac{e^{-x^2/4\sigma^2}}{2\pi\sigma^2} \cdot \sigma\sqrt{\pi} = \frac{1}{\sigma\sqrt{4\pi}} e^{-x^2/4\sigma^2},$$

which is also a normal distribution but with a standard deviation of $\sqrt{2}\,\sigma$. Here we have used

$$\int_{-\infty}^{\infty} e^{-(\xi-x/2)^2/\sigma^2} = \sqrt{\pi}\,\sigma,$$

which independent of any finite x. You can simply confirm this by using

```
n=50000;
u=randn(1,n); v=randn(1,n);
hist(u+v,40);
std(u+v)
```

which should give a standard derivation of about 1.41.

[1] C. M. Grinstead and J. L. Snell, *Introduction to Probability*, AMS, (1997).

In general, if $u \sim N(\mu, \sigma^2)$ and $v \sim N(\nu, \tau^2)$, then $u + v$ is a normal distribution $N(\mu + \nu, \sigma^2 + \tau^2)$.

9.3 The Matlab code can be

```
n=50000; u=rand(1,n);
y=u.^3-u.^2+u/2;
hist(y,40);
mean(y), std(y)
```

You can replace `rand(1,n)` by `randn(1,n)` and do a few runs. In fact, the function $f(u)$ can be any valid combination of basic functions such as $\exp(u)$ and $\sin(u)$.

9.4 As n increase, the distribution should approach to a normal distribution. This is the consequence of the central limit theorem.

SOLUTIONS FOR CHAPTER 10

10.1 For the 2D case, at each step the direction of the walk can be any direction in $[0, 2\pi]$. If the step size of the walk is unity, the location after N steps can be written as

$$d = \sum_{k=1}^{N} e^{i\phi_k}, \tag{D.1}$$

which is conveniently expressed as a sum on a complex plane. Then, that mean distance can be estimated as the expectation $< d^2 >$. From

$$d^2 = d \cdot d^* = \sum_{k=1}^{N} e^{i\phi_k} \sum_{j=1}^{N} e^{-i\phi_j} = \sum_{k=1}^{N}\sum_{j=1}^{N} e^{i(\phi_k - \phi_j)} = N + S_{kj},$$

where

$$S_{kj} = \sum_{k,j=1,k\neq j}^{N} e^{i(\phi_k - \phi_j)}, \tag{D.2}$$

we know that, when averaging over all possible directions, $< S_{kj} > \approx 0$. Therefore, we have

$$< d^2 > = N + < S_{kj} > \approx N, \tag{D.3}$$

which suggests that the mean distance a random walker such as a particle after N steps is approximately $\sqrt{N}$.

10.2 You may observe that small β usually leads to a longer average distance from the origin.

10.3 For example, a few line of Matlab may look like this

```
n=500;
u=randn(1,n);
v=cumsum(u);
plot(v);
```

Obviously, it is easy to extend to higher dimensions.

10.4 It is straightforward to obtain the eigenvalues of

$$\boldsymbol{P} = \begin{pmatrix} 1/4 & 3/4 \\ 3/4 & 1/4 \end{pmatrix},$$

and they are $\lambda_1 = 1$ and $\lambda_2 = -0.5$. In addition, using direct matrix product, we have

$$\boldsymbol{P}^2 = \begin{pmatrix} 5/8 & 3/8 \\ 3/8 & 5/8 \end{pmatrix}, \ \boldsymbol{P}^3 = \boldsymbol{P}^2\boldsymbol{P} = \begin{pmatrix} 7/16 & 9/16 \\ 9/16 & 7/16 \end{pmatrix}, ..., \boldsymbol{P}^\infty \to \begin{pmatrix} 0.5 & 0.5 \\ 0.5 & 0.5 \end{pmatrix}.$$

The rate of convergence of the above sequence is related to the absolute value of λ_2. Here $\boldsymbol{P}$ can be viewed as a simple weather fasting matrix. If it is sunny today, the probability of rain tomorrow is, say, 1/4 and the probability of sunny again is 3/4. So we have $\boldsymbol{\pi}_0 = (0 \quad 1)$. However, when the number of days (n) increases, it is becomes difficult to forecast the weather based on today's condition. For example, for $n = 5$, we have

$$\boldsymbol{P}^5 \approx \begin{pmatrix} 0.48 & 0.52 \\ 0.52 & 0.48 \end{pmatrix},$$

or

$$\boldsymbol{\pi}_5 = \boldsymbol{\pi}_0 \boldsymbol{P}^n \approx \begin{pmatrix} 0.52 & 0.48 \end{pmatrix},$$

which is almost a $50 - 50$ chance. In fact, the long term tendency is

$$\boldsymbol{P}^\infty = \begin{pmatrix} 1/2 & 1/2 \\ 1/2 & 1/2 \end{pmatrix},$$

which gives $\boldsymbol{\pi}^* = \boldsymbol{\pi}_0 \boldsymbol{P}^\infty = \begin{pmatrix} 0.5 & 0.5 \end{pmatrix}$, and this final state is independent of the initial state $\boldsymbol{\pi}_0$. This is an important property of a Markov chain.

10.5 a) It is regular, as all the entries of $\boldsymbol{P}$ are positive. b) Some entries of $\boldsymbol{P}$ are zero, however,

$$\boldsymbol{P}^2 = \begin{pmatrix} 9/20 & 11/40 & 11/40 \\ 12/25 & 7/25 & 6/25 \\ 2/5 & 7/25 & 8/25 \end{pmatrix},$$

has all positive entries, so it is regular. c). There is a zero entry in $\boldsymbol{P}$. If we try to see $\boldsymbol{P}^n$, we have

$$\boldsymbol{P}^2 = \begin{pmatrix} 1 & 0 \\ 3/4 & 1/4 \end{pmatrix}, \boldsymbol{P}^3 = \begin{pmatrix} 1 & 0 \\ 7/8 & 1/8 \end{pmatrix}, ..., \boldsymbol{P}_\infty = \begin{pmatrix} 1 & 0 \\ 1 & 0 \end{pmatrix}.$$

Whatever n is, the top right entry is always zero. So this chain is not regular.

10.6 This is a very famous application of Markov chains. It seems that Google has used $\alpha \approx 0.85$. For more details, please refer to the technical report by Page *et al.* .[2]

SOLUTIONS FOR CHAPTER 11

11.1 This function is part of the Six-hump camel back function if we extend its domain to $-3 \leq x \leq 3$ and $-2 \leq y \leq 2$. Two global minima occur at $(x_*, y_*) = (0.0898, -0.7126)$ and $(-0.0898, 0.7126)$ with $f_* \approx -1.0316$. You can write a simple program in Matlab, or you can modify the Matlab programs in Appendix B.

11.2 This function is the sum of different powers and its global minimum $f_* = 0$ occurs at $(0, 0, ..., 0)$. However, it is not a smooth function, and thus a gradient-free global optimizer should be used.

11.3 This stochastic function has the global minimum $f_* = 0$ at $\boldsymbol{x}_* = (0, 0, ..., 0)$. Interestingly, this global optimum does not depend on any stochastic nature of the coefficient ϵ_i. In addition, this function is singular at the optimum, thus it is impossible to use gradient-based methods directly.

SOLUTIONS FOR CHAPTER 12

12.1 The rate of cooling should be properly controlled. For $T(t) = T_0/(1 + \alpha t)$, α should be chosen so that $T(t) \to 0$ as t approaches the maximum number of iterations $t_{\max}$.

12.2 For some tough multimodal optimization problems, it actually helps to use a non-monotonic function as a cooling schedule. The rise in temperature during iterations increases the system's energy and it is less likely for it to be stuck in local optima. The only requirements for such functions are $T(t) = T_0$ at $t = 0$, and $T(t) \to 0$ as $t \to t_{\max}$ where $t_{\max}$ is the maximum number of iterations. For example, we can even a non-smooth function for cooling

$$T(t) = \frac{1}{2}\Big(\frac{|\sin t|}{t} + e^{-\alpha t}\Big), \qquad \alpha > 0.$$

12.3 This function has the global minimum $f_* = 0$ at $(0, 0, ..., 0)$ in the domain $-2\pi \leq x_i \leq 2\pi$ for $i = 1, 2, ..., n$. However, it has a singularity at this global optimum. Therefore, gradient-based methods are not suitable. In addition,

[2] L. Page, S. Brin, R. Motwani, T. Winograd, "The PageRank citation ranking: bringing order to the web", Technical Report, Stanford InfoLab, (1999).

this function is multimodal, so choose a global optimizer such as simulated annealing and particle swarm optimization.

12.4 This equality-constrained problem has the global minimum $f_* = -1$ at $\boldsymbol{x}_* = (1/\sqrt{n}, ..., 1/\sqrt{n})$. It can be solved directly as a constrained problem or as an unconstrained one using a Lagrange multiplier.

12.5 Simulated annealing is a good global optimizer, but its convergence may be slow for complex optimization problems as it is a trajectory-based algorithm. A potential improvement is to use multiple parallel simulated annealing trajectories, essentially multiple Markov chains in parallel. Information such as the current best solutions may be exchanged among different chains, depending on the variants of the algorithms. For a good start, please refer to Greening's article [3]

SOLUTIONS FOR CHAPTER 13

13.1 You can write the program in any programming language of your choice. When $\gamma = 0$, you will notice the program usually does not converge to the right solution because the solution found will strongly depend on the initial guess. As γ increases slightly, the right optimal solution can be found.

13.2 When n is small, your program will usually be sufficiently quick to find the optimal solution. However, as n increases, it may be very slow or even find a wrong solution. This is because this problem is NP-hard and no efficient algorithms are available at the moment.

13.3 This problem may have many practical applications. For $n = 10$, the number of possible combination is $10! = 3628800$, even with good connections of airlines and/or trains, it may take quite a while to find the optimal solution even if you use a good Internet search engine. To find the optimal solution, it often requires to construct a square matrix of $n \times n$ to store n^2 prices.

SOLUTIONS FOR CHAPTER 14

14.1 The optimal solution should be $f_* = 0$ with $\boldsymbol{x}_* = (0, 0, ..., 0)$. Any approximate solution with a certain accuracy is acceptable. Try to adjust the parameters such as the exploration probability to see how they may affect the rate of convergence.

14.2 There are K work units, and work unit i is assigned to produce x_i parts. The time taken to finish the assignment is x_i/τ_i. The latest time (assuming

[3] D. R. Greening, "Parallel simulated annealing techniques", *Physica D*, **42**, 293-306 (1990).

they start at the same time) to finish its assignment is

$$T = \max(x_1/\tau_1, ..., x_K/\tau_K).$$

In order to finish the overall job as early as possible, we have

$$\text{minimize } T = \min[\max(x_1/\tau_1, ..., x_K/\tau_K)].$$

The constraints are

$$\sum_{i=1}^{K} x_i = N, \qquad x_i \geq 0, \qquad x_i \in \mathcal{I},$$

where $\mathcal{I}$ is the integer set.

14.3 If the overall cost is of concern, the new objective is

$$\text{minimize } \sum_{i=1}^{K} \frac{x_i p_i}{\tau_i},$$

subject to

$$\sum_{i=1}^{K} x_i = N, \;\; 0 \leq x_i \in \mathcal{I}.$$

SOLUTIONS FOR CHAPTER 15

15.1 You can use either a fixed value, say $\theta = 0.9$, or a monotonically decrease function

$$\theta = \gamma + (1 - \gamma)e^{-t/\tau},$$

where $\tau > 0$ is a time constant, t is the iteration index. The constant γ can be taken as $\gamma = 0.8$. By varying the value of θ, you may find that $\theta = 0.4 \sim 0.7$ is a better choice.

15.2 The global minimum $f_* = 0$ is at $\boldsymbol{x}_* = (1, 2, ..., n)$ which is independent of the parameter β. This can be confirmed by using various value of $\beta > 0$ in your program.

15.3 This function has the global minimum $f_* = -1$ at $(\pi, 100\pi)$. If you use the PSO program to find this minimum, you may notice a slightly slow rate of convergence, though not significant. However, the solution quality may be affected. A common way to deal with this is to rescale. For example, use $\zeta = y/100$, we have

$$f(x, \zeta) = -\cos(x)\cos(\zeta)e^{-(x-\pi)^2-(\zeta-\pi)^2/\pi^2},$$

which can be solved using the standard PSO. Here the scales are not too different and thus rescaling does not have much effect. If scales are significantly different, rescaling often helps.

15.4 This function has K^2 local valleys with a global minimum at (π, π), but $f_{\min}$ is random in the range of $(-K^2 - 5, -5)$. For example, you can replace the inline function `f` in the Matlab program by a function

```
function Z=f(u,v)
K=10;
Z=0;
alpha=1; beta=1;
for j=1:K,
  for i=1:K,
    Z=Z-rand*exp(-alpha*((u-i).^2+(v-j).^2));
  end
end
Z=Z-5*exp(-beta*((u-pi).^2+(v-pi).^2));
```

The distinct nature of this function is that its landscape is constantly changing, which usually render many gradient-based or deterministic algorithms useless. However, you may notice that PSO is relatively robust, and will find the optimum easily, though not with a 100% success rate.

SOLUTIONS FOR CHAPTER 16

16.1 This function has the global minimum $f_{\min} = -1$ at $(0, 0, ..., 0)$. It is straightforward to modify the Harmony Search algorithm to find this minimum.

16.2 If you use the simple Matlab code in Appendix B, you will observe the variations of the convergence rates with p_a and r_{pa}. However, in some parameter ranges, the rate of convergence is only weakly dependent on these parameters, and these parameter ranges are a good choice for most problems.

16.3 For some optimization problems, use Lévy flights to adjust pitch may be more efficient. This is still an interesting area of research.

16.4 This is a literature review exercise. New optimization algorithms are often developed by hybridization. Some combinations of existing algorithms may prove to be very fruitful.

SOLUTIONS FOR CHAPTER 17

17.1 Schubert's function has 18 minima with $f_* = -186.7309$ in the domain $x, y \in [-10, 10]$. For a multimodal function with multiple optima, population-based algorithms are a better choice. So particle swarm optimization and firefly algorithm should be used. You can simply modify the Matlab code in this book to complete your task. You may find that the number of fireflies n should be much larger than the number of modes. Try $n = 20, 50, 100$, and see what happens. The initial locations of the fireflies are also important, and they should be distributed relatively uniformly over the entire search space.

17.2 As the solution approaches optimal, it may help to achieve high accuracy by reducing the randomness. However, if the randomness is reduced too early, there is a possibility of premature convergence.

17.3 For this purpose, you can simple replace the last term in (17.11) by an appropriate Lévy flights random number generator. Others distributions with a long tail includes Cauchy distribution, and Student-t distribution. For a detailed study, please refer to a recent paper by the author.[4]

SOLUTIONS FOR CHAPTER 18

18.1 There are many potential ways to extend the standard simulated annealing. A direct and yet simple extension is to modify the acceptance probability as the joint probability

$$p_a = \prod_{j=1}^{m} p_j = \prod_{j=1}^{m} e^{-\Delta f_j / kT_j},$$

where k is Boltzmann constant which can be taken as $k = 1$, and $\Delta f_j = f_j(\boldsymbol{x}_n) - f_j(\boldsymbol{x}_{n-1})$ is the change of each individual objective. For details, please refer to the article by Suppapitnarm *et al.* (2000)[5]

18.2 The methods of least squares requires to minimize

$$\phi = \sum_{i=1}^{n} (y_i - f(x_i))^2,$$

which tends to choose the polynomials of higher orders. A common solution is to impose some penalty on the coefficients so as to choose the simpler models (or lower orders of polynomials). This idea is called Ockham's razor[6], which tries to explain data using the simplest models possible.

[4] X. S. Yang, "Firefly algorithm, Lévy flights and global optimization", in: *Research and Development in Intelligent Systems XXVI* (Eds M. Bramer *et al.*), Springer, London, pp. 209-218 (2010).

[5] A. Suppapitnarm, K. A. Seffen, G. T. Parks and P. J. Clarkson, "A simulated annealing algorithm for multiobjective optimization", *Engineering Optimization*, **33**, 59-85(2000).

[6] William Ockham (1285-1349), *entia non sunt multiplicanda praeter necessitatem* – "Entities must not be multiplied beyond necessity", now referred to as the Ackham's razor or Occam's razor.

As a multiobjective optimization problem, we can write

$$\text{minimize} \quad \phi = \sum_{i=1}^{n} (y_i - f_i)^2, \quad \psi = \sum_{k=0}^{p} \alpha_k^2,$$

where $f_i = f(x_i)$ is the model prediction for y_i at x_i. We can solve this as a multiobjective problem.

However, in data analysis, we often use the penalty least-squares or regularized least squares try to minimize

$$\phi + \lambda\psi = \sum_{i=1}^{n} (y_i - f_i)^2 + \lambda \sum_{k=0}^{p} \alpha_k^2,$$

where $\lambda \gg 0$ is the penalty parameter which has to be determined.

Another more direct approach is to formulate it as an optimization problem with a single objective

$$\text{minimize} \quad \phi = \sum_{i=1}^{n} (y_i - f_i)^2,$$

subject to the constraint

$$\sum_{k=0}^{p} |\alpha_k| \leq Q,$$

where $Q > 0$ is a constant. This method is often referred to as the LASSO method.[7] If you implement this problem properly, you will find that the results are almost independent of Q. That is, any sensible value of Q is acceptable as long as the constraint is valid. In practical, either $Q = 20$ or $Q = 50$ can be used, depending on the scales of your problem of interest.

18.3 This is the test function proposed by Zitzler *et al.* (2000).[8] Its Parato optimal front is convex, and it decreases monotonically when f_2 is plotted against f_1 which varies from 0 to 1. You can write a computer program to visualize this front.

18.4 It is straightforward to verify that this multiobjective optimization problem has the global minimum $\boldsymbol{f}_* = (0, 0, ..., 0)$ at $\boldsymbol{x}_* = (0, 0, ..., 0)$. If you use the weighted sum as a composite objective $\sum_{i=1}^{p} \alpha_i f_i$, you will find that the optimal solution is independent of α_i as long as they are not zero or negative. This is a special case when the Pareto front becomes a single point, which is the global optimum.

[7] R. Ribshirani, "Regression shrinkage and selection via the LASSO", *J. Royal. Statist. Soc. B*, **58**, 267-288 (1996).

[8] E. Zitzler, K. Deb, and L. Thiele, "Comparison of multiobjective evolutionary algorithms: empirical results", *Evolutionary Computation*, **8** (2000) (2), 173-195.

18.5 To achieve this goal, we often choose the most important objective of our preference, say, $f_1(\boldsymbol{x})$ as the main objective, while imposing limits on the other objectives. That is

$$\text{minimize} \quad f_1(\boldsymbol{x}),$$
$$\text{subject to} \quad f_i \leq L_i, \qquad (i = 2, 3, ..., p),$$

where the limits L_i are given. In principle, the problem can be solved using the standard optimization algorithms for single objective optimization. In practice, some prior knowledge is required to impose the correct limits. Otherwise, the solutions obtained may not be the solution to the original problem.

18.6 The main difficulty to extend any single-objective optimization algorithm to accommodate the multiobjectives is that we have to consider the dominance of each candidate solution set. In the standard PSO, we update the velocity

$$\boldsymbol{v}_i^{t+1} = \theta \boldsymbol{v}_i^t + \alpha \epsilon_1 \cdot (\boldsymbol{g}^* - \boldsymbol{x}_i^t) + \beta \epsilon_2 \cdot (\boldsymbol{x}_i^* - \boldsymbol{x}_i^t),$$

where $\boldsymbol{g}^*$ is the current global optimum found so far. A possible extension is to use an external repository or archive to store non-dominated solutions at each iteration. Then, we select such a solution $\boldsymbol{h}^*$ from this repository to replace $\boldsymbol{g}^*$. This idea was first proposed by Coello Coello *et al.* (2004).[9]

SOLUTIONS FOR CHAPTER 19

19.1 The problem can be written as the following optimization problem

$$\text{maximize} \quad S = 2\pi r n,$$
$$\text{subject to} \quad n\pi r^2 \leq A_0, \qquad r \geq a.$$

From the expression $S = 2\pi r n$, it seems that we have to increase r and n so as to increase S; however, if we rewrite the constraint $n\pi r^2 \leq A_0$ as

$$nr \leq \frac{A_0}{\pi r},$$

we have to reduce r in order to increase the product of nr in S. However, r has a minimum value a. Since n must be a positive integer, we have

$$n = \left\lfloor \frac{A_0}{\pi a^2} \right\rfloor,$$

where $\lfloor u \rfloor$ is the greatest integer smaller than u. This leads to the optimum

$$S_{\max} = 2\pi a \left\lfloor \frac{A_0}{a^2} \right\rfloor.$$

[9]C. A. Coello Coello, G. T. Pulido, M. S. Lechuga, "Handling multiple objectives with particle swarm optimization", *IEEE Trans. Evolut. Comp.*, **8** (2004) (3), 256-279.

For example, for $A_0 = 1000$ mm^2 and $a = 5$ mm, we have

$$n = \left\lfloor \frac{1000}{\pi 5^2} \right\rfloor = 12,$$

and

$$S_{\max} = 2\pi a n \approx 377,$$

with a total cross section area of $n\pi a^2 = 942.5$ mm^2. Indeed, both constraints are satisfied as well as the implicit constraints $r > 0$.

19.2 For simplicity, we assume the lighting pole is cylindrical with a radius of r. The total mass of the pole is

$$m = \rho \pi r^2 L,$$

where ρ is the density of the material. For a fixed height, we have to minimize r (and ρ) so as to minimize the total weight.

To maximize the frequencies, we have to consider it as a vibration problem. The vibrations of the pole is governed by

$$EI \frac{\partial^4 u}{\partial z^4} + \lambda \frac{\partial^2 u}{\partial t^2} = 0,$$

where E is Young's modulus, I is the second moment of area. For a cylindrical section, we have $I = \pi r^4/4$. z is the coordinate pointing upwards, and t is time. In addition, $\lambda = \rho A$ is the linear mass density and $A = \pi r^2$ is the area of cross section. This is a fourth-order equation which can be solved by the separation of variables in the case when the pole is considered a simple bending cantilever with one end fixed. Let $u(z, t) = Z(z) \cdot T(t)$, and plug it into the governing equation, we have

$$\frac{\partial^4 Z}{\partial z^4} - \alpha_n^4 Z = 0, \qquad \frac{\partial^2 T}{\partial t^2} + \omega_n^2 T = 0,$$

where α_n is constant related to the angular frequency ω_n

$$\alpha_n^4 = \frac{\omega_n^2 \lambda}{EI}, \qquad \text{or} \qquad \omega_n = \alpha_n^2 \sqrt{\frac{EI}{\rho A}}.$$

Here the subscript n is related to the different modes of vibrations. Using $I = 4r^4/4$ and $A = \pi r^2$, we have $I/\rho A = r/2$. Then, the frequency is simply

$$f_n = \omega_n / 2\pi = \frac{\alpha_n^2 r}{4\pi} \sqrt{\frac{E}{\rho}}.$$

Applying the appropriate boundary conditions, it is straightforward to derive that

$$\cos(\alpha_n L) \cosh(\alpha_n L) + 1 = 0.$$

This equation can be solved numerically, and we have

$$\alpha_n L \approx 1.8751041 \text{ (for } n=1), \qquad 4.6940911 \text{ (for } n=2).$$

The lowest frequency is due to the first mode ($n = 1$), and we have

$$f_1 \approx \frac{1.87510^2 r}{4\pi L^2}\sqrt{\frac{E}{\rho}} \approx 0.2798\frac{r}{L^2}\sqrt{\frac{E}{\rho}}.$$

Clearly, to get a higher frequency, we can either increase r, or select stronger material (higher E) and lighter material (low ρ). However, increase r is conflicting with the minimization of mass m. So the sensible choose is now use stronger and light materials such carbon-fibre reinforced polymer (CFRP). We can also see that it is difficult to achieve two objectives at the same time. For most lighting poles, f_1 is about 1Hz, and f_2 is about 3 to 5 Hz. They may vary in the same range of the wind-induced vortex frequencies, and thus may cause the fatigue failure of the pole. Therefore, we usually have to consider certain damping measures such as the installation of dampers on the poles.

19.3 This is in fact a very tough mathematical problem. To date, the two configurations with the highest packaging factor are the face-centered cubic (FCC) and hexagonal close-packed (HCP) structures. Both have the maximum packaging factor $\pi/\sqrt{18} = \pi/3\sqrt{2} \approx 0.74048$. This problem is related to the Kepler conjecture. Only recently, Thomas C. Hales provided a proof aided by computers using proof by exhaustion. One of the most efficient ways of packing is in a pyramid shape. In engineering materials and metallurgy, atomic packaging factor (APF) is often used. Some real-world materials such as iron and steels are packaged in FCC (with an APF $\approx$ 0.74) and body-centered cubic (BCC) with an APF ≈ 0.68, depends on the temperature and heat treatment history. Interestingly, the diamond cubic has a packaging factor of only about 0.34. Packaging for irregular shapes and sizes is important for many applications, but it still an unsolved mathematical problem.

REFERENCES

1. M. Abramowitz and I. A. Stegun, *Handbook of Mathematical Functions*, Dover Publication, 1965.

2. D. H. Ackley, *A Connectionist Machine for Genetic Hillclimbing*, Kluwer Academic Publishers, 1987.

3. A. Adamatzky, C. Teuscher, *From Utopian to Genuine Unconventional Computers*, Luniver Press, 2006.

4. A. Afshar, O. B. Haddad, M. A. Marino, B. J. Adams, "Honey-bee mating optimization (HBMO) algorithm for optimal reservoir operation", *J. Franklin Institute*, **344**, 452-462 (2007).

5. M. Agrawal, N. Kayal, and N. Saxena, "Primes is in P", *Ann. Mathematics*, **160** (2), 781-793 (2002).

6. G. Arfken, *Mathematical Methods for Physicists*, Academic Press, 1985.

7. J. Arora, *Introduction to Optimum Design*, McGraw-Hill, 1989.

8. S. N. Atluri, *Methods of Computer Modeling in Engineering and the Sciences*, Vol. I, Tech Science Press, 2005.

9. M. Bartholomew-Biggs, B. P. Butler, A. B. Forbes, "Optimization algorithms for generalized regression on metrology", in: *Advanced Mathematica and Computational Tools in Metrology IV* (Eds. P. Ciarlini, A. B. Forbes, F. Pavese, D. Richter), World Scientific, pp. 21-31 (2000).

Engineering Optimization: An Introduction with Metaheuristic Applications.
By Xin-She Yang
Copyright © 2010 John Wiley & Sons, Inc.

10. M. Bartholomew-Biggs, *Nonlinear Optimization with Engineering Applications*, Springer, 2008.

11. B. Basturk and D. Karabogo, "An artificial bee colony (ABC) algorithm for numerical function optimizaton", in: *IEEE Swarm Intelligence Symposium 2006*, May 12-14, Indianapolis, IN, USA, 2006.

12. M. S. Bazaraa, H. D. Serali, and C. M. Shetty, *Nonlinear Programming: Theory and Algorithms*, John Wiley & Sons, 1993.

13. A. Belegundu, *A Study of Mathematical Programming Methods for Structural Optimization*, PhD thesis, University of Iowa, Iowa, 1982.

14. W. J. Bell, *Searching Behaviour: The Behavioural Ecology of Finding Resources*, Chapman & Hall, London, 1991.

15. M. P. Bendsϕe, *Optimization of Structural Topology, Shape and Material*, Springer, Berlin, 1995.

16. M. P. Bendsϕe, O. Sigmund, "Material interpolations in topology optimization", *Arch. Appl. Mech.*, **69**, 635-654 (1999).

17. A. Björck, *Numerical Methods for Least Squares Problems*, SIAM, Philadelphia, 1996.

18. C. Blum and A. Roli, "Metaheuristics in combinatorial optimization: Overview and conceptural comparison", *ACM Comput. Surv.*, **35**, 268-308 (2003).

19. E. Bonabeau, M. Dorigo, G. Theraulaz, *Swarm Intelligence: From Natural to Artificial Systems.* Oxford University Press, 1999.

20. E. Bonabeau and G. Theraulaz, "Swarm smarts", *Scientific Americans.* March, 73-79 (2000).

21. S. P. Boyd and L. Vandenberghe, *Convex Optimization*, Cambridge University Press, (2004). Also http://www.stanford.edu/ boyd

22. D. S. Bridges, *Computatability*, Springer, New York, 1994.

23. C. G. Broyden, "The convergence of a class of double-rank minimization algorithms", *IMA J. Applied Math.*, **6**, 76-90 (1970).

24. L. C. Cagnina, S. C. Esquivel, C. A. Coello, "Solving engineering optimization problems with the simple constrained particle swarm optimizer", *Informatica*, **32**, 319-326 (2008).

25. L. A. Cauchy, "Méthode générale pour la résolution des systeèmes d'équations simulatanées", *Comptes Rendus de l'Académie de Sciences de Paris*, **25**, 3=536-538 (1847).

26. A. Chatterjee and P. Siarry, "Nonlinear inertia variation for dynamic adapation in particle swarm optimization", *Comp. Oper. Research*, **33**, 859-871 (2006).

27. Celis M. Celis, J. E. Dennis, and R. A. Tapia. "A trust region strategy for nonlinear equality constrained optimization", in Numerical Optimization 1994" (Eds. P. Boggs, R. Byrd and R. Schnabel), Philadelphia: SIAM, pp. 71-82 (1985).

28. C. Chong, M. Y. Low, A. I. Sivakumar, K. L. Gay, "A bee colony optimization algorithm to job shop scheduling", *Proc. of 2006 Winter Simulation Conference*, Eds Perrone L. F. et al, (2006).

29. C. A. Coello Coello, "Use of a self-adaptive penalty approach for engineering optimization problems", Computers in Industry, **41**, 113-127 (2000).

30. A. R. Conn, N. I. M. Gould, P. L. Toint, *Trust-region methods*, SIAM & MPS, (2000).

31. B. J. Copeland, *The Essential Turing*, Oxford University Press, 2004.

32. B. J. Copeland, *Alan Turing's Automatic Computing Engine*, Oxford University Press, 2005.

33. R. Courant and D. Hilbert, *Methods of Mathematical Physics*, 2 volumes, Wiley-Interscience, New York, 1962.

34. M. G. Cox, A. B. Forbes, P. M. Harris, I. M. Smith, "The classification and solution of regression problems for calibration", NPL SSfM Technical Report, 2004.

35. B. D. Craven, *Control and Optimization*, Chapman & Hall, 1995.

36. G. B. Dantzig, *Linear Programming and Extensions*, Princeton University Press, 1963.

37. M. Davis, *Computability and Unsolvability*, Dover, (1982).

38. K. De Jong, *Analysis of the Behaviour of a Class of Genetic Adaptive Systems*, PhD thesis, University of Michigan, Ann Arbor, 1975.

39. K. Deb, "An efficient constraint handling method for genetic algorithms", Comput. Methods Appl. Mech. Engrg., **186**, 311-338 (2000).

40. K. Deb., *Optimisation for Engineering Design*: Algorithms and Examples, Prentice-Hall, New Delhi, 1995.

41. J. E. Dennis, "A brief introduction to quasi-Newton methods", in: *Numerical Analysis: Proceedings of Symposia in Applied Mathematics* (Eds G. H. Golub and J. Oliger, pp. 19-52 (1978).

42. R Dorfman, "The discovery of linear programming", *Ann. Hist. Comput.*, **6** (3), 283-295 (1984).

43. M. Dorigo, *Optimization, Learning and Natural Algorithms*, PhD thesis, Politecnico di Milano, Italy, 1992.

44. M. Dorigo and T. Stützle, *Ant Colony Optimization*, MIT Press, Cambridge, 2004.

45. S. Dreyfus, "Richard Bellman on the birth of dynamic programming", *Operations Research*, **50** (1), 48-51 (2002).

46. M. A. El-Beltagy, Keane A. J., "A comparison of various optimization algorithms on a multilevel problem", *Engin. Appl. Art. Intell.*, **12**, 639-654 (1999).

47. A. P. Engelbrecht, *Fundamentals of Computational Swarm Intelligence*, John Wiley & Sons, 2005.

48. L. C. Evans, *An Introduction to Stochastic Differential Equations*, Lecture notes, http://math.berkeley.edu/ evans/SDE.course.pdf

49. M. Fathian, B. Amiri, A. Maroosi, "Application of honey-bee mating optimization algorithm on clustering", *Applied Mathematics and Computation*, **190**, 1502-1513 (2007).

50. T. S. Ferguson, "Who solved the secretary problem?", *Statist. Sci.*, **4** (3), 282-296 (1989).

51. R. P. Feynman, R. B. Leighton, and M. Sands, *The Feynman Lectures on Physics*, vol. 2, Addison-Wesley, Reading, Mass., 1963.

52. A. V. Fiacco and G. P. McCormick, *Nonlinear Porgramming: Sequential Unconstrained Minimization Techniques*, John Wiley & Sons, 1969.

53. S. R. Finch, *Mathematical Constants*, Cambridge University Press, 2003.

54. G. W. Flake, *The Computational Beauty of Nature: Computer Explorations of Fractals, Chaos, Complex Systems, and Adaptation*, Cambridge, Mass.: MIT Press, 1998.

55. R. Fletcher, "A new approach to variable metric algorithms", *Computer Journal*, **13** 317-322 (1970).

56. W. Fleming and R. Rishel, *Deterministic and Stochastic Optimal Control*, Springer, 1975.

57. C. A. Floudas, P. M., Pardalos, C. S. Adjiman, W. R. Esposito, Z. H. Gumus, S. T. Harding, J. L. Klepeis, C. A., Meyer, C. A. Scheiger, *Handbook of Test Problems in Local and Global Optimization*, Springer, (1999).

58. L. J. Fogel, A. J. Owens, and M. J. Walsh, *Artificial Intelligence Through Simulated Evolution*, John Wiley & Sons, 1966.

59. A. R. Forsyth, *Calculus of Variations*, Dover, 1960.

60. A. C. Fowler, *Mathematical Models in the Applied Sciences*, Cambridge University Press, 1997.

61. D. Gamerman, *Markov Chain Monte Carlo*, Chapman & Hall/CRC, 1997.

62. C. W. Gardiner, *Handbook of Stochastic Methods*, Springer, (2004).

63. Z. W. Geem, J. H. Kim, and G. V. Loganathan, "A new heuristic optimization algorithm: Harmony search", *Simulation*, **76**, 60-68 (2001).

64. L. Gerencser, S. D. Hill, Z. Vago, and Z. Vincze, "Discrete optimization, SPSA, and Markov chain Monte Carlo methods", *Proc. 2004 Am. Contr. Conf.*, 3814-3819 (2004).

65. N. Gershenfeld, *The Nature of Mathematical Modeling*, Cambridge University Press, 1998.

66. C. J. Geyer, "Practical Markov Chain Monte Carlo", *Statistical Science*, **7**, 473-511 (1992).

67. A. Ghate and R. Smith, "Adaptive search with stochastic acceptance probabilities for global optimization", *Operations Research Lett.*, **36**, 285-290 (2008).

68. P. E. Gill, W. Murray, and M. H. Wright, *Practical optimization*, Academic Press Inc, 1981.

69. W. R. Gilks, S. Richardson, and D. J. Spiegelhalter, *Markov Chain Monte Carlo in Practice*, Chapman & Hall/CRC, (1996).

70. F. Glover, "Heuristics for Integer Programming Using Surrogate Constraints", *Decision Sciences*, **8**, 156-166 (1977).

71. F. Glover, "Tabu Search - Wellsprings and Challenges", *Euro. J. Operational Research*, **106**, 221-225 (1998).

72. F. Glover, "Future paths for integer programming and links to artificial intelligence", *Comput. Operational Res.*, **13** (5), 533-549 (1986).

73. F. Glover and M. Laguna, *Tabu Search*, Kluwer Academic Publishers, Boston, 1997.

74. Goldberg D. E., *Genetic Algorithms in Search, Optimisation and Machine Learning*, Reading, Mass.: Addison Wesley, 1989.

75. S. M. Goldeldt, R. E. Quandt, and H. F. Trotter, "Maximization by quadratic hill-climbing", *Econometrica*, **34**, 541-551, (1996).

76. D. Goldfarb, "A family of variable metric update derived by variational means", *Mathematics of Computation*, **24**, 23-26 (1970).

77. A. A. Goldstein, "Cauchy's method of minimization", *Numerische Mathematik*, **4**(1), 146-150 (1962).

78. R. Goodman, *Teach Yourself Statistics*, London, 1957.

79. C. M. Grindstead and J. L. Snell, *Introduction to Probability*, American Mathematical Society, 1997.

80. M. Gutowski, "Lévy flights as an underlying mechanism for global optimization algorithms", *ArXiv Mathematical Physics e-Prints*, June, (2001).

81. O. B. Haddad, A. Afshar, M. A. Marino, "Honey bees mating optimization algorithm (HBMO); a new heuristic approach for engineering optimization", in: Proc. of the First Int. Conf. on Modelling, Simulation and Applied Optimization, Sharjah, UAE, (2005).

82. W. K. Hastings, "Monte Carlo sampling methods using Markov chains and their applications", *Biometrika*, **57**, 97-109 (1970).

83. A. Hedar, Test function web pages, http://www-optima.amp.i.kyoto-u.ac.jp /member/student/hedar/Hedar_files/TestGO_files/Page364.htm

84. History of optimization, http://hse-econ.fi/kitti/opthist.html

85. L. Hocking, *Optimal Control: An Introduction to the Theory with Applications*, Oxford University Press, 1991.

86. J. Holland, *Adaptation in Natural and Artificial Systems*, University of Michigan Press, Ann Anbor, 1975.

87. H. H. Holstine, *A History of the Calculus of Variations from the 17th through the 19th Century*, Springer-Verlag, Heidelberg, 1980.

88. D. Jaeggi, G. T. Parks, T. Kipouros, P. J. Clarkson, "A multi-objective Tabu search algorithm for constrained optimization problem", *3rd Int. Conf. Evolutionary Multi-Criterion Optimisation*, Mexico, **3410**, 490-504 (2005).

89. A. Jeffrey, *Advanced Engineering Mathematics*, Academic Press, 2002.

90. J. L. W. V. Jensen, "Sur les fonctions convexes et les inégalités entre les valeurs moyenners", *Acta Mathematica*, **30**, 175-193 (1906).

91. P. Judea, *Heuristics*, Addison-Wesley, 1984.

92. C. L. Kar, I. Yakushin, K. Nicolosi, "Solving inverse initial-value, boundary-value problems via genetic algorithms", *Engineering Applications of Artificial Intelligence*, **13**, 625-633 (2000).

93. D. Karaboga, "An idea based on honey bee swarm for numerical optimization", Technical Report, Erciyes University, 2005.

94. D. Karaboga and B. Basturk, "On the performance of artificial bee colony (ABC) algorithm", *Applied Soft Computing*, **8**, 687-697 (2008).

95. W. Karush, *Minima of Functions of Several Variables with Inequalities as Side Constraints*, MSc Dissertation, Department of Mathematics, University of Chicago, Illinois, 1939.

96. N. Karmarkar, "A new polynomial-time algorithm for linear programming", *Combinatorica*, **4** (4), 373-395 (1984).

97. Keane A. J., "Genetic algorithm optimization of multi-peak problems: studies in convergence and robustness", *Artificial Intelligence in Engineering*, **9**, 75-83 (1995).

98. J. Kennedy and R. C. Eberhart, "Particle swarm optimization", in: *Proc. of IEEE International Conference on Neural Networks*, Piscataway, NJ. pp. 1942-1948 (1995).

99. J. Kennedy, R. C. Eberhart, Y. Shi, *Swarm intelligence*, Academic Press, 2001.

100. S. Kirkpatrick, C. D. Gelatt, and M. P. Vecchi, "Optimization by simulated annealing", *Science*, **220**, No. 4598, 671-680 (1983).

101. G. A. Korn and T. M. Korn, *Mathematical Handbook for Scientists and Engineers*, Dover Publication, 1961.

102. J. R. Koza, *Genetic Programming: One the Programming of Computers by Means of Natural Selection*, MIT Press, 1992.

103. E. Kreyszig, *Advanced Engineering Mathematics*, 6th Edition, Wiley & Sons, New York, 1988.

104. H. W. Kuhn and A. W. Tucker, "Nonlinear programming", *Proc. 2nd Berkeley Symposium*, University of California Press, pp. 481-492 (1951).

105. K. Levenberg, "A method for the solution of certain problems in least squares", *Quart. J. Applied Math.*, **2**, 164-168, (1944).

106. K. S. Lee, Z. W. Geem, "A new meta-heuristic algorithm for continous engineering optimization: harmony search theory and practice", *Comput. Methods Appl. Mech. Engrg.*, **194**, 3902-3933 (2005).

107. J. Macki and A. Strauss, *Introduction to Optimal Control Theory*, Springer, 1982.

108. N. Marco, S. Lanteri, J. A. Desideri, J. Périaux, "A parallel genetic algorithm for multi-objective optimization in CFD", in: *Evolutionary Algorithms in Engineering and Computer Science* (eds Miettinen K. et al), John Wiley & Sons, 1999.

109. D. Marquardt, "An algorithm for least-squares estimation of nonlinear parameters", *SIAM J. Applied Math.*, **11**, 431-441 (1963).

110. E. Marinari and G. Parisi, "Simulated tempering: a new Monte Carlo scheme", *Europhysics Lett.*, **19**, 451-458 (1992).

111. Matlab info, http://www.mathworks.com

112. W. H. McCrea and F. J. Whipple, "Random paths in two and three dimensions", *Proc. Roy. Soc. Edinburgh*, **60**, 281-298 (1940).

113. K. Menger, "Dass botenproblem", in: *Ergebnisse eines Mathematischen Kolloquiums* **2**, Ed. K. Menger), Teubner, Leipzig, p. 11-12 (1932).

114. N. Metropolis, and S. Ulam, "The Monte Carlo method", *J. Amer. Stat. Assoc.*, **44**, 335-341 (1949).

115. N. Metropolis, A. W. Rosenbluth, M. N. Rosenbluth, A. H. Teller, and E. Teller, "Equation of state calculations by fast computing machines", *J. Chem. Phys.*, **21**, 1087-1092 (1953).

116. S. P. Meyn, and R. L. Tweedie, *Markov Chains and Stochastic Stability*, Springer-Verlag, London, 1993.

117. Z. Michaelewicz, *Genetic Algorithm + Data Structure = Evolution Progamming*, New York, Springer, 1996.

118. M. Mitchell, *An Introduction to Genetic Algorithms*, Cambridge, Mass: MIT Press, 1996.

119. M. Molga, C. Smutnicki, "Test functions for optimization needs", http://www.zsd.ict.pwr.wroc.pl/files/docs/functions.pdf

120. L. T. Moore, *Isaac Newton: A Biography*, Dover, New York, 1934.

121. R. F. Moritz and E. E. Southwick, *Bees as superorganisms*, Springer, 1992.

122. H. Murase, "Finite element analysis using a photosynthetic algorithm", *Computers and Electronics in Agriculture*, **29**, 115-123 (2000).

123. D. J. Murdoch and P. J. Green, "Exact sampling from a continuous state space", *Scand. J. Statist.*, **25**, 483-502 (1998).

124. S. Nakrani and C. Tovey, "On honey bees and dynamic server allocation in Internet hosting centers", *Adaptive Behaviour*, **12**, 223-240 (2004).

125. J. A. Nelder and R. Mead, "A simplex method for function optimization", *Computer Journal*, **7**, 308-313 (1965).

126. Y. Nesterov and A. Nemirovskii, *Interior-Point Polynomial Methods in Convex Programming*, Society for Industrial and Applied Mathematics, 1994.

127. Octave, http://www.gnu.org/software/octaive, also J. W. Eaton, *Gnu Octave Manual*, Network Theory Ltd, 2002.

128. I. Pavlyukevich, "Lévy flights, non-local search and simulated annealing", *J. Computational Physics*, **226**, 1830-1844 (2007).

129. I. Pavlyukevich, "Cooling down Lévy flights", *J. Phys. A:Math. Theor.*, **40**, 12299-12313 (2007).

130. C. E. Pearson, *Handbook of Applied Mathematics*, 2nd Ed, Van Nostrand Reinhold, New York, 1983.

131. D. T. Pham, A. Ghanbarzadeh, E. Koc, S. Otri, S. Rahim and M. Zaidi, "The bees algorithm", Technical Note, Manufacturing Engineering Center, Cardiff University, 2005.

132. L. S. Pontryagin, V. G. Boltyanski, R. S. Gamkrelidze and E. F. Mishchenko, *The Mathematical Theory of Optimal Processes*, Interscience, 1962.

133. M. J. D. Powell, "A new algorithm for unconstrained optimization", in: *Nonlinear Programming* (Eds J. B. Rosen, O. L. Mangasarian, and K. Ritter), pp. 31-65 (1970).

134. W. H. Press, S. A. Teukolsky, W. T. Vetterling, B. P. Flannery, *Numerical Recipes in C++: The Art of Scientific Computing*, Cambridge University Press, 2002.

135. N. Quijano, A. E. Gil, and K. M. Passino, "Experimental for dynamic resource allocation, scheduling and control", *IEEE Contrl Systems Magazine*, **25**, 63-79 (2005).

136. S. S. Rao, *Engineering Optimization: Theory and Practice*, 3rd Edition, John Wiley & Sons, 1996.

137. K. F. Riley, M. P. Hobson, and S.J. Bence, *Mathematical Methods for Physics and Engineering*, 3rd Edition, Cambridge University Press, 2006.

138. A. M. Reynolds and M. A. Frye, "Free-flight odor tracking in Drosophila is consistent with an optimal intermittent scale-free search", *PLoS One*, **2**, e354 (2007).

139. G. Ramos-Fernandez, J. L. Mateos, O. Miramontes, G. Cocho, H. Larralde, B. Ayala-Orozco, "Lévy walk patterns in the foraging movements of spider monkeys (*Ateles geoffroyi*)",*Behav. Ecol. Sociobiol.*, **55**, 223-230 (2004).

140. E. Sandgren, "Nonlinear integer and discrete programming in mechanical design optimization", *J. Mech. Des.-T. ASME*, **112** (2), 223-229 (1990).

141. Y. Sawaragi, H. Nakayama, T. Tanino, *Theory of Multiobjective Optimisation*, Academic Press, 1985.

142. A. Schrijver, "On the history of combinatorial optimization (till 1960)", in: *Handbook of Discrete Optimization* (Eds K. Aardal, G. L. Nemhauser, R. Weismantel), Elsevier, Amsterdam, pp.1-68 (2005).

143. D. F. Shanno, "Conditioning of quasi-Newton methods for function minimization", *Mathematics of Computation*, **25**, 647-656 (1970).

144. O. Sheynin, "On the history of the principle of least squares", *Archive for History of Exact Sciences*, **46** (1), 39-54 (1993).

145. P. Sirisalee, M. F. Ashby, G. T. Parks, and P. J. Clarkson, "Multi-criteria material selection in engineering design", *Adv. Eng. Mater.*, **6**, 84-92 (2004).

146. T. D. Seeley, *The Wisdom of the Hive*, Harvard University Press, 1995.

147. T. D. Seeley, S. Camazine, J. Sneyd, "Collective decision-making in honey bees: how colonies choose among nectar sources", *Behavioural Ecoloy and Sociobiology*, **28**, 277-290 (1991).

148. O. Sigmund, "A 99 line topology optimization code written in Matlab", *Struct. Multidisc. Optim.*, **21**, 120-127 (2001).

149. D. R. Smith, *Variation Methods in Optimization*, New York, Dover, 1998.

150. J. C. Spall, *Introduction to Stochastic Search and optimization: Estimation, Simulation, and Control*, Wiley, Hoboken, NJ, 2003.

151. Swarm intelligence, http://www.swarmintelligence.org

152. E.-G. Talbi, *Parallel Combinatorial Optimization*, John Wiley & Sons, 2006.

153. E.-G. Talbi, *Metaheuristics: From Design to Implementation*, John Wiley & Sons, 2009.

154. V. M. Tikhomirov, *Stories about Mixima and Minima*, American Mathematical Society, 1990.

155. A. M. Turing, *Intelligent Machinery*, National Physical Laboratory, 1948.

156. W. T. Tutte, *Graph Theory as I Have Known It*, Oxford Science Publications, 1998.

157. G. M. Viswanathan, S. V. Buldyrev, S. Havlin, M. G. E. da Luz, E. P. Raposo, and H. E. Stanley, "Lévy flight search patterns of wandering albatrosses", *Nature*, **381**, 413-415 (1996).

158. E. W. Weisstein, http://mathworld.wolfram.com

159. D. Wells, *The Penguin Book of Curious and Interesting Geometry*, Penguin, 1991.

160. Wikipedia, http://en.wikipedia.com

161. D. H. Wolpert and W. G. Macready, "No free lunch theorems for optimization", *IEEE Transaction on Evolutionary Computation*, **1**, 67-82 (1997).

162. S. J. Wright, *Primal-Dual Interior-Point Methods*, Society for Industrial and Applied Mathematics, 1997.

163. X. S. Yang, "Engineering optimization via nature-inspired virtual bee algorithms", IWINAC 2005, Lecture Notes in Computer Science, **3562**, 317-323 (2005).

164. X. S. Yang, "Biology-derived algorithms in engineering optimizaton (Chapter 32)", in *Handbook of Bioinspired Algorithms*, edited by Olarius S. and Zomaya A., Chapman & Hall / CRC, 2005.

165. X. S. Yang, "New enzyme algorithm, Tikhonov regularization and inverse parabolic analysis", in: *Advances in Computational Methods in Science and Engineering*, Lecture Series on Computer and Computer Sciences, ICCMSE 2005, Eds T. Simos and G. Maroulis, **4**, 1880-1883 (2005).

166. X. S. Yang, J. M. Lees, C. Morley, "Application of virtual ant algorithms in the otpimization of CFRP shear strengthened precracked structures", *Lecture Notes in Computer Sciences*, **3991**, 834-837 (2006).

167. X. S. Yang, "Firefly algorithms for multimodal optimization", *Proc. 5th Symposium on Stochastic Algorithms, Foundations and Applications, SAGA 2009*, Eds. O. Watanabe and T. Zeugmann, Lecture Notes in Computer Science, **5792**, 169-178 (2009).

168. X. S. Yang, "Harmony search as a metaheuristic algorithm", in: *Music-Inspired Harmony Search Algorithm: Theory and Applications* eds. Z. W. Geem, Springer, p. 1-14 (2009).

169. X. S. Yang, *Nature-Inspired Metaheuristic Algorithms*, Luniver Press, 2008.

170. X. S. Yang, *Introduction to Computational Mathematics*, World Scientific, 2008.

171. X. S. Yang and S. Deb, "Cuckoo search via Lévy flights", *Proceedings of World Congress on Nature & Biologically Inspired Computing* (NaBic 2009, India), IEEE Pulications, USA, 210-214 (2009).

172. X. S. Yang, "Firefly algorithm, Lévy flights and global optimization", in: *Research and Developement in Intelligent Systems XXVI: Incorporating Applications and Innovations in Intelligent Systems XVII* (Eds. M. Barmer, R. Ellis, M. Petridis), Springer-Verlag, London, pp. 209-218 (2010).

INDEX

By Xin-She Yang
Copyright © 2010 John Wiley & Sons, Inc.

CPSIA information can be obtained
at www.ICGtesting.com
Printed in the USA
BVHW042116191119
564303BV00006B/17/P

9 780470 582466